AF873642

Ion Implantation in Semiconductors

Ion Implantation in Semiconductors

SILICON AND GERMANIUM

James W. Mayer
CALIFORNIA INSTITUTE OF TECHNOLOGY
PASADENA, CALIFORNIA

Lennart Eriksson
RESEARCH INSTITUTE FOR PHYSICS
STOCKHOLM, SWEDEN

and John A. Davies
CHALK RIVER NUCLEAR LABORATORIES
CHALK RIVER, CANADA

ACADEMIC PRESS
A Subsidiary of Harcourt Brace Jovanovich, Publishers
New York London Toronto Sydney San Francisco

ACADEMIC PRESS, INC.
111 Fifth Avenue, New York, New York 10003

United Kingdom Edition published by
ACADEMIC PRESS, INC. (LONDON) LTD.
24/28 Oval Road, London NW1

Library of Congress Catalog Card Number: 75–107563

PRINTED IN THE UNITED STATES OF AMERICA
83 84 85 9 8 7 6 5 4 3

Many of the discussions that led to the writing of this book
(and many of our experimental results)
occurred at odd hours, weekends, and holidays.
This book is dedicated, therefore,
to our wives—Betty, Sylvia, and Flo—
for their support, patience, and understanding.

Contents

PREFACE xi

ACKNOWLEDGMENTS xiii

1. **General Features of Ion Implantation**

1.1 Range Distributions *3*
1.2 Lattice Disorder *4*
1.3 Lattice Location and Electrical Properties *6*
1.4 Device Applications *8*

2. **Ranges and Range Distributions of Implanted Atoms**

2.1 Introduction *10*
2.2 Experimental Techniques for Range Distributions *14*

2.2.1 The Radiotracer Method. . . . *14*; 2.2.2 The Scattering Method. . . . *16*; 2.2.3 Hall-Effect and Conductivity Measurements. . . . *16*; 2.2.4 The Junction-Staining Method. . . . *17*; 2.2.5 The Capacitance–Voltage Method. . . . *18*

2.3 Range Distributions in Amorphous Targets *18*

2.3.1 Experimental. . . . *18*; 2.3.2 Theoretical Framework. . . . *21*; 2.3.3 Correction Factors. . . . *27*; 2.3.4 R_p Values—Experiment versus Theory. . . . *31*; 2.3.5 Range Straggling. . . . *32*; 2.3.6 Average Concentration of Implanted Atoms. . . . *36*; 2.3.7 Ranges in Substrates with Oxide Layers. . . . *37*; 2.3.8 Variable-Energy Implants. . . . *37*; 2.3.9 Proton and Helium Ion Ranges. . . . *37*

2.4 Range Distributions in Single Crystals *38*

2.4.1 General Principles. . . .*38*; 2.4.2 Range Distributions of Channeled Beams. . . .*42*; 2.4.3 The Maximum Range R_{max}. . . .*48*; 2.4.4 Z_1 Oscillations in R_{max}. . . .*51*; 2.4.5 Range Distributions in Silicon. . . .*53*

2.5 Enhanced Diffusion *57*

2.5.1 Radiation-Enhanced Diffusion. . . .*59*; 2.5.2 Interstitial Diffusion. . . .*60*; 2.5.3 Enhanced Diffusion during Postbombardment Anneal. . . .*64*

3. Lattice Disorder and Radiation Damage

3.1 Introduction *65*
3.2 Theoretical Considerations *66*

3.2.1 Energy Available for Displacement Processes. . . .*67*; 3.2.2 Average Number of Displaced Atoms. . . .*71*; 3.2.3 Spatial Distribution of Displaced Atoms. . . .*73*

3.3 Experimental Techniques *76*

3.3.1 Optical Effects. . . .*76*; 3.3.2 Electron-Diffraction and Electron-Microscopy Studies. . . .*78*; 3.3.3 X-Ray Transmission Techniques. . . .*85*; 3.3.4 Ion Interactions and Electron Emission. . . .*86*; 3.3.5 Channeling-Effect Measurements. . . .*91*

3.4 Results and Discussion *97*

3.4.1 Lattice Disorder Produced at Room Temperature. . . .*97*; 3.4.2 Depth Distribution of Lattice Disorder. . . .*104*; 3.4.3 Influence of Substrate Temperature on Lattice Disorder. . . .*108*; 3.4.4 Anneal Behavior of Lattice Disorder. . . .*112*

3.5 Comparison with Fast-Neutron Irradiation *118*

3.5.1 Lattice Disorder. . . .*119*; 3.5.2 Electrical and Optical Characteristics. . . .*122*

4. The Lattice Location of Implanted Atoms

4.1 Introduction *125*
4.2 The Channeling Effect Technique *126*

4.2.1 General Principles. . . .*126*; 4.2.2 Alignment of Single Crystals. . . .*135*; 4.2.3 Backscattering Measurements. . . .*136*; 4.2.4 Nuclear-Reaction Techniques. . . .*146*

4.3 Survey of Lattice-Location Experiments *148*

4.3.1 General Observations. . . .*148*; 4.3.2 Summary of Observations. . . .*152*; 4.3.3 Simple Picture of the Behavior of Implanted Atoms. . . .*156*

4.4 Lattice-Location Results in More Detail *158*

4.4.1 Group V Implants in Silicon. . . . *158*; 4.4.2 Group III Implants in Silicon. . . . *161*; 4.4.3 Implantation of Other Ion Species. . . . *165*; 4.4.4 Implantations into Germanium. . . . *167*; 4.4.5 Mixed Group III and Group V Implantations in Silicon and Germanium. . . . *169*; 4.4.6 Postbombardment of Implanted Samples. . . . *170*

4.5 Discussion of Factors Influencing Lattice Location *176*

4.5.1 General Model. . . . *176*; 4.5.2 Implantation Reactions. . . . *177*; 4.5.3 Charge-State Effects. . . . *179*

5. **Hall-Effect and Sheet-Resistivity Measurements in Silicon**

Written in Collaboration with O. J. Marsh

5.1 Introduction *181*
5.2 Interpretation of Hall-Effect Measurements *186*
5.3 Carrier Mobilities and Dopant Ionization Energies *193*
5.4 Anneal Characteristics *198*

5.4.1 General Considerations. . . . *198*; 5.4.2 Temperature Zones in Anneal Characteristics. . . . *200*; 5.4.3 Interstitial Components. . . . *208*; 5.4.4 Influence of Implantation Temperature. . . . *212*; 5.4.5 Influence of Channeling. . . . *215*

5.5 Diffusion Effects *215*
5.6 Limitations on Electrical Behavior *218*
5.7 Analysis of Unconventional Dopant Species *221*

6. **Device Considerations and Applications**

R. W. Bower

6.1 Introduction *224*
6.2 Properties of Doped Layers Produced by Ion Implantation *225*
6.3 Applications to Planar Structures *229*

6.3.1 Discussion of Planar Processing. . . . *229*; 6.3.2 Application of Ion-Implantation Doping to the Planar Process. . . . *231*; 6.3.3 Gate-Masked Ion-Implanted MOSFET. . . . *233*; 6.3.4 Application of the Gate-Masked Ion-Implantation MOS Technique to Integrated Circuits and Large-Scale Arrays. . . . *234*; 6.3.5 Further Application of Ion-Implantation Techniques to the MOSFET. . . . *239*; 6.3.6 Bipolar Transistors. . . . *241*; 6.3.7 Applications of Ion-Implantation Doping to the Planar Integrated-Circuit Technology. . . . *242*

6.4 Application of Ion Implantation to Nonplanar-Process Devices *243*

6.4.1 Point-Contact Diodes. . . . *244*; 6.4.2 Hyperabrupt Controlled Capacitance–Voltage Diodes. . . . *244*; 6.4.3 *p-n* Junction IMPATT Diodes. . . . *245*; 6.4.4 Semiconductor Nuclear-Particle Detectors. . . . *245*

6.5 Analysis of Ion-Implantation Device Applications *247*
6.6 Conclusion *248*

Appendix **Channeling Behavior of Low-Z, MeV Particles in Diamond-Type Lattices**

A.1 General 250
A.2 Experimental 251
A.3 Theory 253
A.4 Comparison between Experiment and Theory 254

References 259

AUTHOR INDEX 269

SUBJECT INDEX 275

Preface

Interest in ion implantation as a method to introduce atoms into the surface layer of a solid has been growing steadily for at least a decade. Until a few years ago this interest was confined to a limited number of research laboratories where implantation was manifesting itself as a useful and versatile tool in many areas of atomic, nuclear, and solid-state physics. The marked upsurge of interest in ion implantation since 1966 can be attributed largely to its emergence as a potentially useful technique to produce electronic components. Indeed, most of the current studies are being made in silicon.

This monograph reviews the recent developments in ion implantation in silicon and germanium and emphasizes the basic aspects of these studies; we feel that experimental work on other semiconductor materials has not been carried through extensively enough to permit a comprehensive picture to emerge, and we have therefore limited our coverage of the subject to the two most studied elements. We have tried to cover in some detail each of the major basic aspects of experimental study: dopant distribution, radiation damage, dopant location, and electrical characteristics.

This is not an historical survey; it is a guide to the more recent developments in the field. Because of this approach, we may not always have given credit to those pioneers in the field of ion-implantation doping who stimulated many of the later investigations. In this regard, the work of M. Bredov, J. O. McCaldin, D. Medved, W. King, and R. Ohl and their colleagues should be noted.

It is our hope that this monograph will serve as a useful summary of the efforts to date in the field of ion implantation of semiconductors. It is aimed at both the specialists in the implantation field and those in other fields who wish to acquaint themselves with the problems and merits of ion implantation.

April 1970

Acknowledgments

The writing of this book has grown out of a joint experimental program between the three authors, extending back to 1965 and involving five different research institutes. We are very deeply indebted to our many colleagues at these institutes for their encouragement, assistance, and criticism and for their stimulating discussions with us during our own investigations. In particular, we wish to mention F. Brown, L. Cheng, J. R. Parsons, D. Marsden, I. Mitchell, and J. L. Whitton at Chalk River Nuclear Laboratories (Canada); J. U. Andersen, E. Bøgh, K. O. Nielsen, H. Schiøtt, and J. Lindhard at the University of Aarhus (Denmark). I. Bergström, K. Björkqvist, B. Domeij, G. Fladda, N. G. E. Johansson, and D. Sigurd at the Research Institute for Physics, Stockholm (Sweden); R. Baron, R. W. Bower, R. R. Hart, O. J. Marsh, and G. A. Shifrin at Hughes Research Laboratories (California); and S. T. Picraux and J. E. Westmoreland at the California Institute of Technology. In writing this book we acknowledge the major contributions of O. J. Marsh who collaborated in writing Chapter 5, R. W. Bower who wrote Chapter 6, and H. Schiøtt who furnished many of the range and straggling data in Chapter 2. The comments of P. Sigmund and F. Eisen on Chapter 3 were very valuable. A special acknowledgment is given to Mrs. N. Kosowicz for her secretarial assistance.

The U.S.A.F. laboratories at Cambridge, Massachusetts and at Wright–Patterson Air Force base, Ohio, are acknowledged for their valuable support in stimulating many of these ion-implantation studies.

Figures 2.3, 2.9a, 2.10, 2.15, 2.17, 2.19, 2.20, 2.23, 2.24, 2.26, 2.27, 2.30, 2.32, 3.14, 3.19, 3.21, 3.27, 3.31, 3.32, 4.9, 4.12, 4.13, 4.17, and 5.18 are reproduced by permission of the National Research Council of Canada from the *Canadian Journal of Physics.*

Figure 2.4 is reproduced by permission of the National Research Council of Canada from the *Canadian Journal of Chemistry.*

Figure 3.34 is reproduced by permission from C. Jech and R. Kelly, *J. Phys. Chem. Solids* **30,** 465 (1969). New York, Pergamon Press.

Figure 5.7 is reproduced from A. H. Clark and K. E. Manchester, Hall measurements of ion-implanted layers in silicon, *Trans. TMS AIME* **242,** 1173–1180 (1968). New York, American Institute of Mining, Metallurgical, and Petroleum Engineers.

Ion Implantation in Semiconductors

1

General Features of Ion Implantation

Ion implantation is the introduction of atoms into the surface layer of a solid substrate by bombardment of the solid with ions in the keV to MeV energy range. The solid-state aspects are particularly broad because of the range of physical properties that are sensitive to the presence of a trace amount of foreign atoms. Mechanical, electrical, optical, magnetic, and superconducting properties are all affected and indeed may even be dominated by the presence of such foreign atoms. Use of implantation techniques affords the possibility of introducing a wide range of atomic species, thus making it possible to obtain impurity concentrations and distributions of particular interest; in many cases, these distributions would not be otherwise attainable. Recent interest in ion implantation has focused on the study of dopant behavior in implanted semiconductors and has been stimulated by the possibilities of fabricating novel device structures in this way. We will therefore direct our attention to those factors which affect the electrical characteristics of implanted layers in silicon and germanium—such factors as range distributions of dopant species, lattice disorder, and location of dopant species on substitutional and interstitial sites in the lattice.

The application of semiconductors in electronic circuitry has been based upon control of the thermal diffusion of dopant elements into semiconducting crystals, normally silicon. These dopants occupy silicon lattice sites and determine the electrical properties of the device. Their concentration is determined by the equilibrium solubility at the process temperature (900–1100° C), and the distribution in depth is given by the diffusion constant and process time.

Ion implantation provides an alternative method of introducing dopant atoms into the lattice. In this case, a beam of dopant ions accelerated through a potential of typically 10–100 kV is allowed to impinge on the semiconductor surface. The implantation system shown in Fig. 1.1 illustrates the basic elements required in this technique. Using different types of available ion sources, a wide variety of beams may be produced with sufficient intensity for implantation purposes: 10^{14}–10^{15} ions/cm^2 (less than a "monolayer") is a representative ion dose. Note that a mass-separating magnet is almost mandatory to eliminate unwanted species that often dominate the extracted beam. Beyond this, however, the basic instrumentation can be quite simple.

An important aspect of the application of implantation to semiconductor technology, in contrast to diffusion processes, is that the number of implanted ions is controlled by the external system, rather than by the physical properties of the substrate. For example, dopants can be implanted at temperatures at which normal diffusion is completely negligible. Also, the dopant concentration is not limited by ordinary solubility considerations, and so a much wider variety of dopant elements may be used. Thus, one potential application of ion implantation is that it might allow the investigation of the properties of species which cannot be introduced into semiconductors by conventional means.

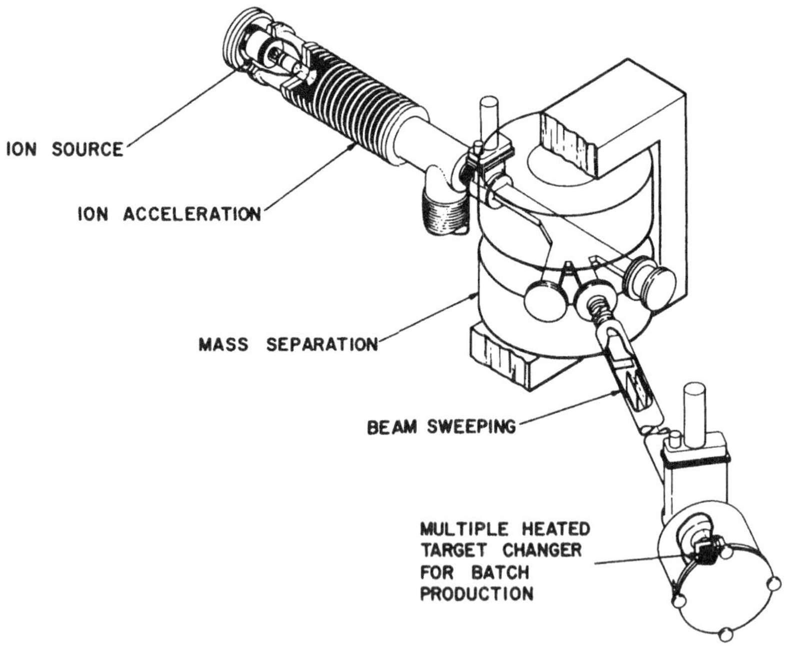

Fig. 1.1. Schematic drawing of an ion-implantation system. A mass-separating magnet is used to select the ion species of interest. Beam-sweeping facilities are provided for large-area uniform implantations.

The major factors governing the successful exploitation of ion implantation are the range distribution of the implanted atoms, the amount and nature of the lattice disorder that is created, the location of the implanted atoms within the unit cell of the crystal, and (ultimately) the electrical characteristics that result from the implantation and subsequent annealing treatment. We will consider all of these factors briefly in the present chapter in order to obtain an overall picture of the problems involved. Subsequent chapters will then treat each one in detail.

1.1 Range Distributions

One of the most important considerations, obviously, in any description of implantation processes is the depth (range) distribution of the implanted ions. In recent years, a large amount of experimental and theoretical work has been devoted to the task of understanding the energy-loss processes that govern the range distribution, and it is now possible to predict fairly accurately most of the factors involved. For example, a typical range distribution in an amorphous substrate is approximately Gaussian in shape, and may therefore be characterized by a mean range and a straggling about this mean value, as depicted in Fig. 1.2. As discussed in Chapter 2, both these quantities depend in a complex but predictable fashion on many variables. It is evident from

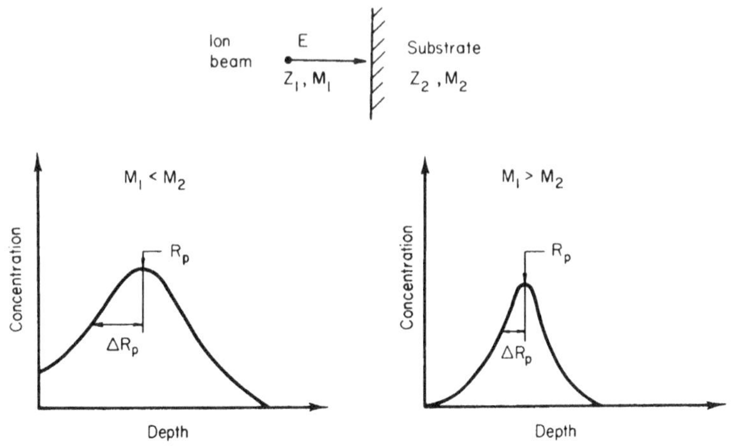

Fig. 1.2. The depth distribution of implanted atoms in an amorphous target for the case in which the ion mass is less than or greater than the mass of the substrate atoms. To a first approximation the mean depth R_p depends on ion mass M_1 and incident energy E, whereas the relative width $\Delta R_p/R_p$ of the distribution depends primarily on the ratio between ion mass and that of the substrate atoms, M_2.

Fig. 1.2 that implanted distributions contrast strongly with the monotonically decreasing profiles that are typical of diffusion processes. Furthermore, by varying the energy continuously during the implantation, one may achieve (in principle) almost any type of dopant profile. Typical values of the mean range for 100-keV ions are ~0.1 micron, whereas diffusion doping usually produces a mean depth of 1–10 microns.

Numerous experiments have shown that in monocrystalline substrates, the range distribution depends strongly on the orientation of the crystal with respect to the implantation direction, i.e., on the "channeling effect." If an ion enters almost parallel to a major axis or plane, then a correlated series of collisions may steer it gently through the lattice, thus reducing its rate of energy loss and increasing its penetration depth. This may result in profiles of the type indicated in Fig. 1.3. In most implantations, only a small fraction of the implanted ions manage to stay channeled throughout their path, and the shape of such a distribution is sensitive to many factors that are difficult to control.

1.2 Lattice Disorder

Other problems inherent in the use of implantation techniques arise from the lattice-disorder (Chapter 3) and radiation-damage effects produced by the incident ion. As an implanted ion slows down and comes to rest, it makes

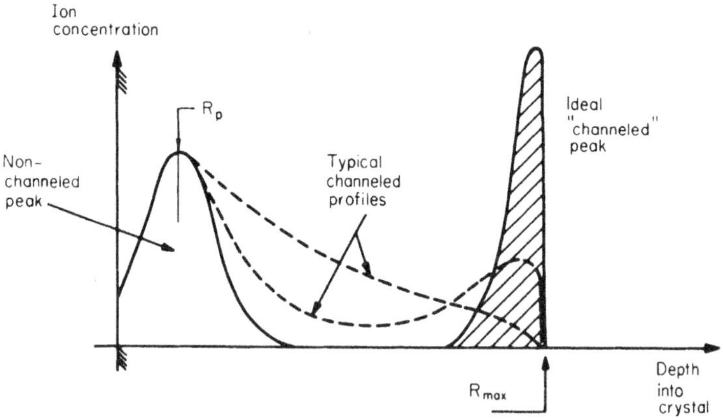

Fig. 1.3. The depth distribution of implanted atoms in a single crystal under conditions such that the beam is aligned with a major crystallographic axis. The shaded portion shows the distribution of perfectly "channeled" ions which penetrate nearly to the maximum channeling range R_{max}. The distribution of atoms is sensitive to many factors, such as beam alignment, lattice vibrations, and surface disorder. The dashed curves indicate the type of distributions that might be obtained under typical implantation conditions in silicon and germanium.

many violent collisions with lattice atoms, displacing them from their lattice sites. These displaced atoms can in turn displace others, and the net result is the production of a highly disordered region around the path of the ion, as shown schematically in Fig. 1.4 for the case of a heavy implanted atom at typically 10–100 keV. At sufficiently high doses, these individual disordered regions may overlap, and a noncrystalline or amorphous layer is formed.

The isolated disordered regions and the amorphous layer have widely different anneal behavior. In the case of germanium and silicon, the isolated disordered regions anneal at moderate temperatures of approximately 200° and 300° C, respectively. The amorphous layers also anneal in a characteristic fashion, but at appreciably higher temperatures, i.e., at approximately 600° C in silicon and 400° C in germanium. It should be noted that even though both types of disorder can be annealed at temperatures well below those where diffusion of the dopant species occurs, there are still defects present. These defects are most evident in the growth of dislocations that occur in silicon at anneal temperatures above 600° C.

If the implantation is performed at a temperature greater than approxi-

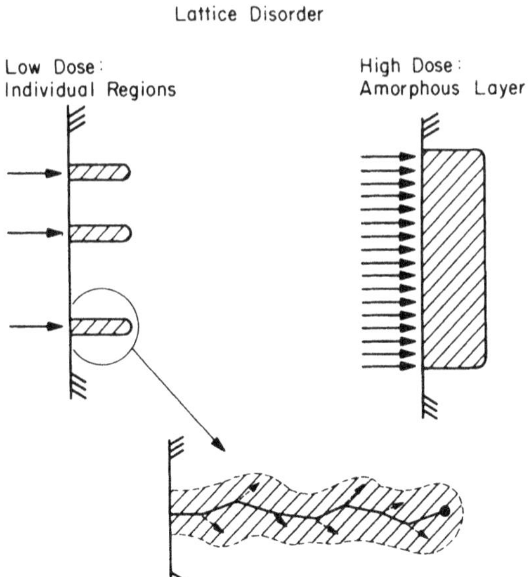

Fig. 1.4. A schematic representation of the disorder produced in room-temperature implantations of heavy ions at energies of 10–100 keV. At low doses, the highly disordered regions around the tracks of the ions are spatially separated from each other. The volume of the disordered region is determined primarily by the stopping point of the ion and the range of the displaced lattice atoms (dashed arrows). At high doses, the disordered regions can overlap to form an amorphous layer.

mately 300° C for silicon and 200° C for germanium, the isolated disordered regions can anneal continuously during the bombardment, thus enabling the implantation to be extended to quite high doses without producing an amorphous layer. Recently, the lattice disorder produced in silicon and germanium has also been investigated for light ions, such as boron. An important feature here is that considerable annealing occurs during implantation even at room temperature, in contrast to the behavior of the heavy elements.

1.3 Lattice Location and Electrical Properties

The overall objective of most ion implantations is to produce a region whose electrical characteristics are determined solely by the dopant atoms located at well-defined positions in the semiconductor lattice. In early studies on implanted layers, it was not at all obvious what type of site an implanted atom would occupy, nor in fact what its dopant nature would be. However, it is now possible to determine the exact lattice location of implanted atoms by channeling-effect measurements (Chapter 4) and their electrical behavior by Hall-effect measurements (Chapter 5). In some cases, a correlation is found

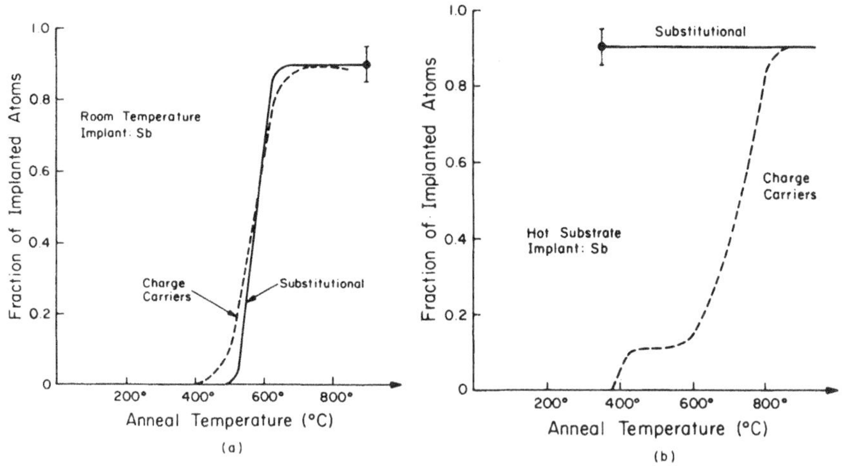

Fig.1.5. The anneal behavior of the number of implanted antimony ions on lattice sites (channeling-effect measurements) and the number of carriers (Hall-effect measurements) for implantation conditions such that the concentration of implanted atoms does not greatly exceed the maximum solubility value: (a) Room-temperature implantations such that an amorphous layer is formed. The steep rise in the curve reflects the anneal of the amorphous layer. (b) Implantation at 350° C substrate temperatures. After implantation approximately 90% of the atoms are on substitutional lattice sites. Anneal temperatures of nearly 800° C are required before the number of carriers approaches the number of implanted atoms.

between the anneal of lattice disorder and the increase in both the atoms on substitutional sites and the number of carriers in the implanted layer. This is illustrated most clearly in room-temperature implantations of Group V elements in silicon. As shown in Fig. 1.5(a), the number of antimony atoms on substitutional sites and the number of carriers are seen to be in excellent agreement. The marked increase at 600° C reflects the epitaxial reordering of the amorphous layer. In implantations at elevated (>300° C) substrate temperatures, this correlation between substitutional and electrical behavior does not occur. As shown in Fig. 1.5(b), implanted antimony atoms attain substitutional sites during implantation and remain there during subsequent anneal, whereas complete electrical activity is not observed until nearly 800° C. The requirement for high anneal temperatures is found in almost all cases where an amorphous layer is not formed during implantation. This is attributed to the presence of electrically active defect centers which tend to compensate the dopant behavior of the implanted atoms. At present, the nature of these centers is not known, and in fact anneal temperatures of 800° C are considerably above those normally used in radiation-damage studies.

A more complex situation is noted in the lattice location of Group III elements in silicon and germanium. Here one often finds *both* an interstitial and a substitutional component. One example of this complex behavior is noted in Fig. 1.6 for thallium implanted in silicon at 450° C. In an "as-implanted" condition (i.e., before annealing) ~30% of the implanted atoms are on regular interstitial sites and another ~30% on substitutional sites. Upon anneal, the substitutional component drops and the interstitial level shows a corresponding increase, indicating that the substitutional atoms are moving to interstitial positions. Finally, at higher temperatures, the interstitial level decreases as the atoms move to nonregular lattice sites such as precipitation centers.

The *electrical* behavior of the Group III dopants in silicon is also complex. However, in room-temperature implantations to sufficiently high dose levels that an amorphous layer is formed, there is the same sharp anneal stage in the electrical behavior at 600° C as was noted in the behavior of Group V dopants in Fig. 1.5(a). Again, as in the Group V case [Fig. 1.5(b)], much higher anneal temperatures (~800° C) are required whenever an amorphous layer is *not* formed. However, for boron at least, the situation is much simpler because the strong increase in carrier concentration around 800° C is closely correlated with a motion of boron atoms onto substitutional sites.

At sufficiently high annealing temperatures, the influence of defects introduced by implantation can be minimized. Under these conditions, the properties of the implanted layer closely resemble those found in heavily doped bulk silicon. Furthermore, at these anneal temperatures, the diffusion constants of the dopant atoms become sufficiently great that the maximum

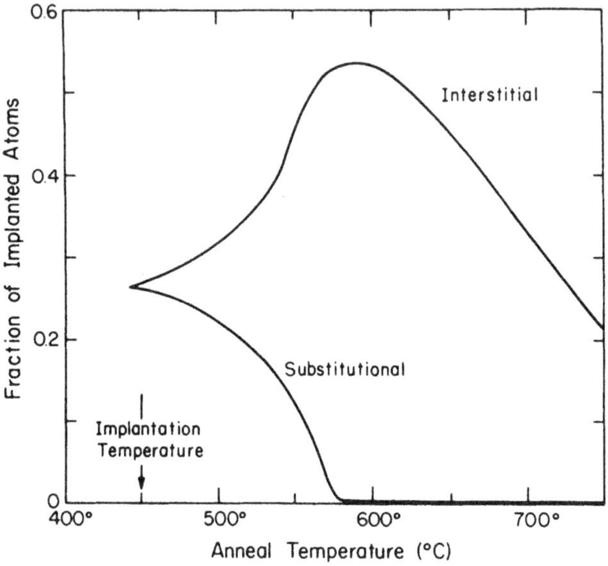

Fig. 1.6. The anneal behavior of the fraction of implanted thallium atoms on substitutional and tetrahedral interstitial sites for concentrations of implanted atoms somewhat greater than $10^{19}/cm^3$. The increase in the number of interstitial atoms around 550° C is reflected in the decrease of the substitutional component.

concentration of implanted atoms on lattice sites should be set by equilibrium-solubility conditions. This suggests the use of implantation techniques to study various properties, such as the solubility, diffusion constant, and dopant energy levels for a wide range of dopant atoms.

1.4 Device Applications

Ion-implantation techniques have been applied to the fabrication of semiconductor devices (Chapter 6). Here the most success has been obtained with single junction devices such as MOSFETs or IMPATT diodes, where the dopant distribution is not a critical factor. The advantages gained in the use of this technique stem from the low process temperatures, ease of masking, and control of the ion dose. To date, two-layer devices, such as bipolar transistors, have not been produced with characteristics superior to those of diffused devices. This appears to be a problem associated with control of the dopant distribution.

Much of this book is devoted to ion-implantation processes in silicon and to a lesser extent germanium. The natural question arises concerning behavior in compound semiconductors. Here, phenomena determined by dynamical events during the stopping of the ions (ranges, production of disorder) should be directly related to these phenomena in silicon. On the other hand, phenomena that occur after the ion comes to rest (such as anneal of disorder, lattice location, and dopant behavior) will be determined by the nature of the semiconductor material. One may expect qualitative differences in the nature of the implanted layers between silicon (covalent bonds) and the more ionic III–V and II–VI compounds. There is some evidence that doping action can be achieved by implanting into these compound semiconductors. However, there have not been sufficient studies as yet to permit an adequate description of the critical factors involved.

2

Ranges and Range Distributions of Implanted Atoms

2.1 Introduction

One of the central considerations in any semiconductor fabrication process is the depth distribution of electrically active dopant species. In thermal diffusion, the surface concentration and the dopant profile are both related to the process temperature. Ion implantation, on the other hand, affords a means of independently controlling these two quantities in that the *profile* of implanted ions is a function of the accelerating voltage, whereas the *number* of implanted ions is determined by the integrated beam current. There are, of course, certain limits in the implantation process: for example, at acceleration voltages greater than a few hundred kV, the implantation system becomes rather large and expensive; and at high ion doses, dopant concentrations greater than a few atomic percent result in excessive lattice strains and formation of precipitates or new phases. For these reasons most implantation studies have been carried out at energies less than 500 keV and at dose levels less than 10^{16} ions/cm^2. Even below this dose level, the *concentration* of the implanted species can often exceed the equilibrium solid-solubility value by quite a large factor; this is another special feature of ion-implantation doping.

To utilize fully the flexibility of ion implantation requires a detailed understanding of the energy-loss processes controlling the depth distribution of implanted ions. In this chapter, therefore, we shall consider the currently available information—experimental and theoretical—on ranges and range distributions.

The energy loss of a heavy charged particle moving through matter at keV energies is determined by screened Coulomb interactions with the substrate atoms. It is customary to distinguish two different mechanisms of energy loss: (1) nuclear collisions, in which energy is transmitted as translatory motion to a target atom as a whole, and (2) electronic collisions, in which the moving particle excites or ejects atomic electrons. For most purposes, this separation into elastic and inelastic collisions is a convenient one and, although not strictly true, it is a good approximation. The first type of collision can involve large discrete energy losses and significant angular deflection of the trajectory; it is also the process responsible for producing displaced lattice atoms (lattice disorder). The second type involves much smaller energy losses per collision, negligible deflection, and negligible lattice damage. The relative importance of the two energy-loss mechanisms changes rapidly with the energy E and atomic number Z_1 of the particle: nuclear stopping predominates for low E and high Z_1, whereas electronic stopping takes over for high E and low Z_1.

A proper understanding of the mechanisms of energy loss is important not only in controlling the depth profile of implanted dopant atoms, but also in determining the nature of the lattice disorder produced during the implantation. In the process of slowing down in the substrate, the implanted ions undergo violent collisions with some of the lattice atoms, thereby displacing them from lattice sites. This problem of lattice disorder is a vital one in most implantation work, and we will return to it again in Chapter 3. The basic principles are, however, treated in the present chapter, since ranges and lattice disorder both involve the same energy-loss mechanisms. Other secondary effects accompanying implantation, such as sputtering of target atoms and back scattering of the incident beam, also depend strongly on the relative importance of nuclear and electronic stopping.

Over the past ten years, there has been extensive experimental and theoretical work on the penetration of keV particles in matter. These studies were given impetus by the development of "unified" range–energy relations by Lindhard *et al.*[1,2] by computer simulations of atomic trajectories (see, for example, Oen *et al.*[3]), and by the innovation of several experimental techniques [4–8] capable of measuring the extremely small penetration depths involved. A review article by Harvey[9] summarizes range–energy theory and measurements prior to 1960. Excellent surveys of the more recent work are contained in the proceedings of the conferences held at Aarhus[10] and Chalk River,[11] and also in review articles by Bergström and Domeij,[12] Dearnaley,[12a] Nichols and Van Lint,[13] and Gibbons.[14] The latter article by Gibbons contains a comprehensive review of ion penetration in silicon.

Up to 1963, range measurements and calculations neglected any possible influence of the ordered arrangement of atoms in a crystal. However, it has now been shown that the crystalline nature of the substrate plays a dominant role in determining the distribution of implanted dopant atoms: An incident ion, having its direction of motion aligned with a low-index crystallographic direction (a channeling direction), undergoes a much lower rate of energy loss in the material, and so penetrates deeper than an ion injected "off-axis." These are not small effects: For example, a 100-keV channeled arsenic ion in silicon can penetrate approximately 50 times deeper than a *non*channeled arsenic ion of similar energy; it even penetrates deeper than would a *channeled* low-Z ion (such as a 100-keV proton or boron ion).

At present, the depth distribution of channeled ions is difficult to characterize, since it depends on many factors that are sometimes difficult to control, such as surface preparation, substrate temperature, beam alignment, and total dose. Consequently, for single crystals only an empirical description of the implanted dopant atom distribution can be given. This should not imply that implantation is too complex a process for the fabrication of semiconductor devices. Similar complexities occur also in conventional diffusion processes where, because of marked deviations from Fick's law, junction depths and diffusion profiles must often be determined empirically. Note also that the distribution of *electrically* active centers may not always coincide with the distribution of implanted atoms; again a similar situation has been found in diffused layers.

The main parameters determining the range of an ion are its energy E and atomic number Z_1, and the atomic number Z_2 of the substrate. In the case of single crystals, the *orientation* of the substrate and the vibrational amplitude (i.e., temperature) of the lattice atoms are also important parameters. In amorphous targets the distance traveled between collisions and the energy transferred per collision are both random variables; hence, all ions of a given type and incident energy do not have the same range. Instead there is a broad distribution in the depth to which individual ions penetrate: this is referred to as the range distribution or range straggling. In the absence of channeling, the range distribution is roughly Gaussian and can be characterized in terms of a mean projected range† R_p and a standard deviation from this mean. However, at depths considerably outside the standard deviation, Sigmund and Sanders[15] and Sanders[16] have shown that the Gaussian approximation is not always valid and that higher moments of the distribution may be necessary in order to predict accurately the profile of the implanted atoms.

When channeling is present, the range distribution contains two distinct components, as shown in Fig. 2.1. Region A illustrates the range distribution of the *non*channeled fraction of the beam. This always involves at least 20% of the dopant atoms (and often considerably more); it is characterized by

† In ion implantation, it is *not* the total distance R traveled by the ion that is of interest, but rather the projection of R along the incident beam direction, i.e., the penetration depth or projected range R_p. The relationship between R_p and R is discussed in Section 2.3.3 (see also Fig. 2.7).

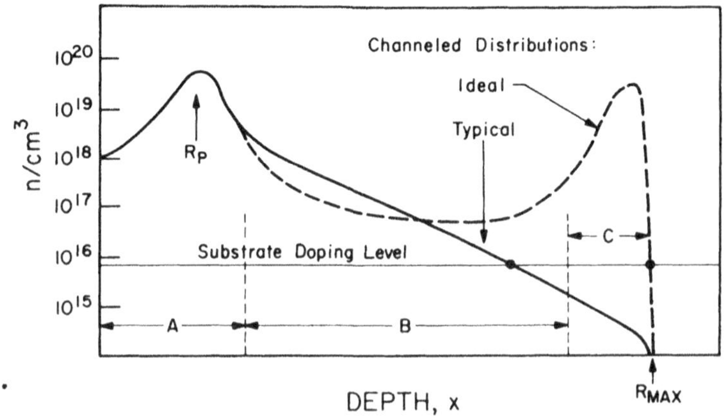

Fig. 2.1. Representative dopant profiles for an initially channeled beam. The dashed curve represents an "ideal" case in which dechanneling is minimized; the solid line represents the case where dechanneling effects are large and only a small fraction of the implanted atoms penetrate to R_{max}. The solid points (●) indicate the depth at which the junction would be located in the two cases.

essentially the same R_p and standard deviation as in an amorphous target. The more penetrating component (regions B and C) is due to the channeled ions; in most cases, it consists of a monotonically decreasing distribution, often an approximately exponential one, which is terminated eventually by a cut-off or maximum value R_{max}. This maximum range R_{max} may be identified with those particles that remain channeled throughout their entire trajectory. Their rate of energy loss is low, and consequently they penetrate much farther than the nonchanneled component. For example, the projected range R_p of 40-keV phosphorous ions in "amorphous" silicon is only 500 Å, whereas the same ions channeled along the $\langle 110 \rangle$ direction in silicon have a maximum penetration of approximately 9000 Å. The depth distribution between R_p and R_{max} arises from particles scattered from the channel direction, i.e., dechanneled, during the slowing-down process. In most lattices, this dechanneling rate is strong enough to dominate the distribution, so that only a very small fraction of the beam penetrates to R_{max}. In such a case, the junction may be located at a depth considerably smaller than R_{max}; its position would then depend strongly on the implantation conditions (cf. Fig. 2.1).

In certain crystals, the dechanneling rate can be reduced to an almost negligible level. Region C then consists of a fairly sharp peak terminated at R_{max}, as indicated by the dotted curve in Fig. 2.1. Examples of a well-defined two-peaked range distribution have been observed in tungsten[17] and occasionally also in silicon.[18]

Our discussion of range distributions is divided into four parts. Part 2.2

deals with the experimental techniques available for range measurements, and the particular advantages and limitations of each. Subsequent Parts 2.3 and 2.4 summarize the existing range data in amorphous and crystalline targets, correlating them wherever possible with theory. Because of the dominant role that channeling plays in monocrystalline targets, we find it convenient at this stage to divide the discussion into two separate sections. Hence, Part 2.3 deals with ranges in amorphous targets, where the distribution of the implanted dopant atoms can often be accurately predicted; and Part 2.4 is devoted to ranges in single crystals, where the distribution is much less predictable. Finally, in Part 2.5, a brief account is presented of the various "enhanced-diffusion" effects that have been observed with implanted dopants.

Measurements in polycrystalline targets are included in Part 2.3; however, their penetration usually exhibits also a significant "channeled" component (cf., for example, Fig. 7 of McCargo *et al.*[19]), due to a small fraction of the crystallites in the bombarded area being sufficiently well aligned with the incident-beam direction.

For many implantation conditions in semiconductor crystals, the amorphous range–energy relations (Part 2.3) are still of value in estimating the location of the peak in the distribution of implanted atoms and the average concentration near that point. This information is particularly useful in interpreting electrical measurements because sheet-resistivity and Hall-effect measurements are strongly influenced by the concentration and mobility of the charge carriers in the region delineated by amorphous range distributions (cf. Chapter 5 on electrical effects). Channeling phenomena (Part 2.4), on the other hand, often determine the junction location and ultimately the current- and capacitance–voltage characteristics of the junction.

2.2 Experimental Techniques for Range Distributions

Several experimental methods have been developed for measuring the small penetration depths involved in ion-implantation studies. Two of the most commonly used ones are the radiotracer method and the elastic scattering method. In addition, in semiconductors where dopant atoms are electrically active, there are at least three other techniques available: the use of Hall-effect or conductivity measurements combined with layer removal, the junction-staining method, and the capacitance-voltage method. A brief description of each is given below.

2.2.1 The Radiotracer Method

The radiotracer method is probably the simplest and most sensitive technique, but requires accelerator facilities capable of handling radioactive projectiles.† First, the target is implanted with a beam of monoenergetic radioactive ions; then, the depth distribution of the embedded radioactivity is determined by removing a series of thin, uniform layers from the surface of the bombarded target, and measuring the residual activity after each layer removal. Several reliable chemical and mechanical techniques for removing sufficiently thin, uniform layers from a variety of targets are now available. The most sensitive is the two-step anodic oxidation method, in which a thin, uniform layer of the target is converted electrochemically to oxide, and this anodic oxide is subsequently dissolved, using a reagent that does not attack the underlying unanodized material. So far, this method has been successfully applied to aluminum,[4] silicon,[5,20]‡ tungsten,[19] and gold [21]; it is probably *not* applicable to many other target systems. Other layer-removal methods, applicable to a wider range of targets, include the vibratory polishing technique developed by Whitton[6] and the sputtering technique[8,22] in which low-energy inert gas ions are used to sputter away the target at a controlled rate. In certain cases, simple chemical etching procedures can even be used, as was recently demonstrated by Andersen and Sørensen[7] for several soft metals, such as copper, silver, and gold; however, etching rates are difficult to control, and are frequently nonuniform over the surface.

One obvious disadvantage of the radiotracer method is that several projectiles of interest in semiconductor doping (such as boron and aluminum) do not have convenient radioactive isotopes.

A useful modification of the anodic oxidation method[23,24] is to form the anodic oxide *before* implanting the radioactive atoms. Subsequent dissolution of the oxide layer then provides information on the fraction of the beam stopped in a known thickness of oxide, instead of in the underlying metal. By using a series of such targets, each having a different thickness of anodic oxide on its surface, it is possible to construct an entire range distribution for the radioactive beam in the anodic oxide. The importance of such measurements lies in the fact that these anodic oxides are amorphous, and so provide a particularly simple method for studying range distributions in the absence of crystal-lattice effects. The method has been used to measure range distributions for a wide variety of projectiles and energies in Al_2O_3,[23,24] in Ta_2O_5,[25] and in WO_3.[23] This modification is not restricted to radiotracer implants, but can also be used (in semiconductors) if the implanted atoms are electrically

† In some cases, the need for *radiotracer* implantation facilities can be avoided by using a subsequent neutron-activation step to convert a small fraction of the implanted dopant atoms into a suitable radioactive isotope.[19a] However, considerably larger implant doses are required, and the problem of competing radioactivity from the substrate atoms must also be considered.

‡ In the case of silicon, two other references on anodic oxidation and stripping have recently been published.[20a,b]

active in the underlying substrate and no compensating centers are present (cf. Section 2.2.3).

2.2.2 The Scattering Method

Another powerful technique for studying the depth distribution of implanted atoms is to bombard the implanted target with a beam of monoenergetic protons (or helium ions), and carry out a careful energy or momentum analysis of the backscattered particles. If the implanted atoms are significantly heavier than the substrate atoms, then the protons scattered off the former will have a somewhat higher energy and can therefore be cleanly resolved. The exact position and shape of this energy spectrum provides quantitative information on the depth distribution of the embedded foreign atoms. This backscattering technique, developed by Powers *et al.*[26–28] and by Bøgh,[29] has provided much of the detailed experimental information on heavy-ion ranges in low-mass targets such as beryllium, boron, carbon and aluminum.

The main limitation of this technique is that the foreign atom should be considerably heavier than the substrate. Also, it has generally required rather high implant doses (typically 10^{15}–10^{17} ions/cm^2), particularly if information about range straggling is desired.

2.2.3 Hall-Effect and Conductivity Measurements

One of the techniques for measuring dopant distributions in semiconductors is to measure the sheet conductivity and the Hall effect (cf. Part 5.2) as successive layers of the semiconductor are removed (using anodic oxidation or etching techniques as described in Section 2.2.1). From such measurements both the mobility and the number of carriers in each layer can be determined independently. One serious limitation of this method (and also of those described in Sections 2.2.4 and 2.2.5) is that it measures the depth distribution of charge carriers associated with the electrically active centers, and not the true range distribution of the implanted atoms. The measured distribution therefore depends on what types of lattice site are available to the impurity atoms, and also on the presence of various radiation-damage centers created by the implantation. Provided the observed number of charge carriers agrees with the total implant dose, the two distributions should be nearly identical. In all other cases, however, it is inadvisable to identify the observed charge-carrier distribution with the true range distribution. In fact, as the electrical measurements in Chapter 5 indicate, this condition is not always fulfilled even after annealing to $>900°$ C (and, by this temperature, diffusion may have broadened the range distribution).

2.2.4 The Junction-Staining Method

The distribution of electrically active centers in an implanted semiconductor can also be obtained by measuring the depths at which *p-n* junctions are formed when ions of one conductivity type are implanted into targets of the opposite conductivity type. The principle of the method is shown in Fig. 2.2. "In the case illustrated, five *p*-type strips of varying resistivities are placed side by side in the target chamber, and implanted under identical conditions. After implantation and annealing, each strip contains a *p-n* junction located at that depth at which the concentration of electrically active implanted ions equals the background concentration. The junction depths can be measured by the conventional lapping and staining technique; hence, each strip gives one point on the implanted-dopant distribution curve. Because a very wide range of background resistivities is available, the profile of electrically active ions can usually be traced over at least six decades in concentration using this technique" (Gibbons[14]). However, it is rather difficult to locate accurately the peak in the range distribution by this method, or to obtain the range straggling ΔR.

A more serious limitation, as in the method of Section 2.2.3, is that the observed distribution does not give the true range distribution, except when all the implanted ions are electrically active.

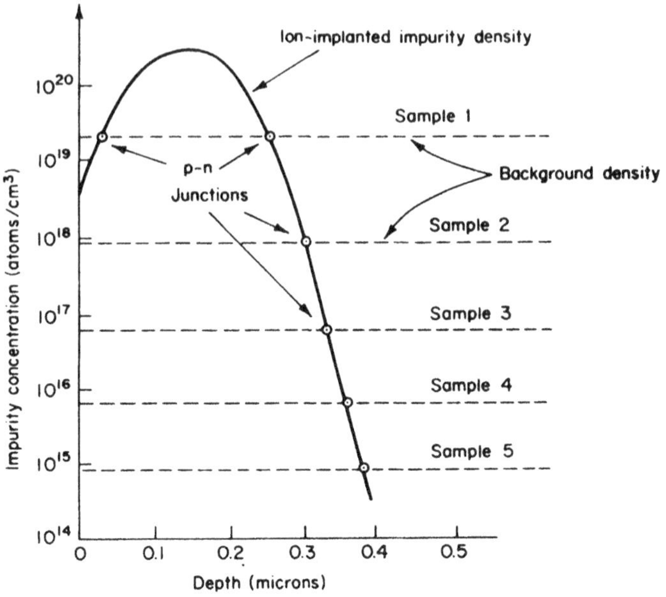

Fig. 2.2. Illustrating the *p-n* junction-location technique for measuring the distribution profile for electrically active impurities (From Gibbons.[14])

2.2.5 The Capacitance–Voltage Method

Another method for determining the distribution of electrically active centers is to measure the capacitance–voltage relation[30] of abrupt, shallow (compared to R_p) *p-n* junctions (or surface barriers) formed on the implanted region under investigation.[31–33] The dopant profile to be evaluated must be of the same conductivity type as the substrate. [If the opposite type of dopant atom is implanted (e.g., *n*-type into *p*-type material) so that a junction is formed, the capacitance–voltage characteristics will provide information about the distribution near the junction[34] rather than that near R_p.] D. E. Davies[31] has used this technique to determine the acceptor distribution in 200- and 400-keV boron implantations into high-resistivity *p*-type silicon. He formed the shallow *n*-type junction required for analysis by implanting phosphorus at a grazing angle.

There are several limitations to the use of this capacitance–voltage technique: (1) low implantation doses must be used to avoid avalanche breakdown; (2) surface effects (such as inversion layers) can contribute to the capacitance; and (3) the free-carrier distribution can also play a role.[35] One advantage is that the distribution of electrically active centers can be found without use of layer-removal techniques.

2.3 Range Distributions in Amorphous Targets

2.3.1 Experimental

A typical set of range distributions in an amorphous target is illustrated in Fig. 2.3, taken from the work of Jespersgård,[24] using the radiotracer method. At each energy, the ^{85}Kr distribution in the Al_2O_3 consists of a broad and rather symmetrical peak, indicating that the individual particle ranges are distributed statistically about a mean value. Both the position and the width of the peak increase almost linearly with implantation energy. Whenever the projectile mass is significantly heavier than that of the target atoms (as in Fig. 2.3), the concentration of implanted atoms in the surface region is seen to be small: i.e., the relative straggling $\Delta R_p/R_p$ is evidently much less than unity. For a light projectile implanted in a heavier substrate, on the other hand, the situation becomes quite different: In such a case, $\Delta R_p/R_p$ may even exceed unity (cf. Fig. 2.11), and the concentration in the surface region is therefore appreciable. An example of such a surface "truncation effect" in the range distribution is shown in Fig. 2.4 for a 10-keV ^{24}Na implantation in polycrystalline aluminum.[36] Note, however, that this truncation effect becomes less significant at higher energies (as shown by the 60-keV ^{24}Na curve in Fig. 2.4) because of the increased contribution of electronic stopping and the corresponding decrease in the relative straggling.

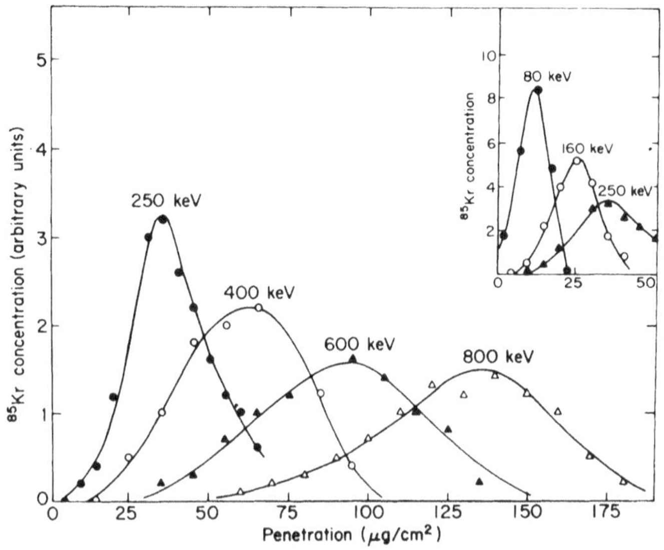

Fig. 2.3. Differential range distributions for ^{85}Kr in amorphous Al_2O_3. (From Jespersgård and Davies.[24]) In Al_2O_3, 1 μg/cm^2 is equivalent to approximately 30 Å.

Each of these distribution curves may be characterized by a mean projected range R_p and a relative range straggling $\Delta R_p/R_p$.† In most cases (for amorphous targets) the range distribution is roughly Gaussian in shape. Hence, R_p corresponds approximately to the position of the peak, and ΔR_p is equal to the full width at half maximum divided by $2(2 \ln 2)^{1/2}$. An average concentration $\bar{n}$ near the maximum can therefore be estimated if ΔR_p is known: i.e., $\bar{n} \simeq N_d/2.5\ \Delta R_p$, where N_d is the number of implanted atoms/cm^2, and $2(2 \ln 2)^{1/2} \simeq 2.5$. Even in single crystals, where channeling phenomena produce a large tail in the distribution, the majority of implanted atoms are often stopped in a region corresponding roughly to the amorphous range. In such cases, the distribution will not be an exact Gaussian; however, the peak position still occurs at approximately R_p, and $\bar{n} = N_d/2.5\ \Delta R_p$ may still be a useful relationship.

It should be noted that experimental data similar to Fig. 2.3 give the mean projected range R_p (along the direction of the incident particles), whereas theoretical calculations often give the mean total range R (i.e., the total path length). This path-length correction factor (see Section 2.3.3) is sometimes quite large; for example, ~30% in the equal-mass case, and as much as a

† Here ΔR_p is defined as the root-mean-square fluctuation in projected range.

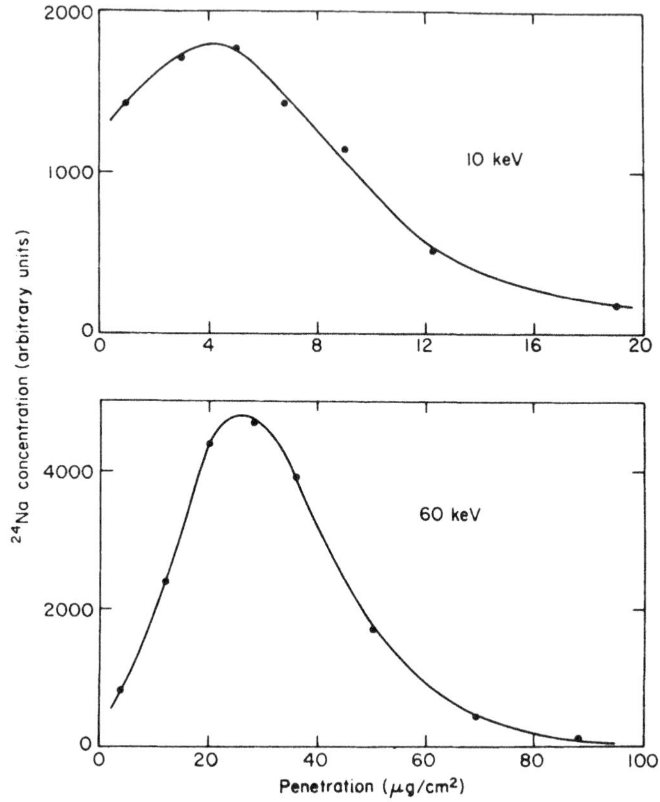

Fig. 2.4. Differential range distributions for ^{24}Na in polycrystalline aluminum. (From Davies and Sims.[36]) Note that in aluminum, 1 μg/cm² is equivalent to 37 Å.

factor of 4 in the case of light ions (boron or phosphorus) implanted into a heavy semiconductor such as germanium or cadmium telluride.†

Experimental range distributions, similar to those in Fig. 2.3, have been measured in several *polycrystalline* targets.[19,27,28,36–39] One such example is seen in Fig. 2.4. In the polycrystalline cases, there is usually a small (5–10%) penetrating tail, due to channeling of the beam that enters the small fraction of crystallites that are by chance aligned with the incident direction. Provided this tail is omitted, the most probable range and the range straggling both agree

† For a *channeled* beam, however, the trajectory is confined within a few degrees (i.e., within the critical channeling angle); thus, there is essentially no difference between the total range and its projected value R_{max} along the direction of incidence. Consequently, unlike the case of amorphous stopping, there is no significant path-length correction factor for channeled ions.

well with the values observed in comparable amorphous targets; if the channeling tail is included, then the range straggling ($\Delta R_p/R_p$) is significantly greater. In certain cases, a significant amount of preferred orientation occurs among individual crystallites, and then the channeling tail may be much larger than a few percent: For example, with rolled tungsten foils, McCargo *et al.*[19] found that more than 50% of the incident beam ended up in the penetrating tail. Under such conditions, meaningful "amorphous" ranges are of course difficult to obtain.

2.3.2 Theoretical Framework

As we have indicated, there are two major types of energy loss to be considered in developing a suitable range–energy relationship for heavy ions: (1) screened Coulomb collisions between the moving ion and the target atoms (nuclear stopping), and (2) interactions between the electrons associated with the moving ion and the various electrons (both bound and free) in the solid (electronic stopping). The basic principles governing both these processes were stated by Bohr[40] almost 25 years ago. More recently, they have been developed into a comprehensive unified theory of atomic stopping by the work of Lindhard and his collaborators[1,2] (referred to subsequently as LSS).

By using a differential scattering cross section based on a Thomas–Fermi potential between the atoms, LSS have derived a universal relationship for the nuclear stopping $(d\varepsilon/d\rho)_n$ in terms of dimensionless length and energy parameters, ρ and ε, defined as

$$\rho = R\pi a^2 N[M_1 M_2/(M_1 + M_2)^2] \tag{2.1}$$

and

$$\varepsilon = E(a/e^2)[M_2/Z_1 Z_2(M_1 + M_2)] \tag{2.2}$$

where

$$a = 0.8853 a_0/(Z_1^{2/3} + Z_2^{2/3})^{1/2} \tag{2.3}$$

a_0 is the Bohr radius (0.529×10^{-8} cm), N is the number of atoms per unit volume, Z and M are the atomic number and mass, and the subscripts 1 and 2 refer to the projectile and target atoms respectively.† The resulting universal relationship between $(-d\varepsilon/d\rho)_n$ and $\varepsilon^{1/2}$ is shown by the solid line in Fig. 2.5 ($\varepsilon^{1/2}$ is proportional to ion velocity). For convenience, we have tabulated in Table 2.1 values of ε/E (keV) and ρ/R (microns) for representative dopants in silicon, germanium, and tin.[41] Table 2.2 lists values of ρ as a function of ε and k.

† Note that the range R in Eq. (2.1) is in units of length (i.e., cm), whereas RNM_2 would be the range in the commonly used unit of gm/cm^2.

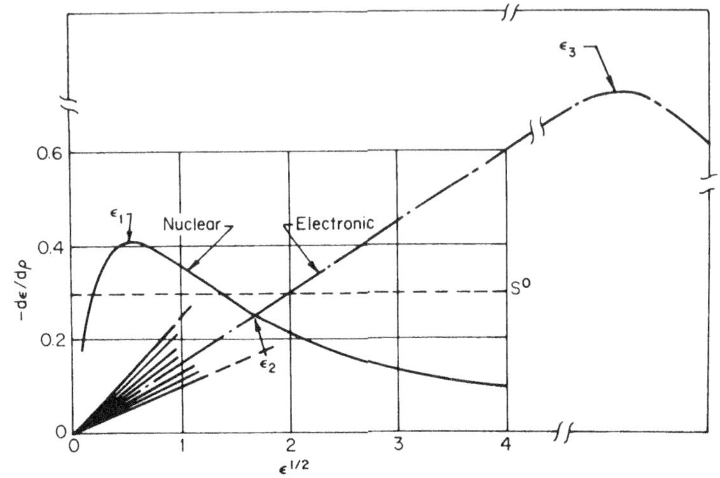

Fig. 2.5. Theoretical nuclear and electronic stopping-power curves, expressed in terms of the reduced variables ρ and ε. (Based on Lindhard *et al.*[2]) For electronic stopping, a family of lines (one for each combination of projectile and target) is obtained; the majority of cases fall within the limits shown. The dot-dash line represents the electronic stopping for $k = 0.15$. The horizontal line labeled S^0 represents the constant-stopping-power approximation suggested by Nielsen.[42] Values of the characteristic energies ε_1, ε_2, and ε_3 are given in Table 2.3.

TABLE 2.1

THE LSS PARAMETERS ε, ρ, AND k FOR VARIOUS DOPANTS IN Si, Ge, AND Sn ($\cong$CdTe)

Ion	ε/E (keV)			ρ/R (microns)[a]			k		
	Si	Ge	Sn (CdTe)	Si	Ge	Sn (CdTe)	Si	Ge	Sn (CdTe)
Li	0.221	0.089	0.052	28.0	8.0	2.7	0.28	0.65	1.04
B	0.113	0.049	0.029	32.2	10.6	3.8	0.22	0.47	0.75
N	0.074	0.033	0.020	32.2	11.8	4.5	0.20	0.42	0.65
Al	0.028	0.015	0.0093	30.5	15.3	6.4	0.14	0.26	0.39
P	0.021	0.012	0.0078	29.0	15.7	6.8	0.14	0.24	0.36
Ga	0.0054	0.0037	0.0027	17.9	15.2	8.1	0.12	0.16	0.21
As	0.0048	0.0034	0.0025	17.0	14.8	8.1	0.12	0.16	0.20
In	0.0021	0.0017	0.00130	11.4	12.2	7.6	0.11	0.14	0.17
Sb	0.0019	0.0015	0.00121	10.7	11.9	7.5	0.11	0.14	0.16
Tl	0.00070	0.00062	0.00052	6.0	8.2	5.9	0.11	0.13	0.14
Bi	0.00066	0.00059	0.00050	5.8	8.0	5.8	0.11	0.13	0.14

[a] Using the following density values (in gm/cm^3): 2.33 for Si; 5.32 for Ge; and 5.84 for Sn (i.e., the density for CdTe).

TABLE 2.2

ρ VALUES AS A FUNCTION OF ε AND k

ϵ	$k = 0.0$	$k = 0.10$	$k = 0.12$	$k = 0.14$	$k = 0.2$	$k = 0.3$	$k = 0.4$	$k = 1.0$
0.01	0.072	0.069	0.069	0.068	0.067	0.064	0.062	0.052
0.02	0.115	0.110	0.109	0.108	0.106	0.102	0.098	0.081
0.05	0.218	0.207	0.205	0.203	0.197	0.188	0.180	0.144
0.10	0.360	0.339	0.335	0.332	0.321	0.304	0.289	0.224
0.20	0.614	0.571	0.563	0.553	0.533	0.501	0.472	0.353
0.50	1.35	1.21	1.19	1.17	1.10	1.01	0.938	0.656
1.0	2.67	2.29	2.22	2.17	2.01	1.80	1.63	1.06
2.0	5.84	4.57	4.39	4.22	3.79	3.26	2.88	1.71
5.0	19.4	11.9	11.1	10.4	8.83	7.11	5.99	3.17
10.0	53.6	23.8	21.6	19.9	16.1	12.3	10.1	4.92

To obtain the stopping contribution due to electronic collisions, LSS have derived a velocity-proportional electronic stopping power $(-d\varepsilon/d\rho)_e$ given by

$$(-d\varepsilon/d\rho)_e = k\varepsilon^{1/2} \tag{2.4}$$

where

$$k = \left[\frac{\xi_e 0.0793 Z_1^{1/2} Z_2^{1/2} (M_1 + M_2)^{3/2}}{(Z_1^{2/3} + Z_2^{2/3})^{3/4} M_1^{3/2} M_2^{1/2}}\right]$$

and ξ_e is of the order of 1 to 2 ($\xi_e \approx Z_1^{1/6}$). The electronic-stopping calculations therefore do *not* produce a universal $(-d\varepsilon/d\rho)_e$ curve, but rather a *set* of curves each characterized by a particular value of k, as show in Fig. 2.5. For most combinations of projectile and target, the appropriate values of k (Table 2.1) fall between 0.1 and 0.25.

Thus, we see that nuclear stopping is the more important process at low energies, that it reaches a maximum value around $\varepsilon = 0.35$ (i.e., at ε_1 in Fig. 2.5), and then falls off. Electronic stopping, on the other hand, increases linearly with velocity over a very wide range, and hence becomes the dominant process for energies greater than $\varepsilon \approx 3$ (i.e., for $\varepsilon > \varepsilon_2$). At much higher energies, $(-d\varepsilon/d\rho)_e$ also passes through a maximum, and subsequently falls off as ε^{-1}. This high-energy region is the familiar "Bethe formula" region where the ion velocity exceeds that of the orbital electrons and hence moves as a fully ionized atom. This energy region has been investigated extensively by nuclear physicists but is far beyond the energy range of interest in most ion implantations. Typical values of the energy (ε_3) at which this maximum in $(-d\varepsilon/d\rho)_e$ occurs are given in Table 2.3 together with the corresponding values of the energy ε_1 at which the maximum in the $(-d\varepsilon/d\rho)_n$ curve occurs, and the transition point ε_2 at which nuclear and electronic stopping become equal.

TABLE 2.3

CHARACTERISTIC ENERGIES (IN keV) CORRESPONDING TO ε_1, ε_2, AND ε_3 IN FIG. 2.5

Ion	ε_1			ε_2			ε_3
	in Si	in Ge	in Sn	in Si	in Ge	in Sn	
B	3	7	12	17	13	10	3×10^3
P	17	29	45	140	140	130	3×10^4
As	73	103	140	800	800	800	2×10^5
Sb	180	230	290	2000	2000	2000	6×10^5
Bi	530	600	700	6000	6000	6000	2×10^6

To obtain a range–energy relationship from these $(-d\varepsilon/d\rho)$ curves in Fig. 2.5, LSS treat the two forms of energy loss as being independent; hence the overall rate of energy loss $(-d\varepsilon/d\rho)_{\text{total}}$ is obtained by adding the appropriate $(-d\varepsilon/d\rho)_e$ curve for electronic stopping to the universal curve $(-d\varepsilon/d\rho)_n$ for nuclear stopping. This can then be integrated from the implantation energy down to zero to give the range as a function of energy. Figure 2.6 shows the resulting ρ versus ε relationship for two limiting cases: (1) where electronic stopping has been completely neglected (Th–F), and (2) where a relatively large electronic stopping term (i.e., $k = 0.4$) is included. More

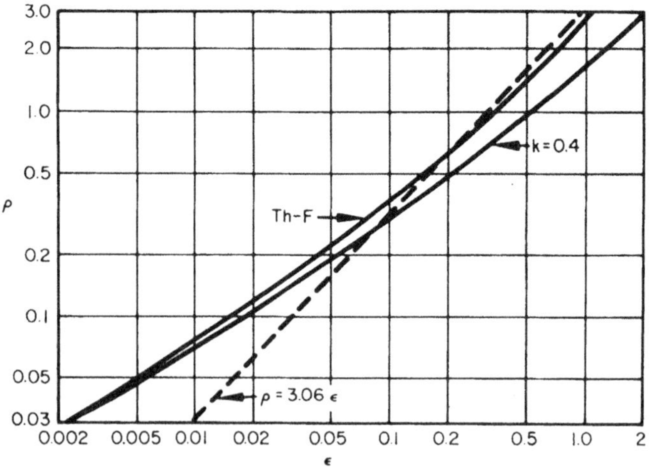

Fig. 2.6. Theoretical range–energy relations for $\varepsilon < 2$. The curve marked Th–F is for $k = 0$ (i.e., neglecting electronic stopping) and the dotted line, $\rho = 3.06\varepsilon$, is derived from the constant stopping power S^0 shown in Fig. 2.5 (from Lindhard *et al.*[2]) Most implantations of interest would fall between the $k = 0$ and $k = 0.4$ curves.

accurate information on this ρ versus ε relationship is contained in Table 2.2. This table in combination with the conversion factors in Table 2.1 provides a quick estimate of the mean range for most cases of interest in ion-implantation studies. Suppose, for example, that we wish to estimate the range of 100-keV arsenic in amorphous silicon. According to Table 2.1, ε/E (keV) = 0.0048, ρ/R (microns) = 17.0, and $k = 0.12$. By suitable interpolation under the $k = 0.12$ column of Table 2.2, we find the appropriate ρ value for $\varepsilon = 0.48$ to be 1.15; hence the range in microns is (1.15/17.0), i.e., 0.067 micron. Note that this is the *total* range R, and not the projected value R_p along the direction of incidence.

For most of the heavier elements in silicon or germanium, we see from Table 2.1 that k is always quite small (less than 0.2), and so nuclear stopping remains the dominant mode of energy loss for ion energies up to several hundred keV (i.e., up to $\varepsilon \approx 3$). For these cases, the Thomas–Fermi curve is quite a useful estimate. On the other hand, for low-Z ions (such as boron in silicon, and boron or phosphorus in germanium) the correction for electronic stopping becomes important at fairly low energies. For example, with 10-keV boron ions in silicon ($\varepsilon \approx 1.1$, $k \approx 0.22$), electronic stopping produces a 25% decrease in ρ as compared to the value calculated from nuclear stopping alone. Even with heavy ions such as gallium, electronic stopping becomes significant if high implantation energies are involved. Table 2.4 lists the *total* range R for several common dopants in silicon, germanium, and tin, calculated by the LSS method, including both nuclear and electronic stopping.

Included in Fig. 2.6 is an empirical range–energy relationship (the $\rho = 3.06\varepsilon$ curve), corresponding to the *constant total* stopping power S^0 in Fig. 2.5. This simple relationship was first published by Nielsen[42] in 1956. At very low energies, it seriously *over*estimates the nuclear stopping (and hence underestimates the range), and at very high energies it *under*estimates the electronic stopping—as can be seen in Fig. 2.5. Nevertheless, at intermediate energies, i.e., $0.05 < \varepsilon < 10$, it provides a rather useful rule of thumb for predicting heavy-ion ranges, usually with an accuracy of ~30%. Using Eqs. (2.1) and (2.2) to transform it into angstroms, the Nielsen equation ($\rho = 3.06\varepsilon$) takes the form

$$R\,(\text{Å}) = \frac{60E\,(\text{keV})}{g}\,\frac{M_2}{Z_2}\,\frac{M_1 + M_2}{M_1}\,\frac{(Z_1^{2/3} + Z_2^{2/3})^{1/2}}{Z_1} \tag{2.5}$$

where g is the density of the substrate in gm/cm^3. By appropriate arithmetical approximations (e.g., $M_1/Z_1 \approx 2.2 \approx M_2/Z_2$), this can be further simplified to

$$R\,(\text{Å}) = \frac{130E\,(\text{keV})}{g}\,\frac{1 + M_2/M_1}{Z_1^{2/3}} \tag{2.6}$$

TABLE 2.4

Total Path Length R (microns) Calculated from the LSS Theory[a]

	^{7}Li	^{11}B	^{14}N	^{27}Al	^{31}P	^{70}Ga	^{75}As	^{115}In	^{204}Tl
a. *In silicon*									
1 keV	0.020	0.011	0.008	0.004	0.004	0.002_5	0.002_5	0.002	0.002
3 keV	0.046	0.025	0.018	0.009	0.008	0.005	0.005	0.004	0.004
10 keV	0.131	0.069	0.048	0.023	0.020	0.012	0.012	0.010	0.009
30 keV	0.334	0.187	0.130	0.058	0.050	0.027	0.026	0.022	0.019
100 keV	0.826	0.527	0.384	0.184	0.155	0.071	0.067	0.052	0.043
300 keV	1.69	1.19	0.920	0.527	0.454	0.199	0.186	0.130	0.098
1 MeV	3.41	2.55	2.08	1.43	1.30	0.675	0.631	0.415	0.269
3 MeV	6.20	4.79	4.10	3.08	2.87	1.88	1.79	1.27	0.777
10 MeV	—	—	7.81	6.39	6.09	4.79	4.67	3.80	2.62
b. *In germanium*									
1 keV	0.030	0.016	0.011	0.005	0.005	0.002	0.002	0.002	0.001
3 keV	0.064	0.034	0.024	0.011	0.010	0.005	0.005	0.003_5	0.003
10 keV	0.153	0.083	0.057	0.026	0.023	0.011	0.010	0.008	0.006
30 keV	0.348	0.196	0.133	0.061	0.051	0.024	0.022	0.017	0.013
100 keV	0.81	0.501	0.353	0.169	0.141	0.059	0.056	0.040	0.029
300 keV	1.63	1.09	0.812	0.443	0.378	0.155	0.144	0.095	0.065
1 MeV	3.27	2.32	1.81	1.16	1.03	0.492	0.457	0.288	0.174
3 MeV	5.94	4.35	3.48	2.48	2.27	1.34	1.27	0.853	0.493
10 MeV	11.17	8.34	6.78	5.15	4.80	3.40	3.28	2.53	1.64
c. *In cadmium telluride*									
1 keV	0.053	0.029	0.020	0.009	0.008	0.003_5	0.003	0.002	0.001_5
3 keV	0.108	0.059	0.041	0.019	0.016	0.007	0.007	0.005	0.003
10 keV	0.241	0.134	0.093	0.043	0.036	0.016	0.015	0.011	0.007
30 keV	0.515	0.295	0.204	0.094	0.080	0.034	0.032	0.022	0.016
100 keV	1.15	0.709	0.501	0.242	0.203	0.083	0.077	0.052	0.036
300 keV	2.28	1.50	1.11	0.595	0.506	0.205	0.190	0.122	0.078
1 MeV	4.55	3.17	2.43	1.51	1.33	0.612	0.567	0.352	0.205
3 MeV	8.26	5.92	4.66	3.19	2.89	1.62	1.53	1.00	0.564
10 MeV	15.5	11.3	9.06	6.62	6.12	4.10	3.94	2.94	1.84

[a] Projected ranges (R_p) may be obtained by combining the above R values with the appropriate R_p/R ratio from Table 2.6.

Considerable care must be exercised in using Eq. (2.5) or (2.6), particularly in cases where the electronic stopping parameter k would be large (i.e., for $Z_1 \ll Z_2$). However, provided $Z_1 > Z_2$, and provided the energy is between 0.3 and 20 keV per atomic mass unit, these simple empirical equations give R values within 30–40% of the LSS total range. These R values may then be divided by $(1 + M_2/3M_1)$ (see Section 2.3.3) to obtain a rough estimate of the mean *projected range* R_p.

2.3.3 Correction Factors

In deriving an accurate range–energy relationship for predicting implantation profiles in amorphous media, there are certain correction factors to be considered:

a. Z_1 *Oscillations in* $(-d\varepsilon/d\rho)_e$. The use of Thomas–Fermi statistical concepts in the LSS treatment leads to an electronic stopping power that increases monotonically with increasing Z_1 (projectile) for a given value of Z_2 and ion velocity. However, experimental studies[43–48] have shown that electronic stopping has a marked periodic dependence on the atomic number of the incident ion. Oscillations of up to 50% on either side of the LSS relationship [Eq. (2.4)] have been observed. Table 2.5 illustrates the magnitude of this

TABLE 2.5

Effect of Z_1 Oscillations in "Amorphous" Silicon

Ion	$(k^1/k)^a$	$R_{corrected}/R_{LSS}$			
		1 keV	10 keV	100 keV	1 MeV
^{11}B	1.20	0.98	0.95	0.89	0.85
	1.5^b	0.95	0.89	0.76	0.69
^{14}N	1.30	0.98	0.95	0.86	0.79
^{23}Na	0.84	1.01	1.02	1.06	1.14
^{27}Al	0.90	1.00	1.01	1.05	1.09
^{31}P	1.14	0.99	0.99	0.96	0.92
^{70}Ga	0.55	1.01	1.02	1.05	1.21
^{75}As	0.60	1.01	1.02	1.05	1.18

[a] k^1/k is the ratio of the *observed* k-value in amorphous carbon targets (Hvelplund and Fastrup[48]) to that predicted by LSS.[2]

[b] Taken from recent experimental data of Eisen.[49]

correction for various dopants in silicon, on the assumption that the Z_1 oscillations in silicon are comparable to those observed in carbon. In the case of boron, an experimentally determined correction factor in "amorphous" silicon is now available from the recent work of Eisen[49]; this has been included in Table 2.5. In the energy region of interest for ion implantation, it is evident that this Z_1-oscillatory effect on the range distribution is usually less than 10%, and hence can often be neglected. In a heavier substrate, such as germanium, the magnitude of this correction should be even smaller. However, for *channeled* ions, the Z_1-oscillation effect is far from negligible, as we shall see in Section 2.4.4.

b. *Projected Range* R_p. The range of an ion, calculated by integrating the basic LSS stopping power curves in Fig. 2.5, is the total distance R that the ion travels in coming to rest. In ion implantation (and in fact in most *experiments*), it is the projection R_p of this total path length on the direction of incidence that is of interest.† The relationship between R_p and R is illustrated in Fig. 2.7. This correction has been discussed in detail by LSS and by Schiøtt.[50,51] They express the correction term in the form $R/R_p = \rho/\rho_p = (1 + bM_2/M_1)$, where b is a slowly varying function of E and R. In the energy region where nuclear stopping dominates and for $M_1 > M_2$, the rule of thumb $R/R_p = 1 + M_2/3M_1$ is a fair approximation (i.e., $b \approx \frac{1}{3}$). Increased electronic stopping at higher energies leads to a smaller value for b. For the case of $M_1 < M_2$, large-angle scattering makes the correction between R_p and R somewhat larger than the value given by the above rule of thumb. On the other hand, electronic stopping is usually appreciable in such cases, and this partially offsets the increase in the correction term. As a consequence, $1 + M_2/3M_1$ is a useful empirical correction for a wide range of implantations as seen in Table 2.6.

Schiøtt[51] discusses the procedure to be followed in obtaining a more quantative correction. His values of R_p/R for representative dopant ions in

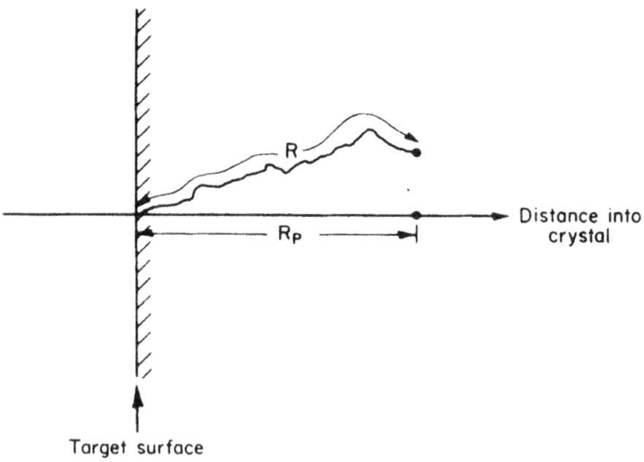

Fig. 2.7. Definition of total range R and projected range R_p. (From Gibbons.[14])

† Strictly speaking, the quantity of interest in implantation is the mean penetration depth x measured perpendicular to the surface, rather than the projected range R_p along the incident-beam direction. In most implantations, however, the angle θ between the incident-beam direction and the surface normal is small ($<5°$), and so R_p and x are almost identical. For oblique incidence, the equation $x = R_p \cos\theta$ may be used to relate these two quantities.

TABLE 2.6

PROJECTED RANGE CORRECTIONS R_p/R

Ion	Substrate	R_p/R values[a]				Rule-of-thumb value
		20 keV	40 keV	100 keV	500 keV	$(1 + M_2/3M_1)^{-1}$
Li	Si	0.54	0.62	0.72	0.86	0.4
B		0.57	0.64	0.73	0.86	0.54
P		0.72	0.75	0.79	0.86	0.77
As		0.83	0.84	0.86	0.89	0.89
Sb		0.88	0.88	0.89	0.91	0.93
Li	Ge	0.33	0.40	0.53	0.74	0.2
B		0.34	0.40	0.50	0.71	0.3
P		0.50	0.52	0.58	0.71	0.56
As		0.67	0.69	0.72	0.77	0.76
Sb		0.76	0.76	0.78	0.81	0.83
Li	Sn	0.22	0.28	0.40	0.63	0.1_5
B		0.24	0.28	0.38	0.60	0.2
P		0.34	0.37	0.43	0.57	0.44
As		0.51	0.53	0.56	0.65	0.65
Sb		0.63	0.64	0.66	0.72	0.75
C	ZnO	0.49	0.54	0.62	0.78	—
Si		0.60	0.63	0.68	0.78	—
Ge		0.75	0.76	0.78	0.83	—
Sn		0.82	0.82	0.83	0.86	—
Pb		0.89	0.89	0.89	0.89	—
C	CdS	0.36	0.41	0.49	0.68	—
Si		0.48	0.51	0.56	0.68	—
Ge		0.65	0.66	0.69	0.75	—
Sn		0.73	0.74	0.75	0.80	—
Pb		0.82	0.82	0.82	0.84	—

[a] These R_p/R ratios have been supplied by Schiøtt.[52]

silicon, germanium, and tin, and also in two "mixed" semiconductors (ZnO and CdS), are included in Table 2.6. For a low-Z ion such as boron, the correction is seen to be strongly energy dependent. These R_p/R ratios, combined with the values of R calculated from the LSS theory (Table 2.4), provide projected ranges R_p for a large variety of projectiles and implantation energies.

Predicted values of the projected range in silicon, germanium, and tin (i.e., cadmium telluride) are illustrated in Fig. 2.8 for three representative ion species. The data are taken from some recent calculations of Schiøtt[52] using the procedure described in his earlier work.[50,51] At constant energy,

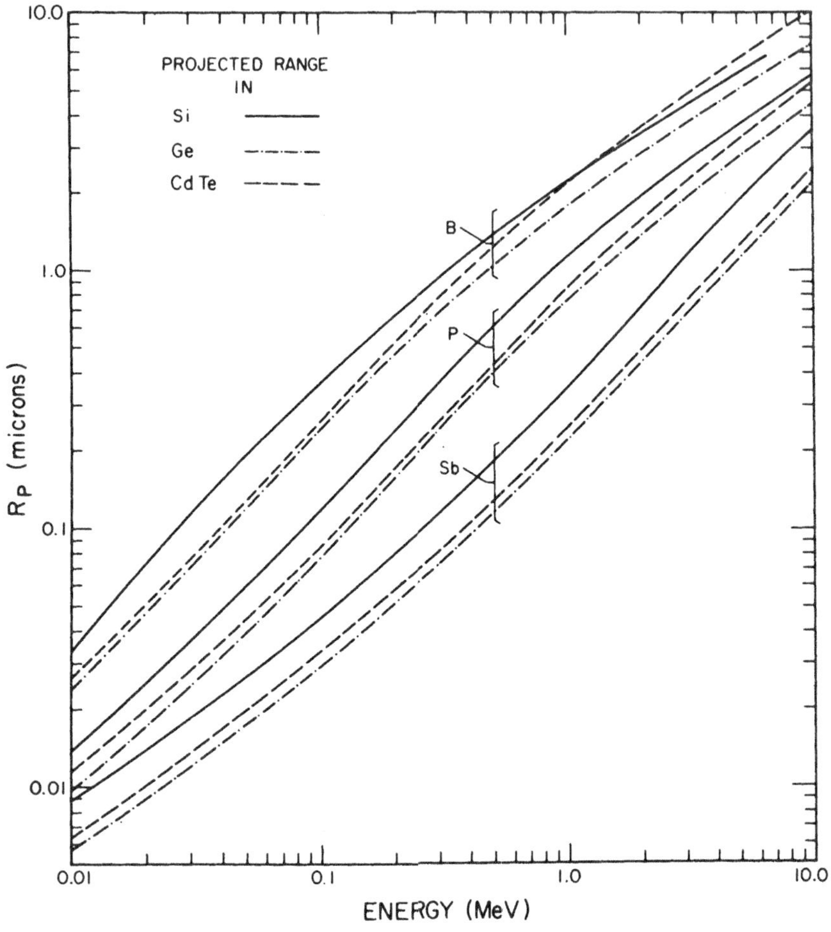

Fig. 2.8. Projected range R_p versus energy for nonchanneled boron, phosphorus, and antimony ions in silicon, germanium, and cadmium telluride. (Based on calculations made by Schiøtt.[52]) Note that the ranges in cadmium telluride fall *between* the corresponding curves in silicon and germanium. This is due to the low value of the density of cadmium telluride, compared to silicon and germanium.

the range is seen to decrease strongly with increasing Z_1. This effect arises from the strong atomic-number dependence of the nuclear stopping. A quite different Z_1 dependence is found for channeled ions (Fig. 2.25) where electronic stopping dominates.

c. *Polyatomic targets.* A further complication arises when the semiconductor substrate consists of two (or more) different atomic species. In many cases, of course, these species are sufficiently close in atomic number—e.g., gallium arsenide and cadmium telluride—that one merely substitutes the mean atomic number (germanium, tin, etc.) into the LSS formulas and proceeds as for a monatomic target. Even in those cases where the two atomic numbers are appreciably different, e.g., Al_2O_3 and GaP, simple range formulas can still be derived, provided that the stopping powers for the two types of target atom have almost the same energy dependence—i.e., that the two ε/E conversion factors are sufficiently similar [Eq. (2.2)] that $(-d\varepsilon/d\rho)$ has almost the same *slope* (Fig. 2.5) for both species. Under these conditions, LSS have shown that the range R in a target composed of two species A and B, having relative abundances X_A and X_B, may be expressed as

$$R = R_A R_B (X_A R_B + X_B R_A)^{-1} \tag{2.7}$$

R_A and R_B are the corresponding ranges in the pure monatomic targets. Jespersgård and Davies[24] and Domeij *et al.*[23] have used Eq. (2.7) to interpret their experimental data in Al_2O_3 and WO_3.

Detailed treatments of projected ranges in polyatomic targets have recently been given by Schiøtt,[51] Sanders,[16] and Winterbon *et al.*[53] Some representative values of R_p/R for various dopants in ZnO and CdS are included in Table 2.6.

2.3.4 R_p Values—Experiment versus Theory

Table 2.7 compares the experimentally observed projected ranges (R_p) in

TABLE 2.7

R_p (in Å) in Amorphous Al_2O_3[23,24] and Ta_2O_5[25]

E	^{24}Na		^{42}K		^{85}Kr		^{133}Xe	
(keV)	Exp.	Theor.	Exp.	Theor.	Exp.	Theor.	Exp.	Theor.
a. *In* Al_2O_3								
10	87	82	62	65	60	58	57	52
20	170	175	130	120	93	88	65	80
40	370	370	232	230	155	150	130	130
80	820	750	450	430	270	260	220	215
160	1870	1500	950	780	570	500	370	365
500	5870	4500	3370	2630	1950	1430	1320	1050
b. *In* Ta_2O_5								
0.5	7	11	9	8	6	6	7	5
1	20	17	13	12	11	9	13	8
2	32	27	21	20	17	16	17	11
5	66	56	44	41	34	29	33	23
10	115	98	72	70	51	49	48	38
20	215	190	130	120	88	81	78	66
40	400	360	235	210	155	145	126	112
80	800	700	435	410	265	245	205	200

amorphous Al_2O_3 and Ta_2O_5 with the values predicted from the LSS treatment. No correction has been made for possible Z_1 oscillations (Section 2.3.3,a). At the higher energies, where electronic stopping becomes appreciable, the theoretical values are ~20% smaller than the experimental data; below 100 keV, however, the agreement is generally within the overall experimental error of ±6%. A similar agreement between experiment and theory is found in polycrystalline targets, as shown for example in Fig. 2.9.

It is evident that in the absence of channeling, the mean penetration depth R_p of an implanted dopant can be estimated with reasonable precision: i.e., usually to within about 10%.

2.3.5 Range Straggling

So far, we have considered only the question of how to derive the mean range R and its projected value R_p along the incident-beam direction. Obviously, in implantation studies, we are interested in the entire depth distribution, and therefore need to know something also about the range straggling. In the absence of channeling, this range distribution is approximately Gaussian in shape, and so can be described by a single extra parameter, namely, the average fluctuation ΔR_p in projected range.†

Again, LSS have developed a useful theoretical framework for treating the problem in terms of the dimensionless energy parameter ε. The solid curves in Fig. 2.10 are the LSS predictions for several projectiles in amorphous Al_2O_3, showing that the straggling is not strongly energy dependent except when electronic stopping becomes dominant. Strictly speaking, these LSS curves represent the relative straggling in the *total* range, i.e., $\Delta R/R$. However, provided $M_1 > M_2$, the difference between $(\Delta R/R)$ and $(\Delta R_p/R_p)$ should be quite small. In such cases, and for $\varepsilon < 3$, LSS find that the predicted range straggling, expressed in terms of the reduced parameter

$$(\Delta R_p/R_p)[(M_1 + M_2)/2(M_1 M_2)^{1/2}]$$

approaches an almost constant value of 0.35. Included in Fig. 2.10 are the experimental values of $\Delta R_p/R_p$ obtained by Jespersgård[24] and by Domeij.[23]

† In most implantations, the incident-beam direction is almost perpendicular to the surface, and hence the depth straggling Δx and the projected range straggling ΔR_p are identical. For oblique incidence, however, Δx involves a contribution also from the transverse straggling $R_\perp$ (i.e., the displacement in the plane perpendicular to the incident-beam direction); in such cases, Schiøtt[50] has shown that the following relationship exists:

$$\Delta x^2 = \Delta R_p^2 \cos^2\theta + \tfrac{1}{2} R_\perp^2 \sin^2\theta$$

where θ is the angle between the incident-beam direction and the surface normal. Sigmund and Sanders[15] have provided tables of ΔR_p and $R_\perp$ for various projectile–target combinations.

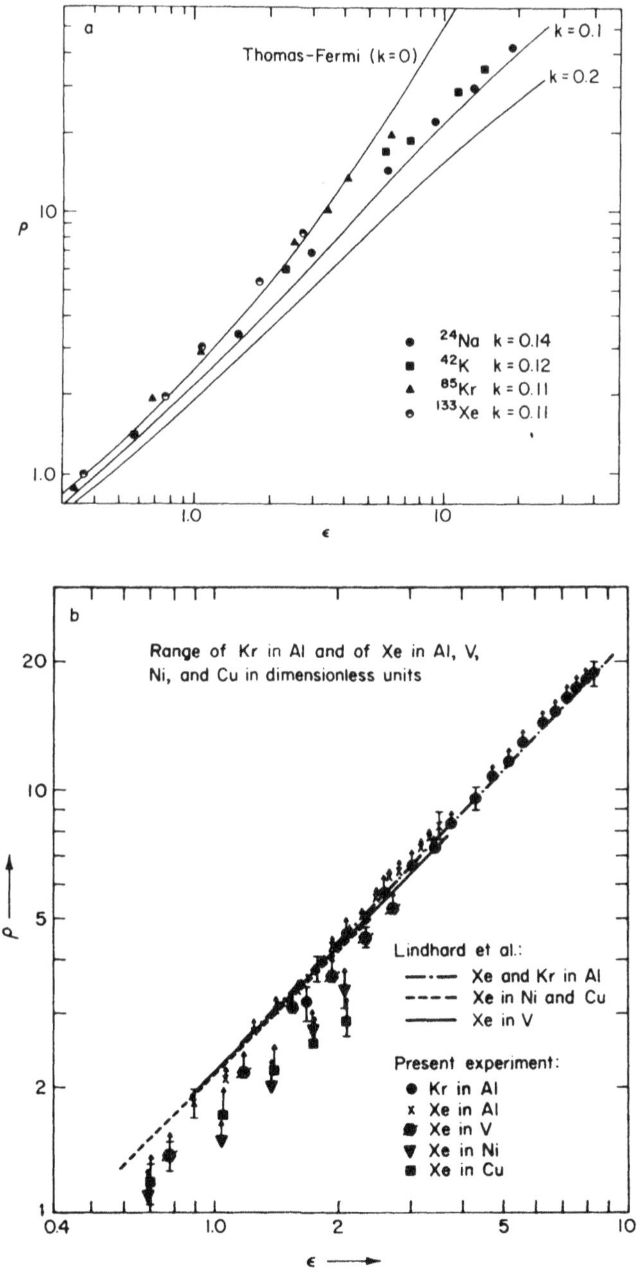

Fig. 2.9. Comparison of experimental results with the theoretical ρ–ε curves given by LSS: (a) Data in amorphous Al_2O_3, converted from projected to total ranges. (From Jespergård and Davies.[24]) (b) Data in several polycrystalline targets. (From Powers *et al.*[27]) The magnitude of the projected-to-total range correction is indicated by vertical arrows on the experimental points.

We see that the experimental points lie somewhat above the theoretical curve, indicating that a better value for the reduced straggling parameter might be ~0.45. Consequently, the full width (2.5 ΔR_p) of the Gaussian distribution measured at half maximum can often be expressed roughly (to within ~25%) as

$$2.5\,\Delta R_p \simeq 1.1 R_p \times 2(M_1 M_2)^{1/2}/(M_1 + M_2) \simeq R_p \quad (2.8)$$

Note that this approximation does not hold for $M_1 < M_2$ because of the dominance of large-angle scattering, nor for $\varepsilon \gg 3$ where electronic stopping becomes dominant. Values of $\Delta R_p/R_p$ versus energy for boron, phosphorus, arsenic, and antimony ions in silicon, germanium, and cadmium telluride are shown in Fig. 2.11, taken from some recent calculations by Schiøtt.[52] The large values of $\Delta R_p/R_p$ for light ions might be anticipated from the fact that R_p is now much less than the total path length R (Table 2.6); the decrease in $\Delta R_p/R_p$ with increasing energy is due to the increased contribution of electronic

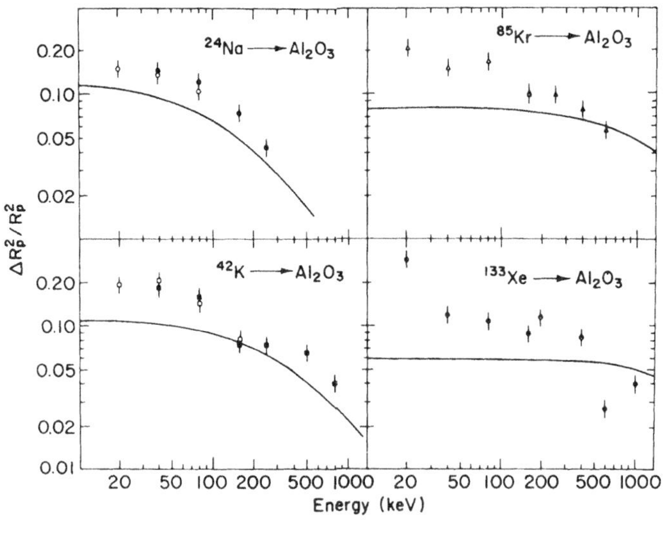

Fig. 2.10. (above) Comparison of the experimental mean *square* range straggling (i.e., $\Delta R_p{}^2/R_p{}^2$) in amorphous Al_2O_3 with the appropriate theoretical curves (solid lines) derived from the LSS treatment.[2] (From Jespersgård and Davies.[24]) Two sets of experimental data are shown for each projectile.

Fig. 2.11. (right) Range straggling $\Delta R_p/R_p$ versus energy for several common dopants, based on the calculations of Schiøtt[52]: (a) boron in silicon, germanium, and cadmium telluride; (b) phosphorus in silicon, germanium, and cadmium telluride and (c) several dopants in cadmium telluride. In (c), the values calculated by Sigmund and Sanders[15] are shown by arrows on the left-hand side; these values do not include electronic stopping.

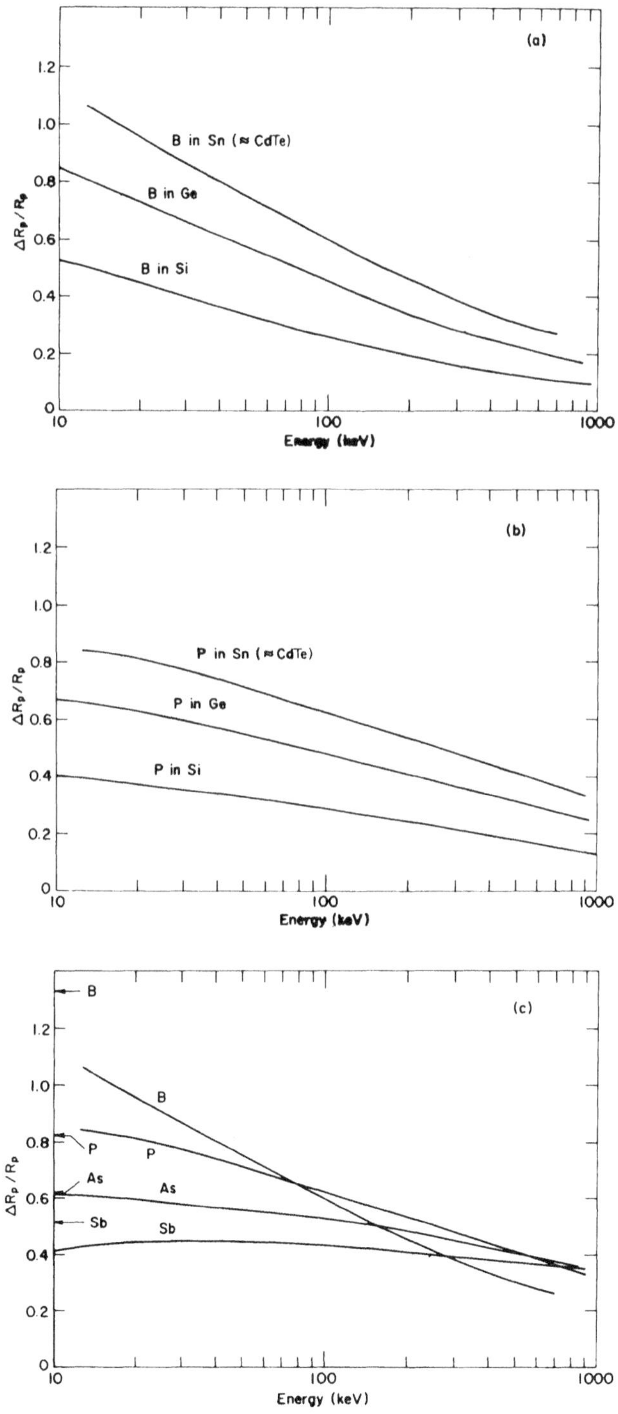

(a)
1.2
1.0
0.8
0.6
0.4
0.2
0
$\Delta R_p / R_p$
B in Sn (≈ CdTe)
B in Ge
B in Si
10
100
1000
Energy (keV)
(b)
$\Delta R_p / R_p$
P in Sn (≈ CdTe)
P in Ge
P in Si
Energy (keV)
(c)
$\Delta R_p / R_p$
B
P
As
Sb
Energy (keV)

Fig. 2.11.

stopping. Schiøtt[51] points out that in the region $M_1 < M_2 < 10\text{–}20M_1$, it is difficult to obtain a one-parameter approximation and therefore each ion-substrate combination must be treated separately. Either one extrapolates from Fig. 2.11, or at low energy uses the calculations of Sigmund and Sanders[15] based on power-law approximations to the nuclear stopping cross section, but neglecting electronic energy loss. Approximate values of $\Delta R_p/R_p$, derived from Sigmund and Sanders' tables for various values of M_2/M_1, are shown by the arrows in Fig. 2.11(c). It is clear that for light ions such as boron and phosphorus, the contribution of electronic stopping usually lowers the value of $\Delta R_p/R_p$ significantly below the Sigmund and Sanders estimate.

2.3.6 Average Concentration of Implanted Atoms

The foregoing calculations provide a means of estimating the depth at which the concentration of implanted dopant reaches its maximum value (viz. $\simeq R_p$), and also the mean straggling ΔR_p about this depth. For many purposes, such as sheet-conductivity measurements, it is desirable to estimate the average dopant concentration $\bar{n}$ in the region around R_p. As noted earlier, provided the distribution is roughly Gaussian,

$$\bar{n} \simeq N_d/2.5\,\Delta R_p \tag{2.9}$$

where N_d is the number of implanted atoms per unit area.

Equation (2.9) is therefore a useful first-order approximation for estimating the *maximum* concentration of the implanted dopants. The concentration $N(x)$ as a function of depth may then be expressed as

$$N(x) = \bar{n}\exp[-(x - R_p)^2/\Delta R_p{}^2] \tag{2.10}$$

i.e., the dopant concentration falls to $\sim\bar{n}/10$ at $x = R_p \pm 2\,\Delta R_p$, and to $\sim\bar{n}/100$ at $x = R_p \pm 3\,\Delta R_p$. These equations must be applied with some caution, however, as they take no account of channeling effects (Part 2.4), nor of possible diffusion effects that may occur during implantation or during any subsequent anneal treatment. Diffusion effects may be particularly serious in the case of high-dose room-temperature implants, where the implanted region becomes converted to an amorphous layer (Chapter 3). During the subsequent reordering (annealing) of this amorphous zone, experimental evidence suggests that for krypton[54] and also for cadmium[55] implants in silicon, considerable motion of the implanted atoms to the surface can occur.† In such cases, the resulting concentration profile differs drastically from the original range distribution. On the other hand, for implant temperatures

† This problem of "out-diffusion" during the annealing of an amorphous layer is discussed further in Section 3.4.4.

or doses where an amorphous layer is *not* formed, subsequent annealing does not ordinarily produce anomalous diffusion effects. In addition, these effects may not be significant for species that occupy substitutional sites (cf. Chapter 4).

2.3.7 Ranges in Substrates with Oxide Layers

For many electronic applications, the ion implantation is carried out through a suitable surface layer or "mask": For example, in the case of silicon, a layer of SiO_2 is often used. In order to obtain the depth distribution in the underlying silicon, the simplest procedure is to calculate the relative stopping powers of SiO_2 and silicon, so that the thickness of the mask may be converted into an "equivalent thickness" of silicon. This equivalent thickness is then subtracted from the total R_p value in silicon to give the residual range of the implanted ions in the underlying semiconductor.

In this regard, it is useful to note that the range of a given projectile, *expressed in units of mass/cm²*, varies only slowly with the atomic number of the target.† Hence, in the case of SiO_2 on silicon, a simple first-order estimate may be obtained by assuming that SiO_2 and Si have the same stopping power per $\mu g/cm^2$.

2.3.8 Variable-Energy Implants

One of the (potentially) attractive features of ion implantation is the possibility of varying the energy *during* the actual implantation in order to achieve any desired dopant profile. So far, this aspect has not been actively explored, although a preliminary study with boron in silicon has been reported by Large *et al.*[56]

From the theoretical standpoint, the overall profile may be calculated in a straightforward manner by summing the individual profiles corresponding to each monoenergetic portion of the total implantation.

2.3.9 Proton and Helium-Ion Ranges

So far, we have omitted any consideration of extremely low-Z ions, such as hydrogen or helium ions. Although such ions have little or no value for semiconductor doping, they are utilized in certain implantation studies because of their large penetration depth. Foyt *et al.*[56a] have used proton bombardment, for example, to create semi-insulating regions in GaAs, thereby isolating various regions of an implanted device from one another. Hence, a brief discussion of the range of these light ions is perhaps justified. In most

† This is especially true if $M_2 < M_1$, as can be seen from Eq. (2.6).

implantation studies, only a rough estimate of the proton or helium ion range is required.

From the Z_1 dependence of the characteristic energies illustrated in Table 2.3, one sees that for proton and helium ions, ε_1 and ε_2 are both less than 1 keV; hence, the total range R is determined almost entirely by electronic stopping. In fact, even ε_3 is now sufficiently small—about 70 keV for protons and 0.5 MeV for helium—that the LSS treatment of electronic stopping [Eq. (2.4)] is not always sufficient.

Whenever the incident energy E is less than ε_3, a reasonable estimate of R_p may still be obtained from the LSS treatment after appropriate allowance for the correction terms discussed in Section 2.3.3. For such light ions, the projected-range correction factor (R_p/R) is much smaller than unity, and is strongly energy-dependent; this problem has been treated in considerable detail, using protons as an example, by Schiøtt.[50]

When $E \approx \varepsilon_3$, the problem becomes more complex, and theoretical range estimates are no longer available. There exists, however, a rather large amount of experimental stopping data, from which semiempirical range-energy relations have been derived. At still higher energies ($E \gg \varepsilon_3$), one enters the "Bethe formula" region, which has been widely investigated, both experimentally and theoretically, by nuclear physicists. Here, theoretical estimates again become relatively simple and accurate, and the correction factors for Z_1 oscillations and for the *projected* range approach unity. Several extensive tabulations of proton and helium-ion ranges as a function of energy are available; see, for example, Williamson and Boujot[56b] and Janni.[56c]

2.4 Range Distributions in Single Crystals

Range distributions in single crystals differ from those in amorphous targets because of the possibility of implanted ions becoming "channeled" along one of the open directions in the lattice. This orientation effect is perhaps best visualized by looking at a model of a silicon crystal (Fig. 2.12), (a) as it appears when viewed along a ⟨110⟩ axis, and (b) the same crystal viewed along an axis tilted 10° with respect to the ⟨110⟩. In the first case, the crystal appears as a honeycombed structure with large open channels extending through the crystal; in the second case, the crystal appears as an almost randomly distributed system of atoms. It is surely not surprising that these two orientations produce very different range distributions.

2.4.1 General Principles

If an energetic ion enters a crystal within a certain critical angle of a major axis [such as the ⟨110⟩ in Fig. 2.12(a)], then each time it approaches one of the aligned rows of lattice atoms, the gradually increasing repulsion between

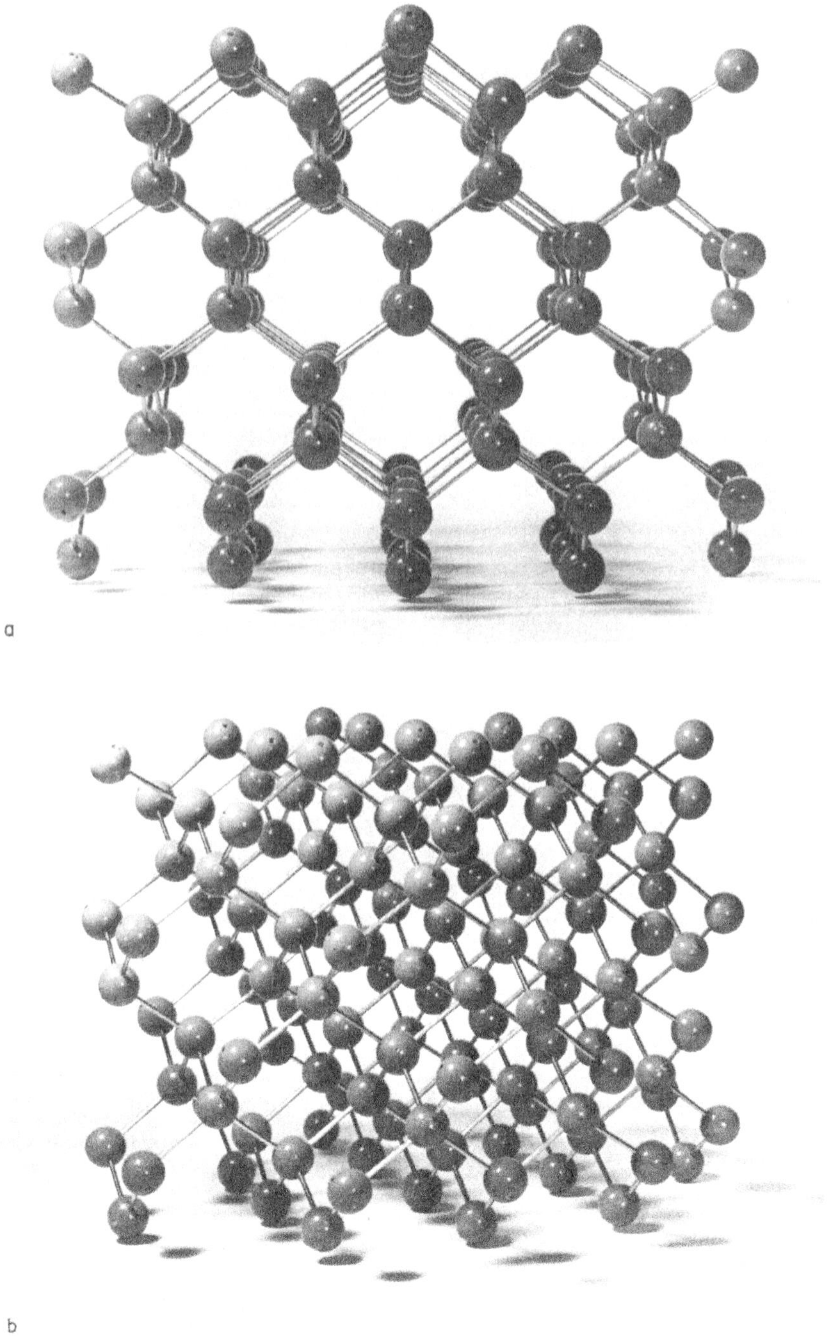

Fig. 2.12. Atomic configuration in the diamond-type lattice (a) viewed along the ⟨110⟩ direction, and (b) viewed along a "random" direction at ~10° from the ⟨110⟩.

the screened Coulomb fields of the projectile and the lattice row is sufficient to steer it away again, thereby preventing violent nuclear collisions from occurring. A reasonable estimate of this critical angle within which channeling occurs can be obtained from the theoretical work of Lindhard[57]; viz. for heavy ions at keV energies, the channeling criterion for a particle of energy E and angle of incidence ψ relative to a row is given by

$$\psi \lesssim \psi_2 \equiv [(a/d)\psi_1]^{1/2} \tag{2.11}$$

where

$$\psi_1 = (2Z_1Z_2e^2/Ed)^{1/2} \tag{2.12}$$

a, the Thomas–Fermi screening distance [Eq. (2.3)], has a magnitude of 0.1–0.2 Å, and d is the atomic spacing along the aligned row. For energies less than a few hundred keV, ψ_2 is between 3° and 5° for standard dopants incident along either the ⟨110⟩ or ⟨111⟩ axes. Equation (2.11) indicates that ψ_2 depends only weakly on energy or atomic number.†

One obvious consequence of this steering mechanism is that the rate of energy loss is greatly reduced and so the ion penetrates more deeply than in an amorphous target. Also, since nuclear stopping depends much more strongly on impact parameter than does electronic stopping, we find that for a channeled beam, (1) the relative importance of nuclear to electronic stopping is much smaller than for a nonchanneled one, and (2) processes related to nuclear stopping, such as radiation damage or sputtering, are also much reduced. As a result, channeling offers at least two potential advantages for the implantation of semiconductors: deeper junctions and less lattice disorder. On the other hand, it has at present one serious disadvantage, viz. the difficulty of obtaining reproducible range profiles. This difficulty arises from the large and often unpredictable role that dechanneling effects can play (cf. Fig. 2.1).

Ion trajectories of interest for axial channeling are shown schematically in Fig. 2.13 in which the crystal lattice is depicted as a set of atomic "strings" of radius a. First, consider particles with oblique incidence [Fig. 2.13(a)]. Particle B, incident at the midchannel with an angle ψ somewhat less than the critical angle ψ_2, will oscillate between strings without undergoing any large-energy-loss nuclear collisions. However, its path has wider-amplitude oscillations than particle C with smaller ψ. Particle A, with $\psi > \psi_2$, will not

† At much higher velocity—e.g., for protons, helium ions, etc. at MeV energies—a quite different critical-angle relationship is applicable (viz. $\psi \lesssim \psi_1$). This velocity region is discussed in more detail in Chapters 3 and 4 and the appendix, since it forms the basis of the channeling-effect technique for studying lattice disorder and for locating dopant atoms in the unit cell of a crystal.

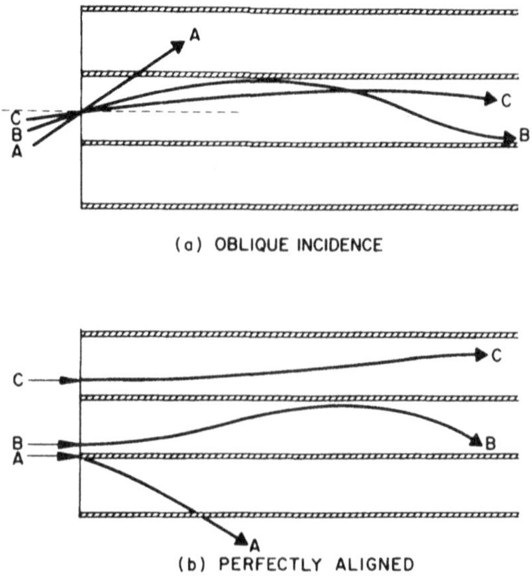

Fig. 2.13. Schematic representation of ion trajectories for axial channeling. (From Mayer and Marsh.[58]) The crystal lattice is depicted as a set of atomic "strings" of radius a, i.e., the shaded rows. (a) Trajectories for various angles of incidence ψ relative to the lattice row: B and C represent trajectories for ψ values less than the critical angle, and A for values greater than the critical angle. (b) Trajectories for parallel incidence as a function of impact position.

be steered by the lattice rows in Fig. 2.13, and so has a trajectory similar to that in an amorphous solid—unless, of course, it happens to be aligned with some other low-index direction.

The impact position of a particle in the plane perpendicular to the channel axis also influences its trajectory. Consider ions incident parallel to the channel direction, as shown in Fig. 2.13(b). Those entering the crystal close to an atomic row are scattered immediately through an angle sufficiently large that they do not become channeled (trajectory A). Particles entering at a position slightly farther away are channeled (trajectory B), but with larger oscillations than particles entering near the center of the channel (trajectory C).

For convenience, therefore, the incident particles may be divided into three broad categories, corresponding to the trajectories shown in Fig. 2.13:

1. Group A, those particles that do not "feel" the lattice and so have a range distribution similar to those in amorphous material.
2. Group B, those particles that start out with large oscillations in the channel. Such particles are probably scattered out of the preferred direction—

i.e., become *dechanneled*—long before they become stopped, and so do not penetrate as deeply as those in Group C.

3. Group C, those particles that start out "well channeled" and so have a better chance of remaining channeled throughout the slowing-down process.

2.4.2 Range Distributions of Channeled Beams

Figure 2.14 illustrates the type of range distribution that is observed when a heavy-ion beam is implanted into a well-aligned tungsten crystal. Tungsten has been chosen as a rather "ideal" case in that its lattice vibrations at room temperature are small, and so dechanneling effects are minimized. We see that the distribution curve consists of three clearly divided regions, which may be identified roughly with the three types of trajectory shown in Fig. 2.13(b). The two peaks, A and C, correspond to the amorphous and well-channeled ranges respectively; the latter is due to those particles that remain channeled throughout their entire path, and is characterized by a sharp cut-off or maximum range, R_{max}. This is true to a lesser extent in silicon,[5,18] and perhaps also in other crystals if the implantation temperature is low enough. Whitton,[59]

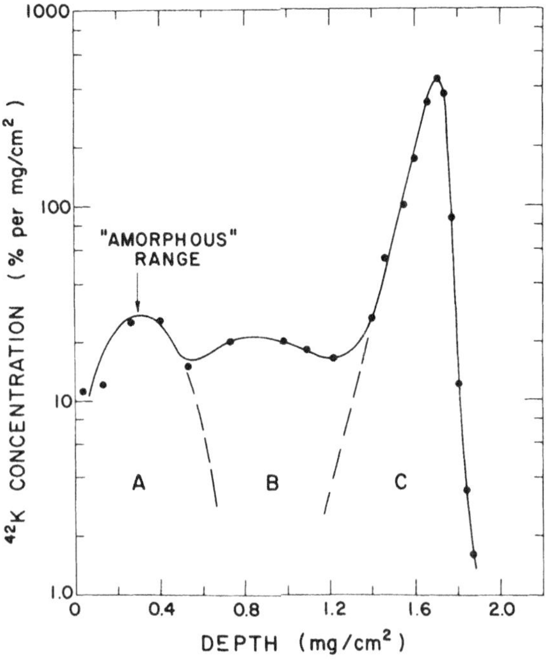

Fig. 2.14. The observed range distribution of 500-keV ^{42}K ions injected at 25° C along the ⟨111⟩ direction in tungsten. (From Eriksson *et al.*[17]) Note that in tungsten, 1 mg/cm^2 is equivalent to 0.52 microns.

for example, observed a measurable R_{max} in gold at 30° K, but not at room temperature.

In well-aligned (i.e., within 0.1°) tungsten crystals,[17,60] region C is usually the dominant feature of the range distribution, particularly at energies greater than ~100 keV. In most crystals, however, dechanneling is the dominant effect at room temperature even for particles starting out in category C; it causes most of them to end up somewhere in the intermediate region B between the two peaks of the range distribution (cf. Fig. 2.1). The resulting distribution then appears as a roughly exponential tail with an eventual cut-off at R_{max}; in some cases, the distribution falls to an immeasurably low level long before reaching R_{max}, so that the cut-off is not actually observed. Several such distributions are illustrated in Fig. 2.15.

Even in an "ideal" crystal such as tungsten, there is a significant amount of dechanneling involved, and the observed distribution is quite sensitive to many factors, such as lattice temperature, surface contamination, small misalignment, or angular divergence—some of which are often difficult to control. The amount of dechanneling also depends very much on how far the ions must travel to reach R_{max}—i.e., on the magnitude of the electronic

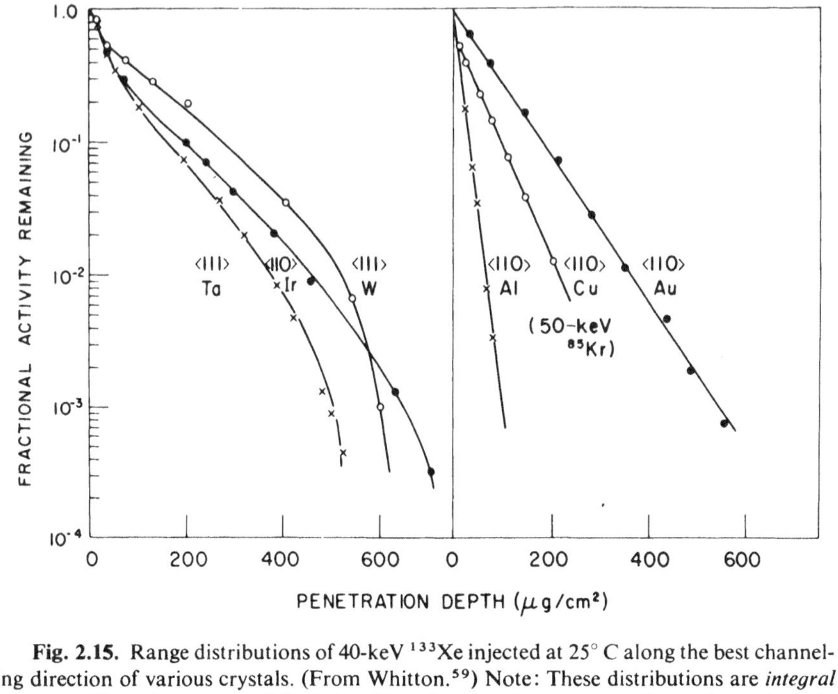

Fig. 2.15. Range distributions of 40-keV ^{133}Xe injected at 25° C along the best channeling direction of various crystals. (From Whitton.[59]) Note: These distributions are *integral* range profiles and must be differentiated to give concentration profiles such as those in Figs. 2.14 and 2.16.

stopping power. Thus ^{42}K, for example, exhibits a much more pronounced channeled peak in tungsten than does ^{64}Cu. The latter ion has a much smaller $-dE/dx$ value [cf. Fig. 2.22(a)], hence a larger R_{max} than ^{42}K, and so exhibits a greater amount of dechanneling.[17]

Let us look briefly at the more important factors governing the range distribution of channeled ions.

a. *Dependence on Increased Temperature.* The effect of lattice vibration is illustrated in Fig. 2.16 for 500-keV ^{42}K ions injected into tungsten along the ⟨111⟩. At 250° C, the number of particles appearing in the channeled peak C is a factor of 2–3 less than at 25° C, presumably because of the increased scattering arising from the larger mean vibrational amplitude (0.115 Å versus 0.085 Å, respectively). A similar temperature effect has also been reported in silicon by Dearnaley *et al.* [Fig. 2.17(a)]. As one might expect, there is no significant change in the fraction of particles in the amorphous peak A.

b. *Dependence on Bombardment Dose.* The total ion dose can strongly influence the concentration profile in semiconductor implantations made at *room temperature.* The incident ions create a large amount of disorder (cf.

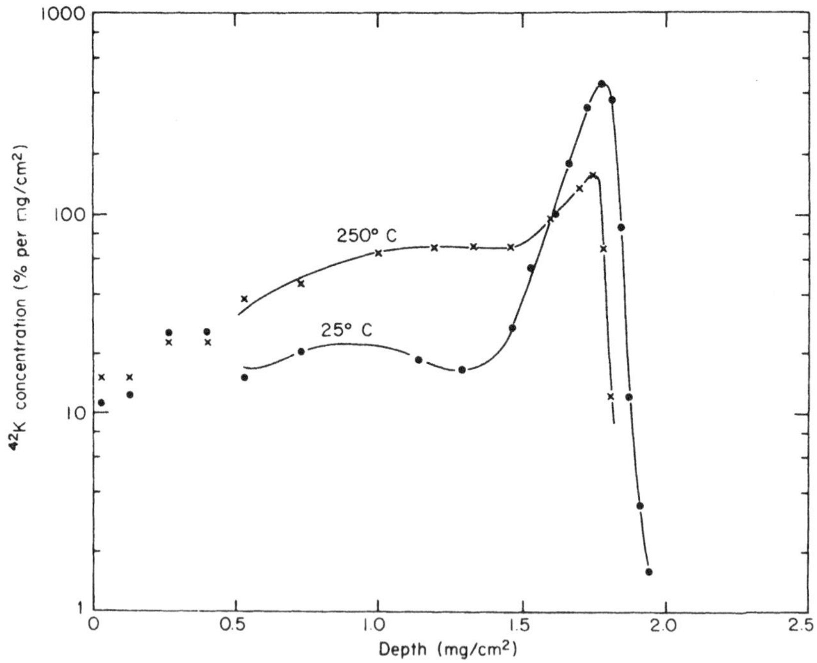

Fig. 2.16. The influence of target temperature on the range distribution of 500-keV ^{42}K injected along the ⟨111⟩ direction in tungsten. (From Eriksson.[60])

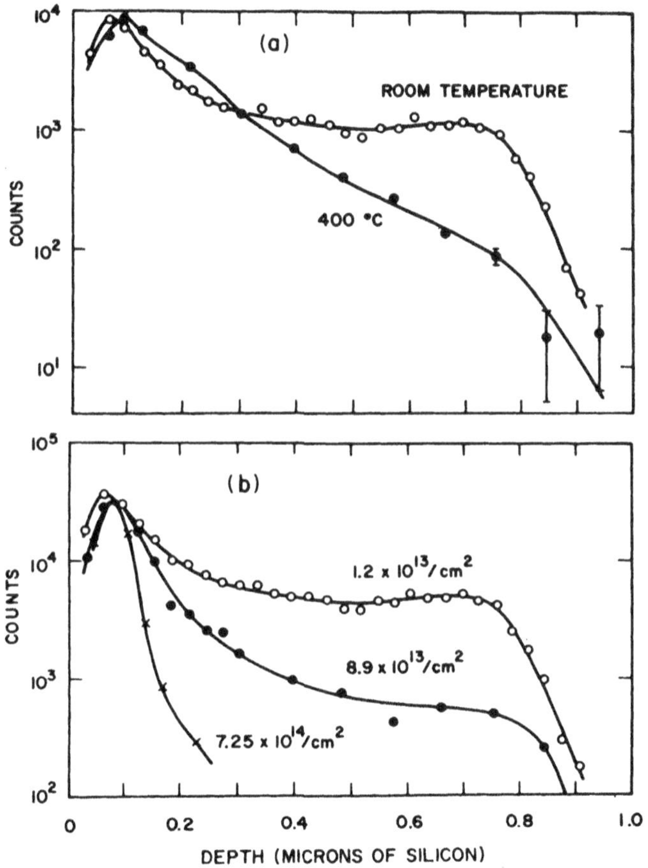

Fig. 2.17. Concentration profiles of ^{32}P implanted in silicon at 40 keV along the ⟨110⟩ direction (from Dearnaley *et al.*[18]): (a) distribution as a function of substrate temperature during implantation for the low-dose (1.2×10^{13} ions/cm^2) implant; (b) distribution as a function of total dose, normalized to the concentration maximum.

Chapter 3) and hence destroy the lattice structure necessary for channeling. The earlier work of Davies *et al.*[5] showed that doses above $\sim 10^{13}$ xenon ions/cm^2 produced a drastic reduction in the fraction of channeled ions; this can also be clearly seen in the recent work of Gibbons[61,62] and Dearnaley.[18] For example, Fig. 2.17(b) shows the influence of dose on the distribution of 40-keV ^{32}P ions in silicon for implantation along the ⟨110⟩ direction. The concentration profiles have been normalized to the same peak height in order to show how the *relative* number of channeled particles decreases with increasing dose. The *total* number of particles penetrating beyond the amorphous range obviously cannot decrease, but at doses greater than $\sim 10^{13}$

ions/cm^2 it levels off toward an almost constant value, indicating that the lattice structure has been essentially destroyed. Whitton *et al.*[63] have recently observed a similar dose dependence in the range profile of ^{85}Kr in gallium arsenide.

It should be noted that this sensitivity to bombardment dose does not occur in all crystals. In tungsten, for example, dose levels in excess of 10^{16} ions/cm^2 at 25° C are required before drastic changes in the range distribution are observed (Fig. 2.18). It does, however, seem to be a characteristic feature of all semiconductor lattices studied to date.

c. *Dependence on Misalignment and Surface Disorder.* Figure 2.19 shows the effect of a 7° misalignment and also of a surface oxide layer (~200 Å thick) on the measured implantation profile of ^{125}Xe in a ⟨110⟩ silicon crystal. In both cases, the number of well-channeled particles is reduced by some orders of magnitude. It should be noted that in this experiment, we are not introducing new effects, but are merely increasing (by about a factor of 10) two effects

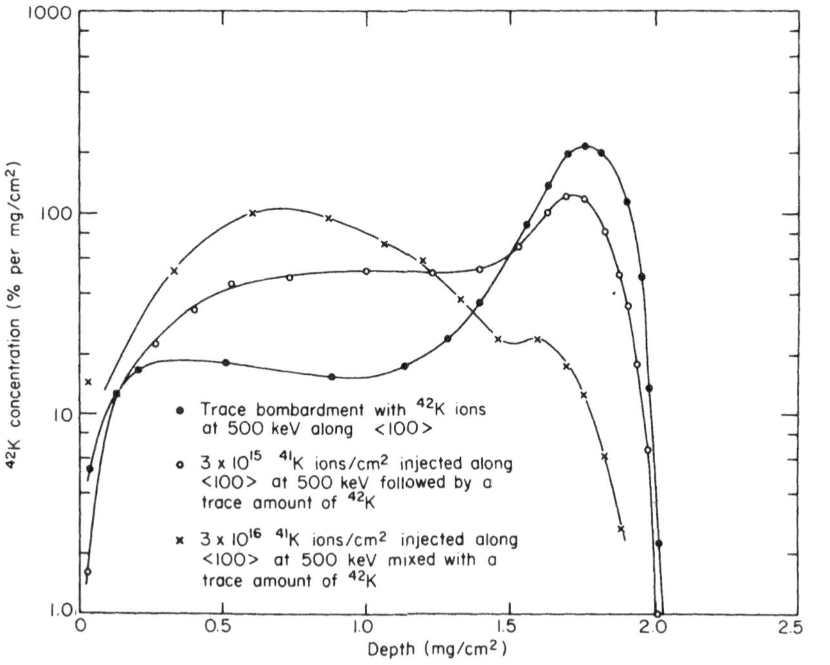

Fig. 2.18. The influence of pre-bombardment with a stable isotope (^{41}K) on the distribution of radiotracer ^{42}K ions in tungsten. (From Eriksson.[60]) Both bombardments were made at 500 keV along the ⟨100⟩ direction.

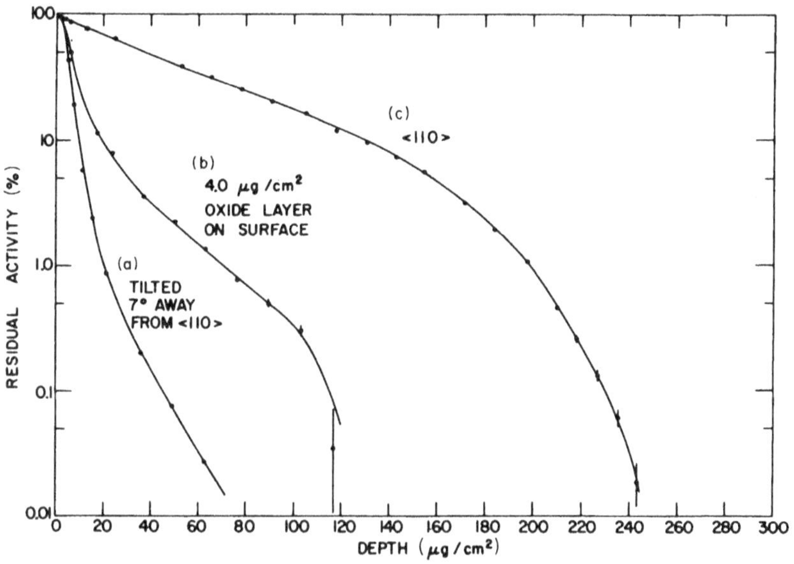

Fig. 2.19. Effect of misalignment and surface disorder on the range distribution of 40-keV ^{125}Xe injected into ⟨110⟩ silicon at 25° C. (From Davies *et al.*[5]) (a) The crystal was tilted so that the beam entered at 7° from the ⟨110⟩ direction. (b) The crystal had an anodic oxide layer 4.0 μg/cm² (~200 Å) on its surface. (c) The profile obtained in a well-aligned *un*anodized crystal is included for comparison. Note: All of these curves are *integral* range profiles and must be differentiated to obtain *concentration* profiles such as those in Figs. 2.14, 2.16–2.18, and 2.20.

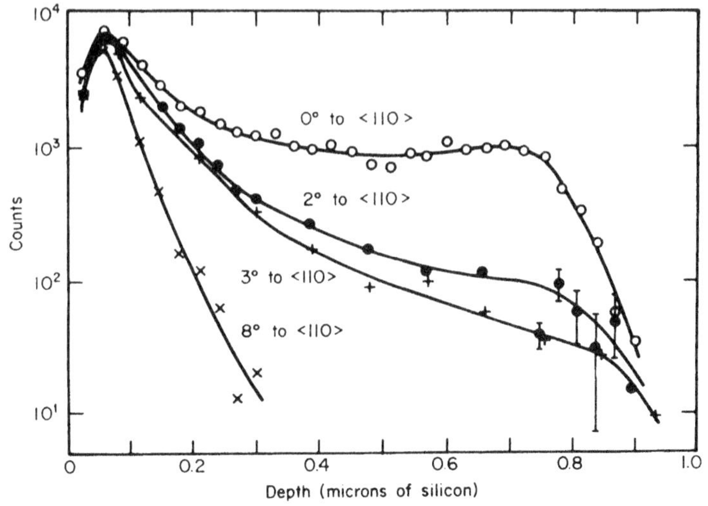

Fig. 2.20. Concentration profiles of ^{32}P implanted in silicon at room temperature as a function of beam alignment with the ⟨110⟩ direction. (From Dearnaley *et al.*[18]) Implantations were carried out at 40 keV and for doses $\lesssim 10^{13}$ ions/cm².

that are present in most implantations. The drastic change in the observed penetration profile indicates that even an angular spread (or misalignment) of 0.5°, or the presence of some tens of angstroms of surface oxide, would affect significantly the shape of a *channeled* range distribution. Dearnaley *et al.*[18] found it desirable to maintain beam divergence and beam-to-substrate alignment at less than 0.1° in order to obtain reproducible profiles. To orient a crystal this accurately with respect to the beam direction, there are several simple techniques available based on the well-defined channeling behavior of high-energy protons.[64,65] With some of these, it is even possible to use the same accelerator as in the subsequent ion implantation, thereby permitting the crystal to be oriented *in situ* (cf. Section 4.2.2).

It might seem comparatively easy to avoid any influence of channeling in semiconductors, viz. by orienting the beam several degrees away from any low-index axial direction. However, this is not always the case. First, planar channeling (or higher-order axial channeling) cannot be neglected; therefore, one must be very careful in choosing a "random" alignment. Second, the electrical characteristics of implanted *p-n* junctions are determined by the implantation profile in the region where the dopant concentration is comparable to the background dopant level in the substrate. In this region, the concentration levels involved may be several orders of magnitude below the maximum concentration of implanted atoms, unless low-dose implants or low-resistivity (high-dopant-level) substrates are used. Hence, even a small fraction of the beam scattered *into* a channeling direction might suffice to determine the junction depth. For example, Dearnaley *et al.*[18] found that, by carefully tilting out 8° in a direction that did not correspond with any major channeling plane, they were able to suppress most of the influence of channeling (Fig. 2.20). However, there was still a *tail* in the distribution that in high-resistivity material might establish the junction depth at 0.2–0.4 micron, instead of near the amorphous range of 0.05 micron.

2.4.3 The Maximum Range R_{max}

At present, the most reproducible experimental parameter for characterizing range distributions in single crystals is the maximum range R_{max}. Studies in tungsten[17,60,66,67] over a wide range of experimental conditions have established that the measured value of R_{max} depends only on the choice of crystal, the crystal direction, the projectile, and its energy. It does *not* vary significantly with bombardment dose, lattice temperature, surface contamination, small misalignment, or beam divergence (cf., for example, Figs. 2.16 and 2.18). On the other hand, the *number* of particles approaching R_{max}—i.e., the overall *shape* of the range distribution—is quite sensitive to all these factors.

A detailed investigation of the energy dependence of R_{max} for various projectiles in tungsten[17] has shown that R_{max} approximates very closely the $E^{0.5}$-dependence characteristic of electronic stopping (cf. Fig. 2.5) down to energies of a few keV, thus confirming that for a well-channeled ion, electronic stopping is normally the dominant mechanism of energy loss. This is illustrated in Fig. 2.21, in which the effective stopping cross section S of a well-

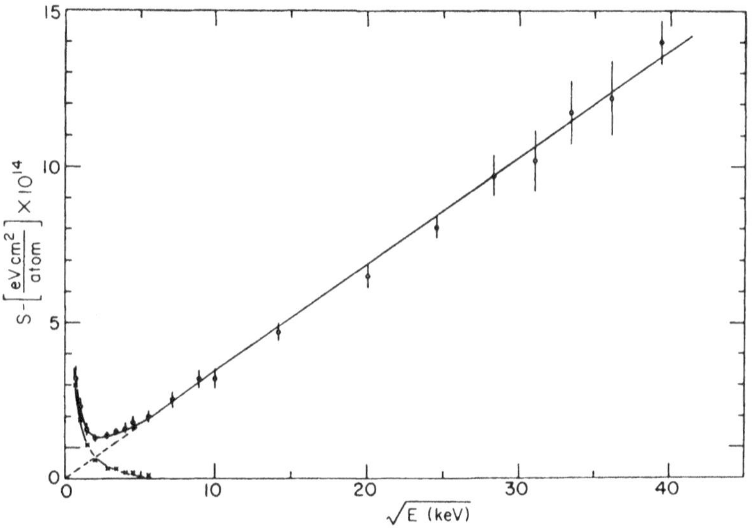

Fig. 2.21. Experimentally derived values of the total stopping cross section S versus $E^{1/2}$ for well-channeled xenon ions injected into tungsten along the ⟨100⟩ direction. (From Eriksson *et al.*[17]) The dotted line is an extrapolation of the electronic-stopping contribution S_e to lower energies. Crosses indicate the nuclear stopping S_n obtained by subtracting the extrapolated electronic stopping from the measured total values.

channeled beam, obtained by differentiating the experimental R_{max} versus E curve, is plotted as a function of $E^{0.5}$. For such a channeled beam, nuclear stopping is evidently negligible, except at energies below ~5 keV. For a *non*channeled xenon beam in tungsten, on the other hand, the LSS theory predicts that the transition energy ε_2 (Fig. 2.5) between nuclear and electronic stopping occurs at 2.7 MeV—i.e., almost a factor of 10^3 higher in energy. This leads to the observation that for well-channeled ions of equal *energy*, the higher the mass the lower the velocity, and hence the lower the stopping power. Therefore, contrary to the case for amorphous ranges where nuclear stopping usually dominates, channeled particles of high mass penetrate deeper than lower-mass particles (cf. Fig. 2.25).

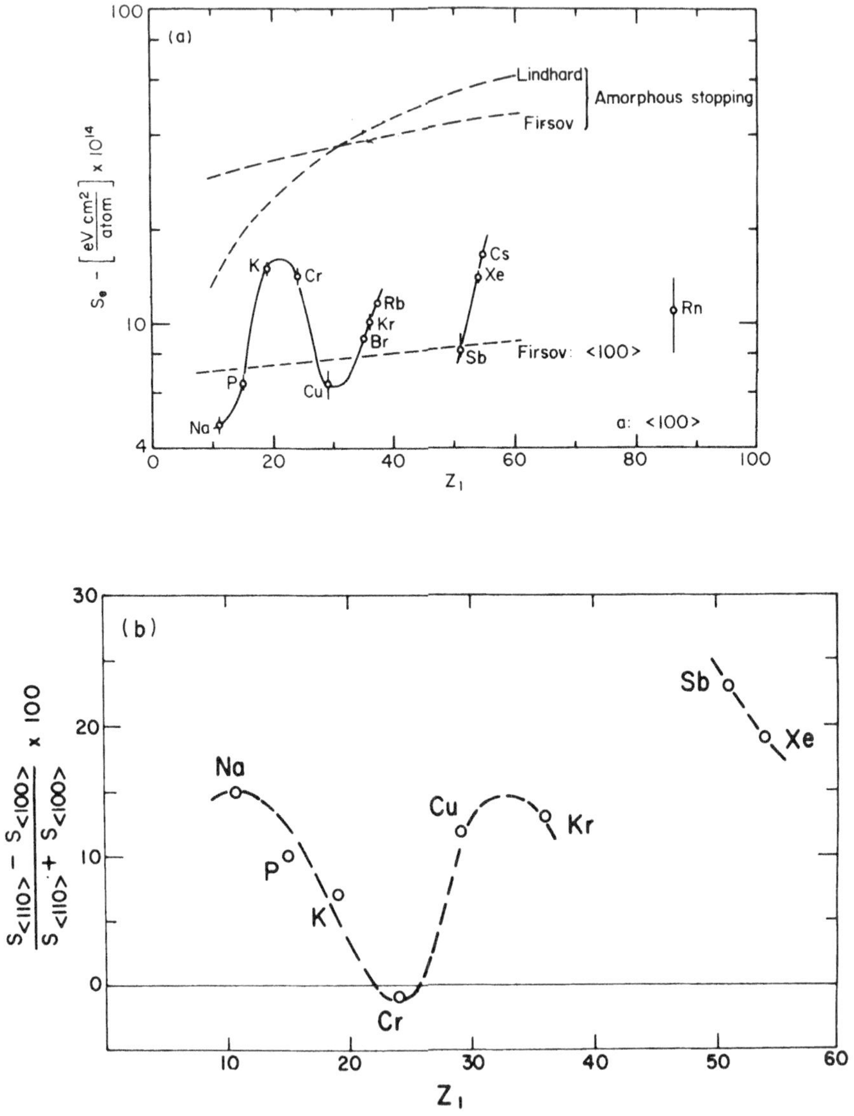

Fig. 2.22. Z_1-oscillation effect in tungsten crystals: (a) The electronic stopping cross section S_e (measured at a constant velocity $v = 1.5 \times 10^8$ cm/sec) versus the atomic number Z_1 of the projectile for the $\langle 100 \rangle$ direction in tungsten. (From Eriksson *et al.*[17]) The curve labeled "Firsov: $\langle 100 \rangle$" is calculated from the electronic-stopping equation given by Firsov[69]. (b) Relative difference in S_e values between the $\langle 100 \rangle$ and $\langle 110 \rangle$ directions in tungsten. (Derived from the data of Eriksson *et al.*[17])

2.4.4 Z_1 Oscillations in R_{max}

There is one serious difficulty in attempting to predict R_{max}, as can be seen from Fig. 2.22(a), in which the electronic stopping cross section S_e of a well-channeled ion is plotted at constant velocity as a function of the atomic number Z_1 of the implanted atom. Not only are the experimentally observed values of S_e significantly smaller than those for an amorphous stopping medium, but also they exhibit very large oscillations about a smooth Z_1 dependence. These oscillations, which in amorphous targets never exceeded $\pm 50\%$ (Section 2.3.3), are evidently much enhanced for a channeled beam. Eisen's recent measurements in silicon,[68] for example, show peak-to-valley oscillations greater than a factor of 10. He has carried out a series of measurements (Fig. 2.23) of the minimum energy loss of channeled ions in passing through thin (0.2–1.5-micron) silicon wafers. The positions of the maxima and minima in stopping power are close to those observed in tungsten [Fig. 2.22(a)] and in amorphous carbon (Hvelplund and Fastrup[48]); hence, the nature of the oscillations does not seem to depend strongly on the target material.

Large Z_1 oscillations are also noted in the relative difference of stopping powers between the various low-index directions [Fig. 2.22(b)]. A similar effect was seen by Eisen in silicon (Fig 2.23). He also found that the energy loss of well-channeled ions is not always proportional to $E^{0.5}$. By fitting his stopping-power data to a relation of the type $S_e = kE^p$, Eisen observed marked Z_1 oscillations in the energy exponent p around the predicted value of 0.5, as shown in Fig. 2.24. The maxima and minima in the p versus Z_1 curve correspond rather closely to the minima and maxima respectively in the S_e versus Z_1 relation.

For nonchanneled beams, the effect of such stopping-power oscillations on the *range* distribution is markedly reduced (Table 2.5) by the dominant contribution of nuclear stopping over most of the energy region of interest. For channeled beams, however, nuclear stopping is negligible down to very low energies (Fig. 2.21), and so the R_{max} values will exhibit Z_1 oscillations comparable in magnitude to those in Figs. 2.22 and 2.23—i.e., oscillations of up to a factor of 10.

The theoretical curves labeled "Firsov" in Fig. 2.22(a) are calculated from a velocity-proportional stopping-power equation derived by Firsov[69]; his treatment differs considerably from that used by LSS,[2] but has the advantage of predicting the dependence of S_e on impact parameter; hence, it can be used to estimate S_e for a well-channeled ion. As can be seen in Fig. 2.22, Firsov's treatment predicts resonably well the "average" attenuation in S_e in going from the amorphous to the well-channeled case. Such a treatment, however, is unable to predict the Z_1 oscillations.

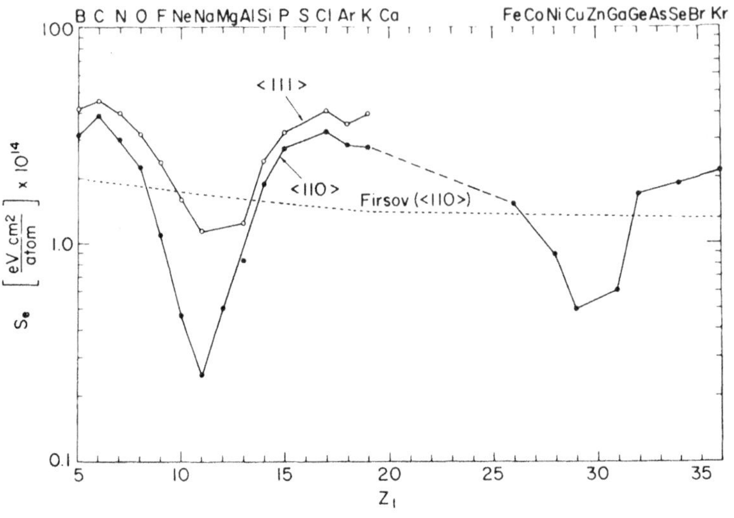

Fig. 2.23. Z_1 dependence of S_e for well-channeled ions injected at 100 keV along the ⟨110⟩ in silicon. (From Eisen.[68])

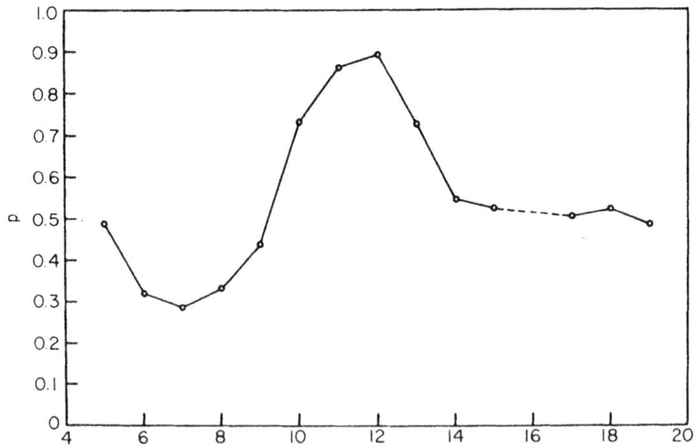

Fig. 2.24. Values of p obtained by Eisen[68] in fitting his observed S_e versus E data to an equation of the form $S_e = kE^p$. (From Eisen.[68])

Recently, several theoretical groups[70–73] have tackled the question of these Z_1 oscillations and have shown that the positions of the maxima and minima can be correlated roughly with ion size. At the same time, Lindhard and Finnemann[74] have developed a quite different theoretical approach to the question of the Z_1 oscillations that also appears to fit reasonably well the observed positions of the maxima and minima. Unfortunately, none of the theoretical treatments is yet able to predict accurately the magnitude of the oscillations, nor their dependence on energy or on channel size [cf. Fig. 2.22(b)]. It is hoped that further theoretical work will clarify the situation. In the meantime, the available evidence suggests that R_{max} is best obtained from experimental measurements.

2.4.5 Range Distributions in Silicon†

Experimental determinations of R_{max} versus energy in silicon are shown in Fig. 2.25. The measurements were obtained from radiotracer stripping profiles by Dearnaley *et al.*[18] and from measurements of the junction depth by Eisen,[68] Ruth and Eisen,[75] and Galaktionova.[76] The phosphorus R_{max} versus E curve follows an $E^{0.5}$ dependence, in agreement with Eisen's earlier $-dE/dx$ result that for phosphorus $S_e \propto E^{0.5}$. It is also clear that the range of 200-keV antimony is greater than that of arsenic, which in turn is greater than that of phosphorus. This is consistent with the nature of electronic stopping, in that slower ions of the same energy have a lower stopping power. The fact that the boron ranges lie between the ⟨110⟩ and ⟨111⟩ phosphorus ranges is attributable to the oscillatory nature of the stopping powers, since boron lies approximately at the first *maximum* in S_e (Fig. 2.23). The R_{max} values for ⟨110⟩ phosphorus and for ⟨110⟩ and ⟨111⟩ boron are in agreement with those derived by Eisen[49] from his experimental values of S_e. However, the ⟨111⟩ phosphorus values shown in Fig. 2.25 are ~40% less than those derived from the earlier stopping-power data.[49] The origin of this discrepancy is not known.

Dearnaley *et al.*[18] have studied the effect of misalignment, ion dose, energy, and lattice temperature on the depth distribution of channeled phosphorus ions in silicon (for example, Figs. 2.17, 2.20). Their differential range distributions were obtained by anodically oxidizing the implanted samples and then measuring the ^{32}P activity of each layer of oxide removed (Section 2.2.1). The depth distributions for 40-keV room-temperature implantations of $\lesssim 10^{13}$ ions/cm^2 along the ⟨110⟩ and ⟨111⟩ directions are shown in Fig. 2.26. Note that there is *not* a well-defined channeled peak in the

† At present, silicon is the only semiconductor lattice that has been extensively studied. Whitton *et al.*[63] have recently measured ^{85}Kr range profiles in gallium arsenide and are currently extending this investigation to other implanted ions.

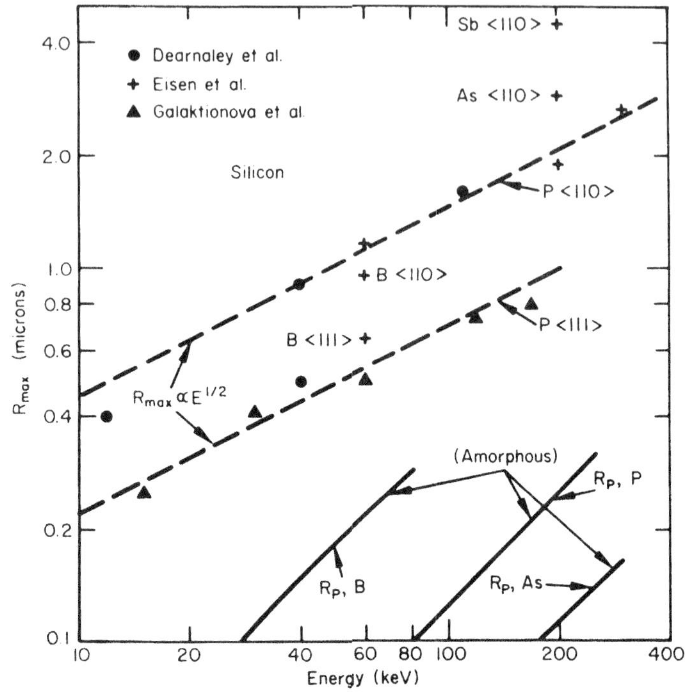

Fig. 2.25. Experimental measurements of R_{max} versus E for well-channeled ions in silicon. (From Mayer and Marsh.[58]) The dashed lines, fitted to the phosphorus data, represent an $E^{0.5}$ dependence. Projected ranges R_p of boron, phosphorus, and arsenic in amorphous silicon are shown by solid lines. (Data from Dearnaley *et al.*[18] Eisen,[68] Ruth and Eisen,[75] and Galaktionova.[76])

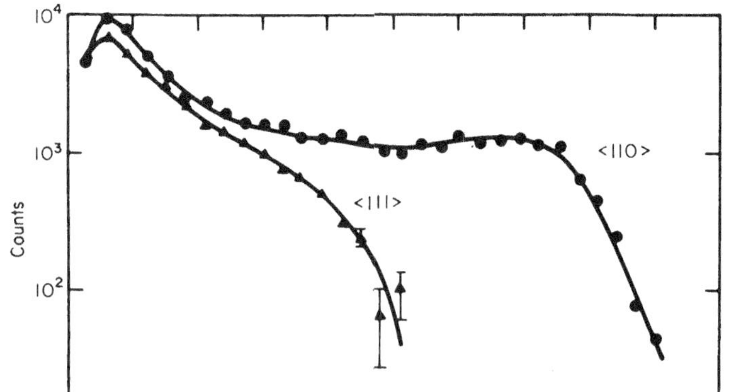

Fig. 2.26. Concentration profiles of ^{32}P implanted in silicon at room temperature for doses $\lesssim 10^{13}$ ions/cm^2 at 40 keV. (From Dearnaley *et al.*[18]) Beam alignment relative to the $\langle 110 \rangle$ direction was $\pm 0.1°$.

distribution, nor are the values of R_{max} as well defined as in tungsten (Figs. 2.14 and 2.16). Presumably, this difference can be attributed at least in part to the larger mean vibrational amplitude (and hence to a larger rate of dechanneling) in the silicon case. In a later study,[76a] Dearnaley *et al.* have extended these range investigations to several other projectiles—^{24}Na, ^{35}S, ^{64}Cu, and ^{85}Kr. They observe that the maximum range is strongly influenced by the ion size, thus providing further evidence for the oscillatory Z_1 dependence of electronic stopping.

With poorer alignment and at higher doses, the channeling effect becomes much less pronounced. This is clearly illustrated in the implantation profiles in Fig. 2.27† taken from the recent work of Kleinfelder *et al.*[62] These implantations were performed with the substrate heated to 625° C in order to minimize the accumulation of lattice disorder (Chapter 3); at the same time, the increased amplitude of the lattice vibrations increases markedly the rate of dechanneling [as seen in Figs. 2.16, 2.17(a)]. Note how widely the implantation profiles in Fig. 2.27 differ for the various dopants. Boron and nitrogen fit the LSS *amorphous* stopping curve reasonably well, although definite indications of a channeled *tail* may be seen in the boron profile. Phosphorus, however, gives a completely different profile,‡ viz. an approximately exponential distribution over several decades in concentration, with an immeasurably small fraction reaching the R_{max} region. This type of distribution is the one most commonly observed in other crystals—for example, in aluminum,[78] copper,[79] or gold.[80] The arsenic case [Fig. 2.27(d)] is particularly puzzling in that the observed tail appears to exhibit an easily recognizable maximum range around 0.25 microns, whereas the value calculated from Eisen's $-dE/dx$ measurements is much larger, viz. ~ 1.5 microns. One possible explanation for this discrepancy, and also for differences in profile between the various dopants, may lie in the fact that Kleinfelder's measurements were made by the junction-staining technique outlined in Section 2.2.4. As noted earlier, this technique measures the depth distribution of the charge carriers associated with the *electrically active* dopant atoms, rather than the depth distribution of all the implanted atoms. These two distributions are not necessarily identical. Furthermore, the electrical measurements discussed in Chapter 5 show that in hot-substrate implantations, anneal temperatures $\gtrsim 800°$ C are sometimes required to achieve the maximum number of carriers per cm^2.

Roosild *et al.*[81] have obtained range distributions of B, N, and P ions in silicon at much higher energies, viz. 1–2.5 MeV, using the junction-staining

† This is also illustrated in the profiles obtained recently by Crowder and Fairfield[76] for 280-keV phosphorus, arsenic, and antimony ions in silicon, using the technique of neutron activation followed by anodic oxidation and stripping.

‡ A similar range profile for ^{32}P in silicon has been observed by Manchester.[77] He has also investigated the dose dependence of the exponential tail.

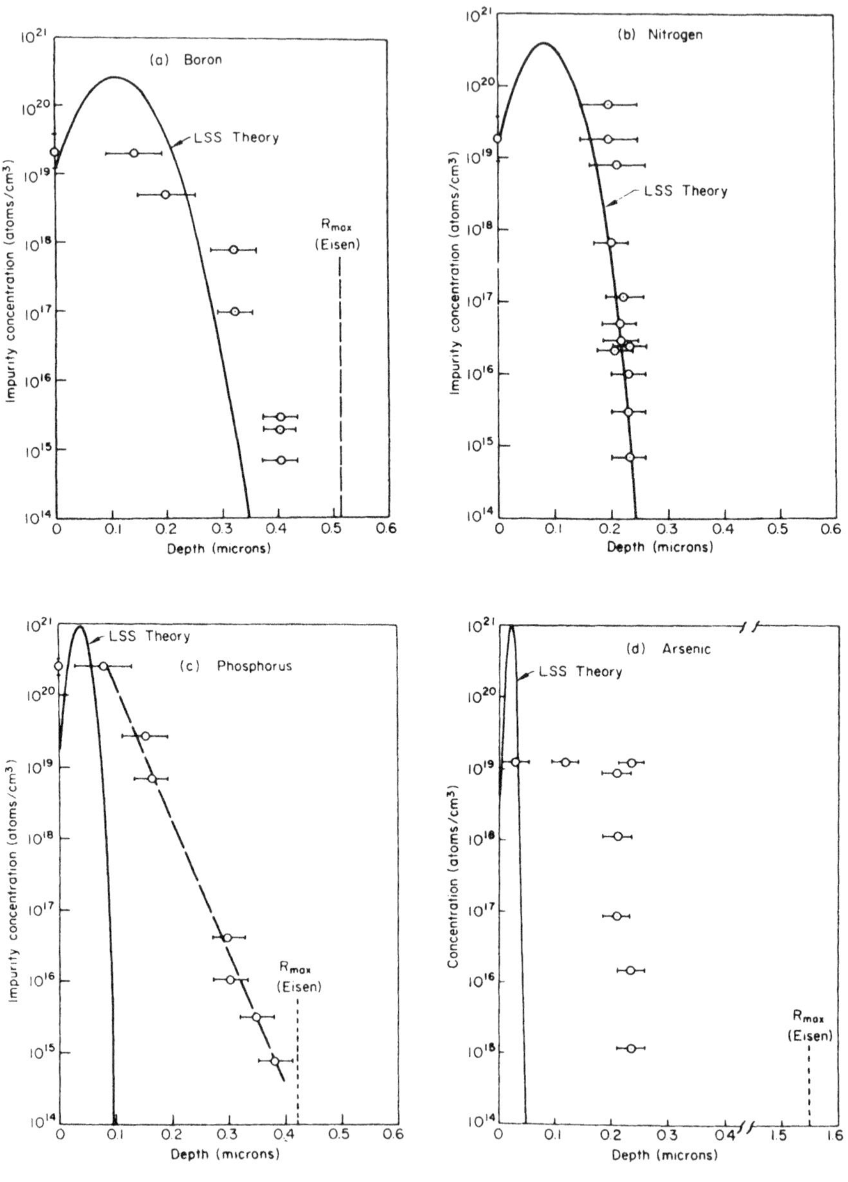

Fig. 2.27. Distribution of electrically active dopant ions implanted at 30-keV and 625° C in silicon within ±2° of the ⟨111⟩ direction: (a) boron, (b) nitrogen, (c) phosphorous, (d) arsenic. (Based on the data of Kleinfelder *et al.*[62]) Implantation dose: 3×10^{15} ions/cm^2, except in (d) where the dose was 1.5×10^{15} ions/cm^2. In each case, the solid curve is the LSS predicted distribution for amorphous stopping, and the dotted line indicates the R_{max} value calculated from Eisen's[49] stopping-power measurements.

technique. Their range profiles agree reasonably well with those predicted for amorphous targets by the LSS theory, particularly if the Z_1-oscillation effect (Table 2.5) is included. No attempt was made to orient their crystals better than "within a few degrees of the ⟨111⟩"; hence, channeling effects were not observed. A similar study of the range of high-energy implants (0.2–1.8-MeV boron, phosphorus, and arsenic ions) has recently been reported by D. E. Davies,[31, 82] using a differential-capacitance technique. The crystals were deliberately *mis*aligned to minimize channeling effects; again, the observed ranges agree reasonably well with the LSS predictions for amorphous targets. Profiles obtained[83] by junction staining of boron and phosphorus implantations at 78° K in silicon samples tilted $\sim 7°$ from the ⟨111⟩ are shown in Fig. 2.28. Again, the peak positions agree well with the R_p values predicted by LSS for amorphous targets. In the phosphorus implants, the small tail in the observed distribution was attributed to enhanced diffusion; however, these tails do not extend beyond the R_{max} values for channeled phosphorus atoms (Fig. 2.25).

In summary, it is clear that channeling phenomena can have a dominant influence on the distribution of implanted ions in semiconductors. In such cases, channeling may affect not only the location of the junction in an implanted *p-n* device, but also the electrical characteristics of the implanted layer. Unfortunately, the channeled part of a range profile is sensitive to many factors that may be difficult to control: For example, in most implantations a dose in excess of 10^{13} ions/cm^2 is desired; at room temperature this dose causes enough lattice disorder to markedly reduce the channeled distribution [Fig. 2.17(b)], whereas at elevated temperature (where the disorder can be continuously annealed during the implantation) the enhanced vibrational amplitude of the lattice atoms has an appreciable effect [Figure 2.17(a)]. Hence, the observed distribution is not easily reproducible, nor can it be accurately predicted. In many applications, it is therefore desirable to suppress the channeled component as much as possible. This may be achieved by carefully aligning the crystal so that the beam enters several degrees off-axis (cf. Figs. 2.19, 2.20). In semiconductors, if the dose is large enough, it may become suppressed still further by the accumulation of lattice disorder during the implantation [Fig. 2.17(b)]. One can indeed suppress the channeled *tail* completely by prebombarding the crystal with some electrically inactive ion species (such as neon or, better still, silicon) in order to destroy the crystal structure before injecting the dopant atoms.[61]

2.5 Enhanced Diffusion

A number of experiments indicate that enhanced diffusion effects can sometimes influence the depth profile of implanted dopant species. Two mechanisms have been clearly identified. One is radiation enhancement of the

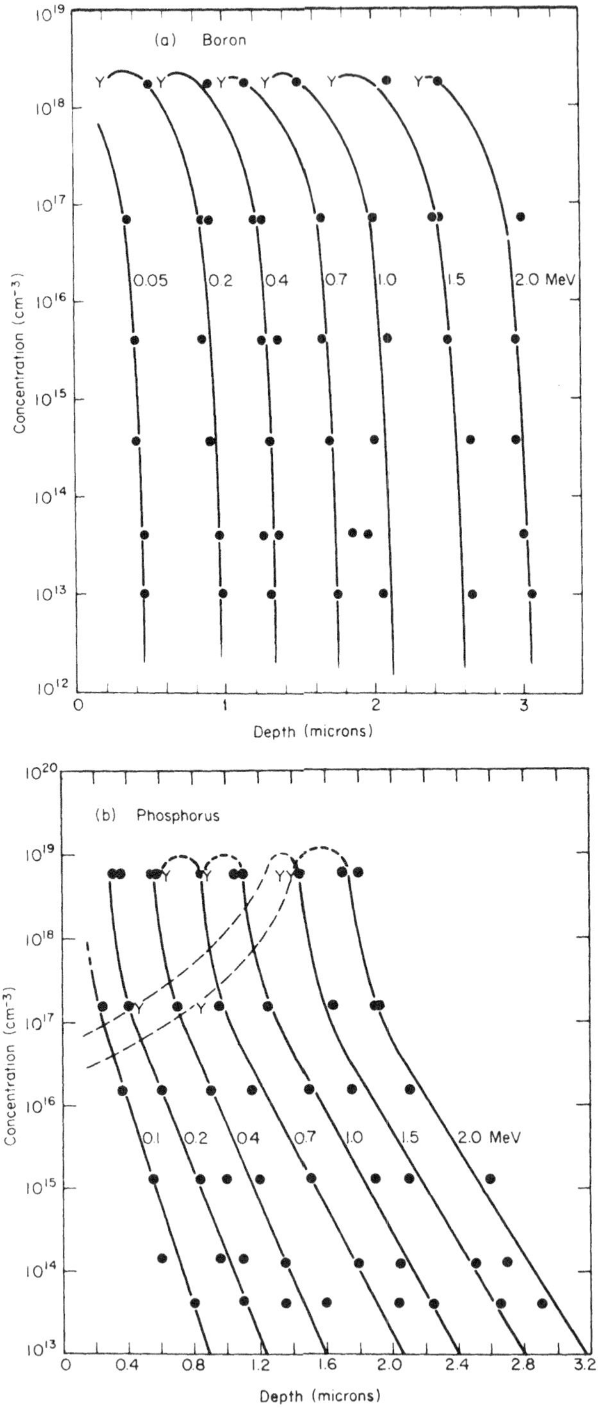

Fig. 2.28. Profiles for (a) boron and (b) phosphorus ions implanted at various energies into silicon at $-190°$ C, and subsequently annealed at $650°$ C. (From D. E. Davies.[83]) Implant conditions: 10^{15} ions/cm^2 at 7° from the $\langle 111 \rangle$ direction.

diffusion constant due to the increase in the vacancy concentration. This effect can be seen at temperatures a few hundred degrees below the normal thermal-equilibrium diffusion temperature regime. The other mechanism is attributed to the diffusion of an interstitial component of the implanted species, and occurs even at much lower temperatures. The magnitude of this effect is sensitive to the concentration of defect centers which could "trap" the interstitial component.

2.5.1 Radiation-Enhanced Diffusion

Several groups[84–87] have indicated that a radiation-enhanced diffusion may be explained by the generation and diffusion of vacancies created during irradiation with energetic protons. A most striking indication of this effect is shown schematically in Fig. 2.29. Silicon (or germanium) slices containing a thermally diffused junction are irradiated at temperatures between 600° and 1200° C with protons of 0.2–1.0-MeV energy. The penetration depth of the protons is less than the original junction depth. In the portion of the sample immediately beneath the irradiated region, a significant increase in junction depth occurs, as can be seen in Fig. 2.29. For example, in germanium[87] an irradiation of $\sim 10^{18}$ protons/cm^2 at 1.5 MeV and 600° C resulted in a 0.77-micron increase in junction depth. The movement of the junction is dose-dependent, and the detailed behavior depends on the irradiation temperature. In the high-temperature range (above 600° C in germanium) and for low dose rates, the increase ΔD in the impurity diffusion coefficient is proportional to the total radiation dose and is independent of temperature. At lower irradiation temperatures, ΔD is dependent on both total dose and temperature. The physical mechanisms responsible for the observed enhanced diffusion are not simple. Pfister[86] proposes that the creation of point defects leads to an enhanced substitutional diffusion via a vacancy mechanism.

Glotin[88] finds a similar enhanced diffusion for phosphorus and boron implantations in silicon at substrate temperatures above 650° C. Figure 2.30

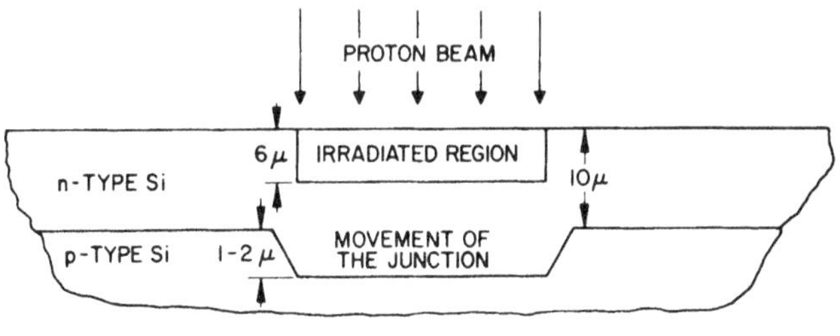

Fig. 2.29. Schematic representation of the movement of a *p-n* juction following proton irradiation at elevated temperature. (From Pfister.[86])

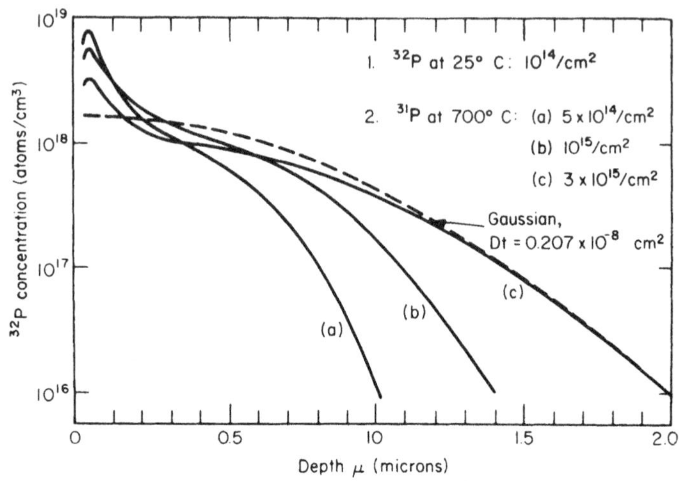

Fig. 2.30. Influence of the dose on the enhanced diffusion of ^{32}P in silicon when irradiated with ^{31}P at 700° C. (From Glotin.[88]) The procedure was to implant ^{32}P at room temperature, and then to implant the stable isotope at substrate temperatures of 700° C.

(curve b) illustrates the influence of radiation-enhanced diffusion on the measured radiotracer profile. First, Glotin implanted ^{32}P (a β-emitter) at room temperature to a dose of 10^{14} ions/cm^2; he then increased the substrate temperature to 700° C and implanted the stable isotope ^{31}P to dose levels between 5×10^{14} and 5×10^{15} ions/cm^2. He found that this second implantation (into the heated substrate) caused an increase in the depth of the first implantation. The radiotracer profile gradually approached that of a Gaussian, with a diffusion constant D of the order of 2.7×10^{-13} cm^2/sec. He noted that this effect is negligible if the target temperature is only 550° C. Obviously, the onset of radiation-enhanced diffusion sets an upper limit on the implantation temperature, if one wishes to have the dopant depth distribution determined by the dynamics of the slowing-down process. It is possible that this temperature limit will depend on the dopant species.

2.5.2 Interstitial Diffusion

A characteristic feature of experimental range distributions in tungsten has been the existence of a deeply penetrating component of the implanted species.[66,89–92] This component involves only ~0.1% of the total dose in room-temperature implants, but increases to the 5–10% level in low-temperature (30° K) implantations (Figs. 2.31, 2.32); typically, it extends at least 5–10 microns into the crystal. Davies and Jespersgård[91] established that this penetration (commonly referred to as a "supertail") is caused by a rapid

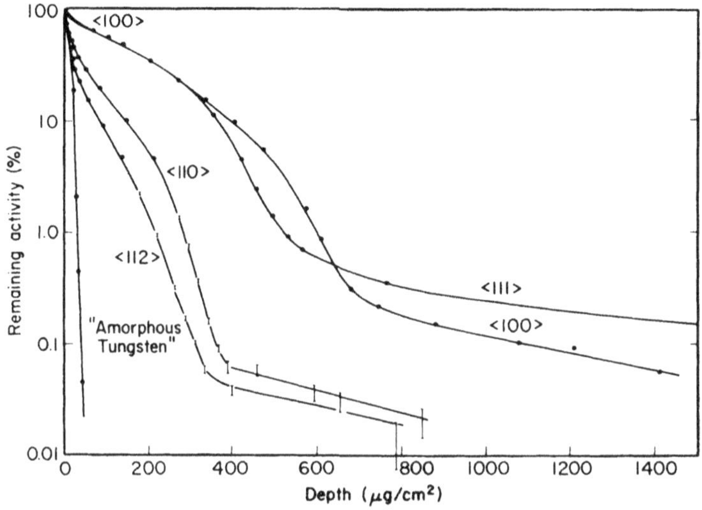

Fig. 2.31. Integral penetration distributions for 40-keV ^{125}Xe injected parallel to the indicated directions in monocrystalline tungsten. (From Domeij *et al.*[89]) Note that in tungsten, 1 mg/cm^2 is equivalent to 0.52 micron.

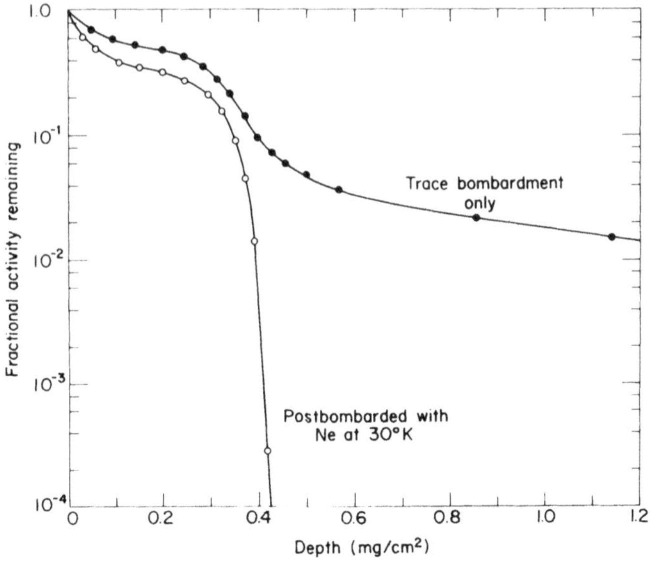

Fig. 2.32. Integral penetration of 40-keV ^{42}K injected into tungsten along the ⟨100⟩ direction at 30° K: ●—^{42}K (trace amount) injected without postbombardment; O—^{42}K (trace amount), followed by 3×10^{15} ^{20}Ne ions/cm^2 to introduce trapping centers. (From Davies *et al.*[66]) The range distributions were measured *after* the crystal had warmed to room temperature.

diffusion process occurring after the particles have lost their incident energy; they interpret this rapid diffusion process as an *interstitial* diffusion involving almost perfectly channeled particles that have come to rest without producing damage in the surrounding lattice. Such atoms must be (at least initially) in interstitial sites, and hence must diffuse interstitially until they encounter a suitable trapping center such as a vacancy, a surface, a dislocation, or perhaps another impurity atom.

Figure 2.32 shows the integral range distribution of 40-keV ^{42}K ions injected at 30° K along the ⟨100⟩ in tungsten. The range distributions were measured by the anodic oxidation and stripping technique described in Section 2.2.1, after the crystal had warmed to room temperature. It had earlier been shown by Hermann *et al.*[92] that the supertail does not occur if enough trapping centers are introduced into the crystal before allowing it to warm up. This suppression of the supertail by lattice defects is clearly illustrated in Fig. 2.32. In this case, immediately after one of the low-temperature ^{42}K bombardments, the crystal was subjected to a second bombardment at 30° K with 3×10^{15} neon ions/cm^2 to introduce a large concentration of trapping centers. Obviously, the interstitial diffusion had not occurred while the crystal was at 30° K. For the room-temperature ^{42}K bombardment, on the other hand, the diffusion process is so rapid that one obtains the "infinite-time" distribution, extending to depths of several microns, in less than one hour. Consequently, the activation energy for this interstitial motion in tungsten is between 0.5 and 0.8 eV[91]; this is almost a factor of 10 smaller than the corresponding activation energy for normal diffusion in tungsten—i.e., 4–5 eV.

In silicon, McCaldin[93] reports that a similar interstitial diffusion of ion-implanted ^{24}Na occurs at implantation temperatures above 450° C, but not at room temperature—suggesting that the activation energy for interstitial dopant motion in silicon is much higher than in tungsten. Moreover, the thallium implantation studies[94] discussed in Chapter 4 have shown that interstitial thallium atoms are not mobile in silicon below at least 650° C [Fig. 4.18(b)]. On the other hand, in the range distribution of ^{32}P implanted into silicon along the ⟨110⟩ direction, Dearnaley *et al.*[76a] observe a significant "supertail" even at room temperature. Furthermore, they find that at 77° K, the supertail can be largely suppressed by postbombardment with a large dose of ^{20}Ne ions (cf. Fig. 2.32), thus confirming that it is *not* a channeling effect.

McCaldin[95] has suggested several criteria that should be fulfilled in order to observe such an interstitial diffusion process:

1. The damage ratio (i.e., the number of vacancies produced per injected ion) must be less than unity; otherwise each injected atom has a high probability of becoming trapped in one of the vacancies it has itself created.

2. The interstitial atoms must have a reasonable lifetime in the crystal, and must not interchange readily with the lattice atoms.

3. The activation energy for interstitial *motion* must be small enough (or the lattice temperature high enough) to permit measurable diffusion of the injected interstitials to occur.

In most lattices, a negligible fraction of the beam remains sufficiently well channeled throughout the slowing-down process to fulfill condition (1). Furthermore, Andersen and Sigmund[96] have pointed out (from a theoretical consideration of the slowing-down behavior of channeled particles) that when $Z_1 > Z_2$, even perfectly channeled ions can initiate collision cascades and thus become trapped. In many implantation studies, Z_1 is considerably greater than Z_2, and hence this type of interstitial diffusion should not occur.

An interesting theoretical treatment of these interstitial diffusion effects has been advanced by Sparks.[96a] He has developed various multistream steady-state-diffusion models, in which several species—such as vacancies,

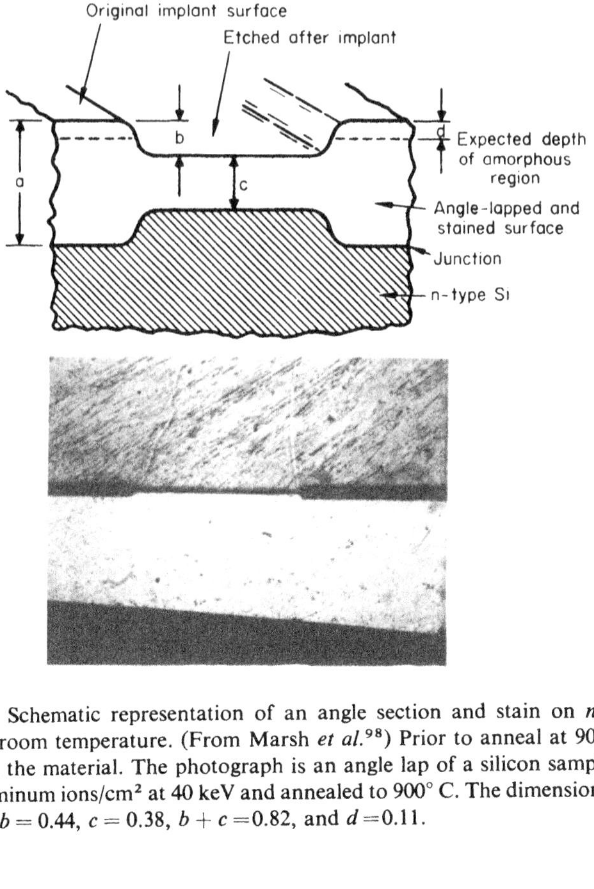

Fig. 2.33. Schematic representation of an angle section and stain on *n*-type silicon implanted at room temperature. (From Marsh *et al.*[98]) Prior to anneal at 900° C, a strip was etched in the material. The photograph is an angle lap of a silicon sample implanted with 10^{15} aluminum ions/cm^2 at 40 keV and annealed to 900° C. The dimensions in microns are $a = 1.03$, $b = 0.44$, $c = 0.38$, $b + c = 0.82$, and $d = 0.11$.

self-interstitials, and implanted ions—all diffuse rapidly enough to maintain a steady state during the implantation. He finds that the resulting ion distributions can be fitted to the experimental supertail data reasonably well. Sparks suggests that a similar multistream diffusion mechanism may also be involved in radiation-enhanced diffusion.

2.5.3 Enhanced Diffusion during Postbombardment Anneal

In aluminum implants in silicon at room temperature, a pronounced increase in junction depth has been reported at annealing temperatures well below that at which normal diffusion occurs.[97–98] Figure 2.33 shows an angle section and stain of *n*-type silicon implanted at room temperature with 10^{14} aluminum ions/cm^2 at 40 keV. Prior to anneal at 900° C for 10 min, a strip of the material was etched to a depth greater than the amorphous range R_p, but not nearly as great as R_{max}. After anneal, the junction beneath the etched region was not as deep as that beneath the *un*etched region.

This experiment shows that the increased junction depth cannot be attributed to the penetration of channeled ions to that depth. The experiments do not distinguish, however, between the diffusion of an interstitial component originating near the surface and an enhanced substitutional diffusion resulting from the release of vacancies from the heavily disordered surface region. Further work is required to clarify the nature of the physical mechanisms involved.

3

Lattice Disorder and Radiation Damage

3.1 Introduction

It has been known for many years that bombardment of a crystal with energetic (keV to MeV) heavy ions produces regions of lattice disorder. An extensive review of work in this field up to 1966 has been given by Carter and Colligon.[99] More recently, ion implantation in silicon and germanium has received increased attention. Consequently, this discussion will be based primarily upon results obtained in these two semiconductors. Although qualitative aspects of these results may be extended to the III–V semiconductors such as gallium arsenide or to the II–VI semiconductors such as cadmium telluride, there is insufficient experimental evidence to justify a unified treatment of all diamond-lattice semiconductors.

The disorder can be directly observed by techniques sensitive to lattice structure, such as electron-transmission microscopy, MeV-particle channeling, and electron diffraction. The use of these and other techniques, along with the theoretical treatment of ion interactions in solids, has provided a basis for evaluation of implantation processes. A general description can be given, even though many of the details have not yet been resolved. In addition, it has now been shown that information on specific defects produced by ion implantation can be provided by optical absorption and electron-resonance studies. These defects are found in both ion-implanted and neutron-irradiated silicon.

As an ion slows down and comes to rest in a crystal, it makes a number of collisions with the lattice atoms (Chapter 2). In these collisions, sufficient energy may be transferred from the ion to displace an atom from its lattice site. The displaced atom can in turn displace other atoms, and so on—thus creating a cascade of atomic collisions. This leads to a distribution of vacancies, interstitial atoms, and other types of lattice disorder in the region around the ion track. The precise nature of these disordered regions is not known in detail at present. As the number of ions incident on the crystal increases, the individual disordered regions begin to overlap. At some point an amorphous layer is formed.† The total amount of disorder and the distribution in depth depend on ion species, temperature, energy, total dose, and channeling effects.

There is a marked difference between the anneal behavior of the amorphous layer and that of the isolated disordered regions around the track of each particle. The isolated disordered regions anneal at markedly lower temperatures than the epitaxial reordering of the amorphous layer. If the implantations are made in substrates at elevated temperatures, then the individual disordered regions can anneal before the next ion strikes in their vicinity, thus preventing formation of the amorphous layer.

In the following sections, we will present a general picture of the theoretical developments (Part 3.2) and the experimental techniques (Part 3.3) that have been used to investigate implantation processes. The experimental results (Part 3.4) are discussed from the viewpoint of an attempt to correlate the diverse experimental evidence and also to indicate gaps in the data. In the last part (Part 3.5), a comparison is made between fast-neutron irradiation and ion implantation, since there are striking similarities in the anneal characteristics of the disorder produced by the two processes.

3.2 Theoretical Considerations

The approach we have taken is to develop a simple qualitative estimate of the disorder produced by the incident ion, with particular attention paid to those parameters that are amenable to experimental measurement. In general, we will deal with *average* quantities resulting from implantation by many atoms, i.e., the average number of displaced lattice atoms, their spatial distribution, etc. The more general aspects of radiation damage in solids have been discussed in several reviews.[99–101]

† In keeping with common usage in implantation studies, we will use the word *amorphous* to signify a noncrystalline region in which there is no long-range crystallographic ordering. However, this region may contain close-range order similar to that in crystalline silicon (i.e., tetrahedral bonding of the atoms).

3.2.1 Energy Available for Displacement Processes

An ion loses energy in both electronic and nuclear collisions as it slows down and comes to rest in a crystal (cf. Chapter 2). Only the latter process creates lattice disorder around the ion track and is responsible for radiation-damage effects. Therefore, in considering the disorder created by an incident ion, one must first determine the partition of energy between electronic and nuclear processes. A similar procedure was used in determining the range distribution of implanted atoms in Chapter 2. The difference between range and disorder calculations is that in the latter case, the energy partition of the displaced atoms must be considered also. As in the case of ion-range distributions, the crystal structure can influence the amount of energy lost in nuclear collisions. For example, a well-channeled particle loses more of its energy in electronic processes, and so creates less disorder than a particle whose initial direction of motion is not aligned with any low-order lattice axis or

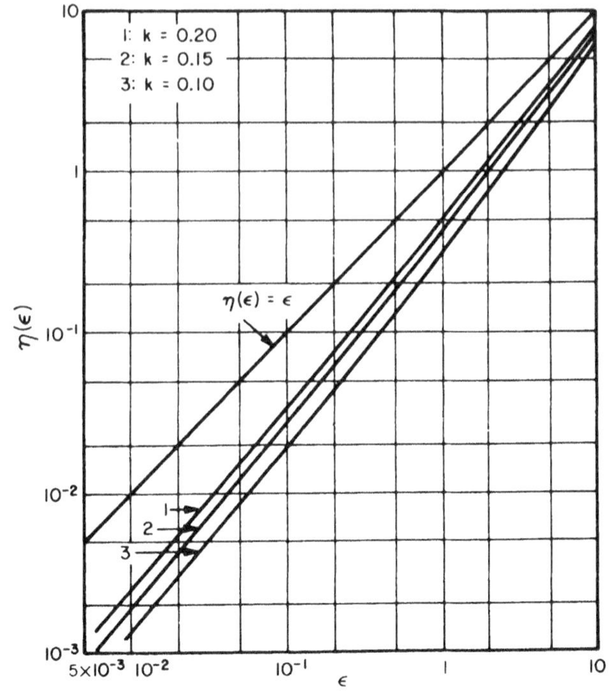

Fig. 3.1. The function $\eta(\varepsilon)$ versus ε for $Z_1 = Z_2$ and for three different values of the electronic stopping parameter k [Eq. (2.4)]. The upper line corresponds to $\eta(\varepsilon) = \varepsilon$ (i.e., pure electronic stopping). Note that both η and ε are plotted in dimensionless units. (From Lindhard *et al.*[102])

plane. However, for the moment we will exclude consideration of the influence of channeling effects.

Lindhard *et al.*[102] have derived a theoretical treatment for the distribution of energy between electronic and nuclear processes for both the primary (incident) particle and the secondary (knocked-on) particles. They consider η as the sum total of the energy given to electrons, ν as the total energy ending up in atomic motion, and $\eta + \nu = E$ the energy of the incoming particle. They present analytical calculations for the case $M_1 = M_2$ (Fig. 3.1) for $\eta(\varepsilon)$ versus ε in dimensionless units [Eq. (2.2)]. The unequal-mass case must be calculated numerically (program available at Aarhus University, Denmark). The explicit case of alpha particles in silicon is given by Lindhard *et al.*[102]

Table 3.1A gives values for the total amount of energy ν lost in nuclear collisions by energetic Group IV ions incident on silicon and germanium. These energy-loss values include the nuclear-collision contributions from the whole cascade, and not merely from the primary particle. The values for silicon in silicon and for germanium in germanium could be obtained directly from Fig. 3.1 by converting energy from keV to ε units (cf. Table 2.1). These ν values for Group IV elements can be used for the adjacent Group III or V elements without introducing significant error.

A straightforward measurement of the energy partition has been obtained by studying the response of a silicon or germanium nuclear-particle detector

TABLE 3.1A

TOTAL AMOUNT OF ENERGY AVAILABLE FOR NUCLEAR COLLISIONS[a,b]

Incident ion	Incident energy (keV)						
	1	3	10	30	100	300	1000
In silicon							
C	0.80	2.2	5.9	14	27	41	54
Si	0.83	2.4	7.3	19	51	100	170
Ge	0.84	2.5	7.7	21	63	160	370
Sn	0.85	2.5	7.9	22	68	180	460
Pb	0.86	2.5	8.0	23	70	190	530
In germanium							
C	0.74	2.0	5.8	14	30	48	65
Si	0.80	2.3	7.2	20	54	110	200
Ge	0.85	2.5	7.9	23	69	175	440
Sn	0.86	2.5	8.1	24	72	195	530
Pb	0.87	2.6	8.2	24	74	210	580

[a] Data supplied by P. V. Thomsen (Aarhus University).
[b] Z_1 *oscillations* in electronic stopping are neglected.

to energetic ions. A typical experiment is to produce energetic recoil ions in the active volume of a detector by irradiation with fast neutrons, and then compare the pulse height produced by the ion with that produced by an electron or X ray of the same energy. The difference in pulse height, commonly called the pulse-height defect, is attributed to the energy lost in nuclear collisions. Such measurements have been summarized in a review by Sattler.[103] In Fig. 3.2, Sattler plots the amount of ionization produced in silicon and germanium versus the recoil energy of ions whose mass is comparable to that of the lattice atoms. These data show that for equal ion masses of energies less than 100 keV, most of the energy loss is available for displacements.

Haines and Whitehead[104] have calculated the fraction of energy expended in nuclear collisions for energetic ions in silicon and germanium, using a simpler approach based on Bohr's treatment.[40] Their results are shown in Fig. 3.3 (solid curve) along with experimental data in terms of a pulse-height defect $\nu(\varepsilon)$ and reduced energy ε in the Lindhard notation [Eq. (2.2)].

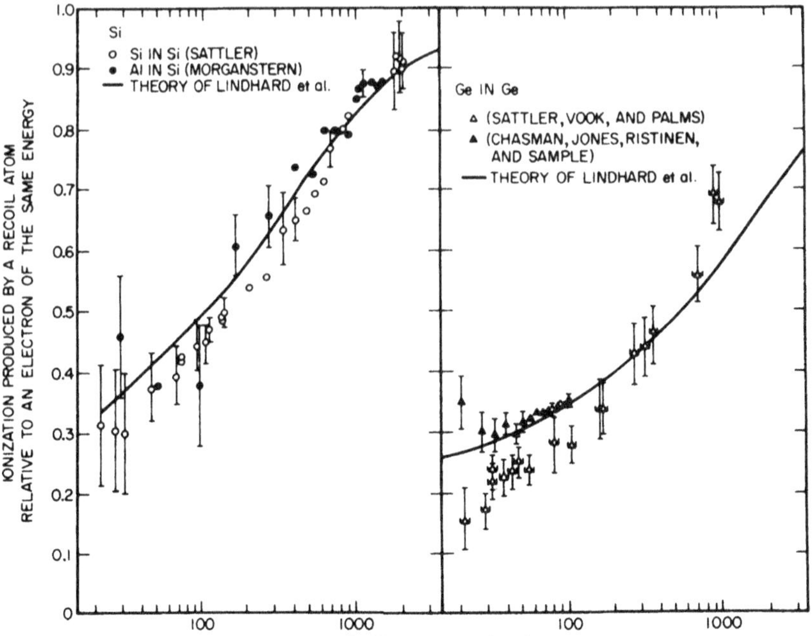

Fig. 3.2. Ionization produced by energetic ions in silicon and germanium relative to that produced by an electron of the same energy. (From Sattler.[103]) Data on Al in Si from H. H. Morganstern, Hahn Meitner Institut für Kernforschung, Berlin, 1965; Si in Si, A. R. Sattler, *Phys. Rev.* **138**, A1815 (1965); Ge in Ge, Sattler *et al.*, *Phys. Rev.* **143**, 538 (1966); and Chasman *et al.*, *Phys. Rev. Letters* **15**, 245 (1965).

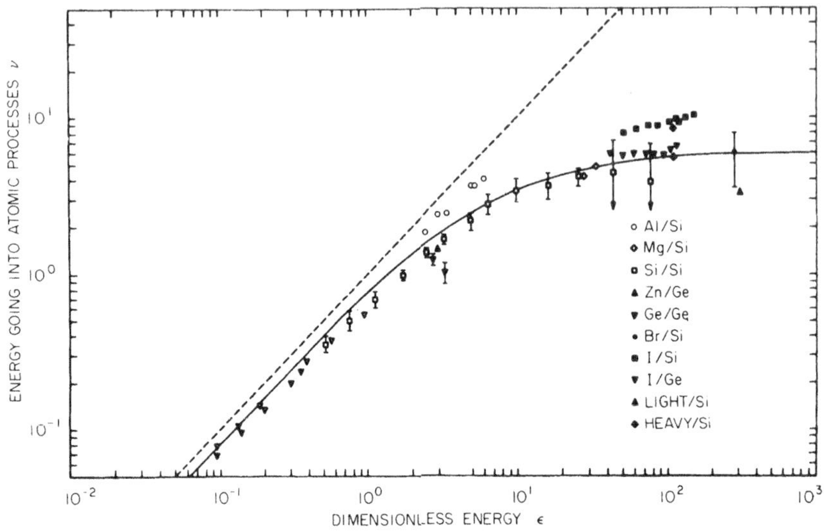

Fig. 3.3. Energy lost in atomic collisions ν versus particle energy ε. The dashed line represents the limit where all the energy is lost in nonionizing events. Solid line is the calculation of Haines and Whitehead. (From Haines and Whitehead.[104])

The dashed line corresponds to the situation in which all the incident energy is going into nonionizing processes. Comparison of these data and the calculations of Lindhard *et al.* indicate that the $\nu(\varepsilon)$ relation derived for $M_1 = M_2$ holds approximately for $M_1 \neq M_2$.

Note that, below $\varepsilon \simeq 3$ and for $M_1 \geq M_2$, more than one-half of the incident energy is available for displacement processes. Qualitatively this is what would be predicted from Fig. 2.5. Table 3.1B gives values of ion energy

TABLE 3.1B

INTERSECTION BETWEEN ELECTRONIC AND NUCLEAR STOPPING IN SILICON AND GERMANIUM[a]

		$S_n(E_0) = S_e(E_0)$			
		B	P	As	Sb
Si	ε_0	1.9	3.1	3.8	3.8
	E_0 (keV)	17	140	800	2000
Ge	ε_0	0.66	1.6	2.8	3.2
	E_0 (keV)	13	140	800	2000

[a] Taken from Table 2.3.

and ε at the intersection between electronic and nuclear stopping (Fig. 2.5) for various ions incident on germanium and silicon. It can be seen that except for low-Z ions or high implantation energies, most of the ion energy is available for displacement processes.

3.2.2 Average Number of Displaced Atoms

From a knowledge of the amount of energy available for nuclear collisions, it should be possible to estimate the number of displaced atoms. However, there are complications in any such calculation.[99,100] The use of a well-defined displacement threshold energy implies that the displacement processes are spatially well separated. In a heavily disordered region where several neighboring atoms have been displaced, the use of a displacement energy may therefore not be justified. If the displacement events are sufficiently concentrated, they might even produce a highly disordered or amorphous region. Several possible mechanisms for producing highly disordered regions have been suggested.† Another complication is the fact that one is interested in the residual number of displaced atoms after the initial transient displacement process occurs. At room temperature, a considerable amount of vacancy–interstitial recombination is expected to occur. At present, there is not a good estimate as to the extent of this recombination in ion-implanted semiconductors. However, recent experimental work shows that for a low-mass species, such as boron, this effect is much more significant than for the heavier ions (cf. Section 3.4.3).

Usually in calculating the number of displaced atoms, the possibility of channeling of the incident ion is neglected. Collisions are treated as isolated two-body elastic scattering events, and both atoms are available to make further collisions after the displacement event. Although electronic energy loss terms could be added, this correction is often not included in the calculation. Different forms of the interaction potential have been used; however, the number of displaced atoms is not strongly dependent on the form of the potential.

The usual estimate of the number $N(E)$ of displaced atoms in a solid is based on the relationship derived by Kinchin and Pease[107]:

$$N(E) = E/2E_d \tag{3.1}$$

† According to Seitz and Koehler,[105] a large portion of the energy of the incident particle could be transferred to the lattice in the form of lattice vibration. In these regions, where the temperature may rise above the melting point during a very short time (of the order of 10^{-11} sec), it is possible that the disorder will be retained as a region with a frozen-in-liquid structure. Brinkman[106] considers a displacement spike which, in its initial unstable stage, consists of a void region around which associated interstitials form a shell. Because of the high local temperature, this configuration will collapse to form a disordered region.

where E is the ion energy and E_d an effective displacement threshold energy. This was derived for the equal-mass case; however, it also holds[108] for unequal masses of ion and target. Sigmund[108] points out that the Kinchin and Pease assumptions of elastic collisions and hard-sphere scattering may be eliminated by modifying the equation to

$$N(E) = \xi \, \nu(E)/2E_d \tag{3.2}$$

where $\nu(E)$ is the amount of energy not lost to electronic excitation and ξ is a factor (<1) dependent on the form of the scattering potential. Sigmund's calculations indicate that ξ is ~ 0.8.

By using a power-law approximation to the Thomas–Fermi cross section and neglecting the loss of defects by replacement collisions, Sigmund[108] has calculated an *upper limit* for the number of displaced atoms:

$$N(E) = \frac{6}{\pi^2}\frac{E}{U}\ln\left(1 + \frac{U}{E_d}\right) \tag{3.3}$$

where U is the binding energy lost by an atom when leaving a lattice site. The displacement model of Bauerlein[109] gives $E_d = 4E_b$, where E_b is the bond energy. With $U = E_d$, Eq. (3.3) reduces to

$$N(E) = 0.42E/E_d \tag{3.4}$$

This equation is essentially the same as Eq. (3.2), despite rather different underlying assumptions. It is based on the fact that each displaced atom had to break its four nearest-neighbor bonds. However, if all the atoms within a localized region become displaced, then only two bonds on the average would have to be disrupted per atom. This would increase $N(E)$ by about a factor of 2.

The essential feature of the different theories on defect production by heavy ions is that the number of displaced atoms is linearly proportional to the energy of the incident ion in the region where electronic collisions can be neglected. For a heavy ion such as antimony in silicon or germanium at an energy of about 40 keV, the different theories would suggest that between 1200 and 2400 atoms would be displaced (depending on whether isolated defects or clusters are formed). It assumes a displacement threshold energy E_d of about 14 eV in the semiconductors.[100] Strictly speaking, this value of 14 eV is the *threshold value* of the displacement energy, i.e., the *minimum* recoil energy at which observable displacements are detected. Since the displacement energy should depend significantly on the recoil direction, the mean value of E_d will obviously be somewhat larger than the threshold value, and so the number of displaced atoms will be less than calculated. The magnitude of this effect is difficult to estimate. Also, no allowance has been made for any vacancy–interstitial recombination that occurs during the implantation. Hence, the predicted number of displaced atoms must be considered only an upper limit.

3.2.3 Spatial Distribution of Displaced Atoms

In discussing the spatial distribution of atoms displaced by energetic ions, one should distinguish between the displacement events produced by a single ion and the average distribution of displaced atoms produced by many ions. An example of the defects produced along the track of one ion is given in the Monte Carlo calculations of Yoshida,[110] who found that the distribution of displaced atoms is not uniform; instead, the vacancies and interstitials are distributed in several groups along the path of the incident atoms. The model was not very realistic in that hard-sphere collisions were assumed. Better potentials were used by Beeler[111] in computer simulation of collision cascades in cubic metals. His results also indicated the nonuniform distribution of vacancies and interstitials. Part of this nonuniformity can be attributed to the channeling of recoil atoms.

Calculations of the disorder produced by *many* ions have been made by Sigmund and Sanders[15] and Pavlov *et al.*[112] In general they find that the characteristics of the depth distribution depend on the ion energy and on the relative masses of the ion and substrate atoms. Pavlov *et al.*[112] included electronic stopping, but instead of the more realistic Thomas–Fermi potential, they used the simpler Bohr potential in their Monte Carlo calculations of the trajectories in silicon. They determined the distribution in depth for boron, aluminum, phosphorus, and arsenic implantations and the resultant defect distributions for boron and aluminum. The influence of the crystal structure was not included. In the calculation of disorder, they neglected recombination of point defects and assumed a threshold displacement energy E_d of 30 eV. Calculations carried out for $E_d = 13$ eV showed that the total number of defects increased but their spatial distribution was influenced only slightly. Figure 3.4 shows the distribution of impurity atoms (solid line) and of vacancies (dashed line) produced by boron ions of energies of 20, 40, and 60 keV. The data indicate that the maximum in the distribution of implanted ions lies somewhat deeper than the maximum in the defect distribution. This same general result was found for aluminum ions in silicon; however, the number of defects is greater for aluminum than for boron implantations. Since the range of the heavier ions is less than that of the lighter ions (Fig. 2.8), this leads to an increase in the concentration of defects with an increase in ion mass—i.e., the heavier the ion, the larger the concentration of defects.

Sigmund and Sanders[15] have calculated the spatial distribution of defects

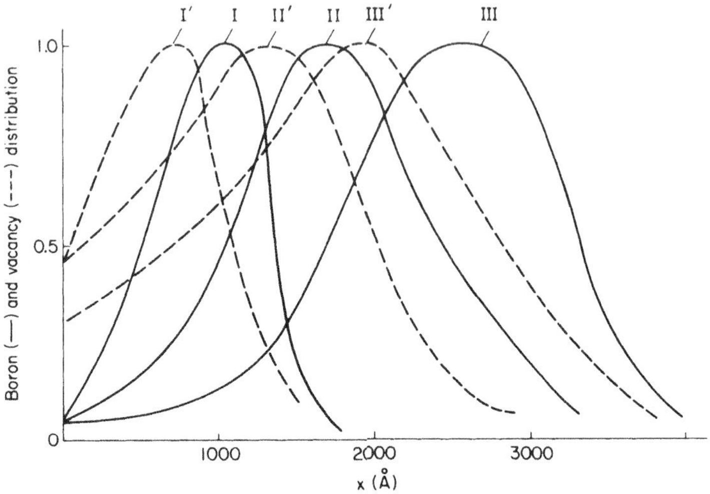

Fig. 3.4. Distribution of implanted boron atoms (solid lines, I–III) and of vacancies (dashed lines, I′–III′) in silicon. The maximum concentrations are normalized to unity. Energy in keV: I and I′, 20; II and II′, 40; III and III′, 60. The maximum concentrations per cm^3 of vacancies for 1000 ions/cm^2 are: I′, 11.95×10^9; II′ 11.8×10^9; III′, $11.9 \times 10.^9$ (From Pavlov *et al.*[112])

and of implanted atoms by using a power-law approximation to the Thomas–Fermi potential and neglecting the electronic stopping component. A displacement threshold energy is not included, which means that in effect, they calculate the spatial distribution of *energy* available to displace atoms. Numerical values for the range (Chapter 2, Part 2.3) and damage distribution have been computed for power cross sections of the form

$$d\sigma \propto E^{-m} T^{-1-m} \, dT,$$

where T is the recoil energy ($m = \frac{1}{3}$ for $\varepsilon \lesssim 0.2$, $m = \frac{1}{2}$ for $0.08 \lesssim \varepsilon \lesssim 2.0$). With this cross section, there is one single length unit proportional to E^{2m} characterizing both range and damage distribution.[2] Their curves for the ratio between average projected range $\langle x \rangle$ (i.e., R_p) and damage depth $\langle x \rangle_D$ versus M_2/M_1 are shown in Fig. 3.5(a), and the ratio between straggling in projected range $\langle \Delta x^2 \rangle$ and damage $\langle \Delta x^2 \rangle_D$ in Fig. 3.5(b) in which

$$\langle x \rangle = \frac{\int x f(\bar{r}, \bar{v}) \, d^3\bar{r}}{\int f(\bar{r}, \bar{v}) \, d^3\bar{r}} \tag{3.5}$$

$$\langle \Delta x^2 \rangle = \langle (x - \langle x \rangle)^2 \rangle$$

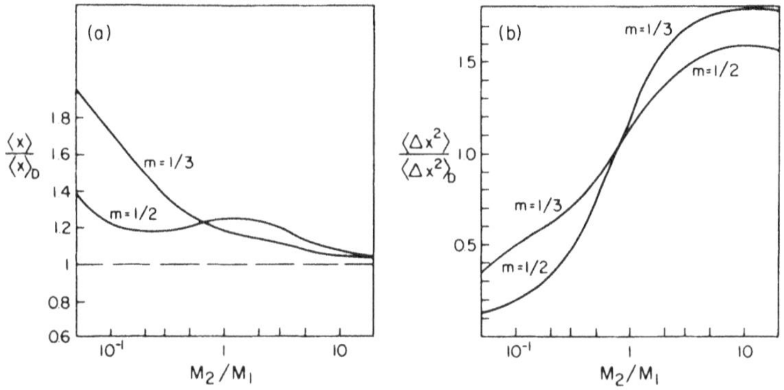

Fig. 3.5. Proportion between projected range and damage parameters versus M_2/M_1: (a) ratio of average projected range to damage depth; (b) ratio between range and damage straggling. (Modified from Sigmund and Sanders.[15])

and $f(\bar{r}, \bar{v})\, d^3\bar{r}$ is the probability that a projectile starting with velocity $\bar{v}$ at the origin ($\bar{r} = 0$) comes to rest in the volume element ($\bar{r}$, $d^3\bar{r}$). For disorder, $f_D(\bar{r}, \bar{v})\, d^3\bar{r}$ is the average amount of energy deposited in the volume element ($\bar{r}$, $d^3\bar{r}$). Normalizing conditions are such that the integral of the range distribution is unity and that of the distribution f_D of energy deposited in nuclear collisions equals $\nu(E)$. As indicated in Fig. 3.5(a), they find that the incident ion comes to rest at a position that is *deeper* than the center of the disorder cascade. For light ions ($M_1 < M_2$), large-angle scattering of the primary beam can occur, and so both the defect and range distributions averaged for many ions, are comparable in shape. The peak in the defect distribution occurs at almost the same depth as the peak in the range distribution. The width of the defect distribution is somewhat narrower than the range straggling. For the case of equal masses, the ratio of the first moment of the range distribution to that of the defect distribution is about 1.2. For extremely heavy projectiles at low energies (i.e., $m = \frac{1}{3}$), the range becomes almost twice the average depth of disorder, as indicated in Fig. 3.5(a). The latter is a consequence of the fact that for $M_1 \gg M_2$ the incident-particle track is relatively straight (large-angle scattering is an unlikely process) and the range straggling is small. The damage profile extends over almost the entire particle range and follows the variation of nuclear stopping with depth.

For most cases the distribution of damage is slightly cigar-shaped, with the transverse extent of the defects typically 60% of that along the x direction. It should be emphasized again that these calculations give the *mean* damage distribution for many ions with the same initial conditions; they do *not* describe the disorder around a single track.

Brice[112a] has developed a method of calculating the energy deposited per unit depth by determining the rate at which the energy is transferred from the incident ion to the host atoms. The fraction of this energy which is lost in nuclear collisions can be determined from data similar to those in Figs. 3.1 or 3.2. The calculation does not take into account the spread in deposited energy due to the finite range distribution of the recoiling host atoms. This approximation is valid in cases where the average energy transfers are small. The disorder calculations agree well with Monte Carlo calculations of Pavlov *et al.*[112] for boron in silicon (Fig. 3.4).

3.3 Experimental Techniques

A number of different experimental techniques have been used to characterize the lattice disorder in semiconductors. In general, each of these techniques measures a different property of the lattice disorder. In this section, descriptions of experimental methods and the nature of the observations will be presented so as to compare the different types of information that have been obtained. The electron-paramagnetic-resonance and optical-absorption experimental techniques that have recently been applied to the study of implanted layers are not described. However, the data obtained from these measurements are discussed in Parts 3.4 and 3.5.

3.3.1 Optical Effects

One of the most striking features of lattice disorder is the appearance of a hazy or milky color on the surface of ion-implanted silicon.† Similar, but less pronounced, color changes occur on germanium. This effect seems to be a general one, and occurs in varying degree for all implanted species once a certain dose is exceeded. This critical dose is specific to each ion type, ion energy, and substrate temperature.[113–116] Figure 3.6 shows the change in the visual appearance of implanted silicon (the cross corresponds to an area shielded from the ion beam).

Two mechanisms have been proposed to account for these visual effects: Rayleigh scattering by small disordered regions[114,115] and a change in reflectivity due to a change in the average dielectric constant.[116a] Nelson and Mazey[114,115] suggest that at low ion doses the individual disordered zones act as isolated scattering centers, and if their radii are very much less than the wavelength of the illumination, the scattered energy per unit volume can be described by the Rayleigh relation. Then, as the ion dose increases, the density

† Equally striking is the "breath test." By breathing on a sample, one can see the implanted area in the vapor-condensation pattern. This is true even for low-dose or hot-substrate implantations where little visible coloration is present.

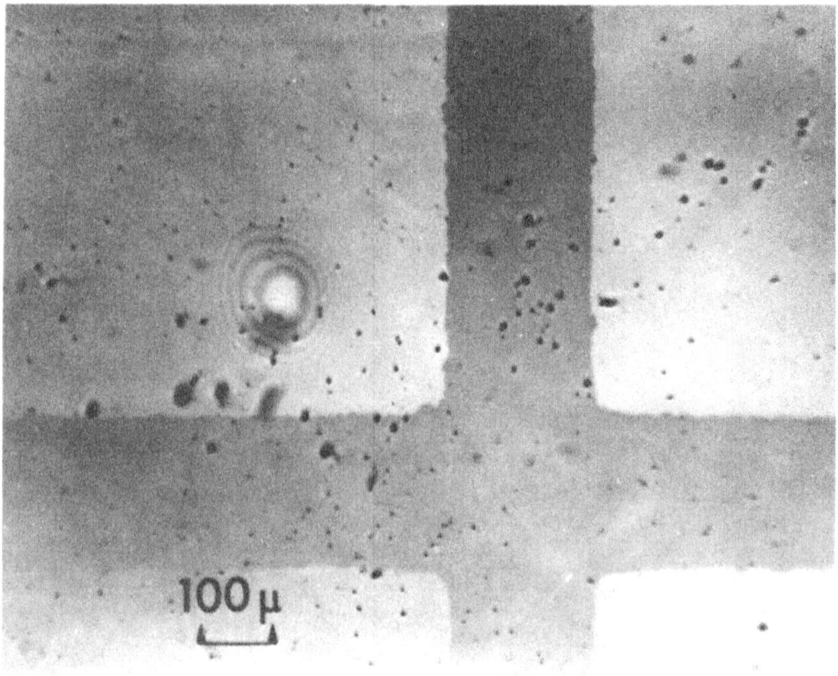

Fig. 3.6. Photograph of a silicon slice bombarded with 80-keV neon ions to a dose of $\sim 10^{15}$ ions/cm^2. The cross corresponds to a part of the surface that was shielded from the ion beam. (From Nelson and Mazey.[114])

of scattering centers will also increase and, because of multiple scattering, the end result will be a milky appearance typical of diffuse scattering.

However, recent measurements of the optical reflectivity suggest that the milky appearance is associated with the properties of amorphous material. A number of optical and photoemission measurements have been made on vacuum-deposited amorphous germanium thin films.[117,118] A distinct change in the optical properties is observed as one goes from a crystalline to an amorphous film. The work by Donovan and Spicer[118] on photoemission indicates that there is a change in the valence-band and conduction-band density of states and also a change in the optical selection rules.

A direct comparison of the optical reflectivity of crystalline silicon, sputtered silicon (amorphous), and ion-implanted silicon has been made by Kurtin *et al.*[116] A salient feature of their data (Fig. 3.7) is that the optical reflectivity peaks characteristic of crystalline silicon are absent in the amorphous material; in implanted silicon, these peaks diminish monotonically with increasing dose. For a dose of approximately 10^{15} antimony ions/cm^2,

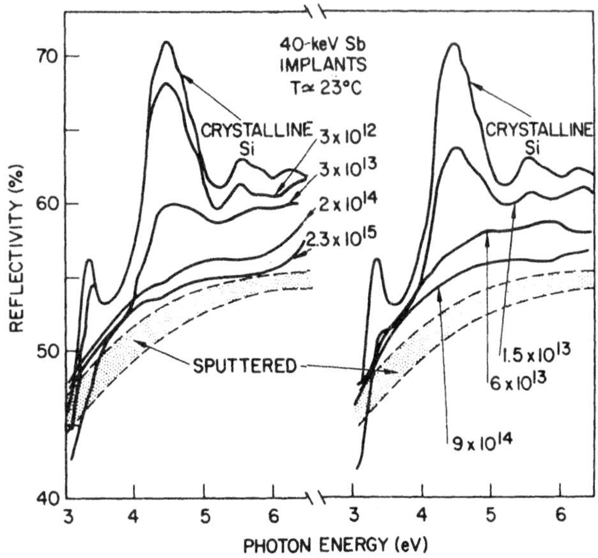

Fig. 3.7. Reflectivity spectra for crystalline, sputtered, and antimony-implanted silicon (40 keV at room temperature) as a function of ion dose. The left- and right-hand portions were separated to show the dose dependence more clearly. (From Kurtin *et al.*[116])

the reflectivity curve is quite similar to that observed in vacuum-deposited amorphous silicon. The measured change of reflectivity with implantation dose throughout most of the visible region is consistent with observations of milkiness of silicon crystals.

3.3.2 Electron-Diffraction and Electron-Microscopy Studies

Both neutron-irradiated and ion-implanted silicon and germanium have been studied by a number of techniques involving electron beams. Although these methods do not permit a quantitative determination of the amount of disorder, as is possible in channeling-effect measurements (Section 3.3.5), they do provide insight that is not obtained by other methods. For example, transmission electron microscopy has been used to determine the growth of dislocation loops upon anneal. We will discuss briefly three of the different methods that have been used; a comprehensive survey of the techniques is given in the *Proceedings of the Sixth International Congress for Electron Microscopy.*[119]

a. *Low-Energy Electron Diffraction.* Low-energy electron-diffraction (LEED) studies are highly sensitive to the structural arrangement of the atoms in the outermost few atomic layers. Any disturbance in the structural arrangement can be observed in the LEED response. For implantation studies, however, the shallow depth that can be probed by low-velocity electrons is a definite limitation unless layer-removal techniques are used. The basic experimental details have been described by many authors.[120–123]

Jacobson and Wehner[120] studied the degradation in intensity of the diffraction patterns observed on a germanium surface bombarded at room temperature with argon ions. Figure 3.8 shows the degradation of (10) beam intensity versus the integrated number of ions for argon-ion energies ranging from 10 to 1000 eV. In these experiments the intensity was measured for the surface in the initial ordered state and again after each of a number of low-dose bombardments. In Fig. 3.8 a highly ordered surface corresponds to $I/I_0 = 1$, a highly disordered surface to $I/I_0 = 0$. It can be seen that for bombardment energies greater than 100 eV, degradation goes to completion for sufficiently large doses of argon ions.

The extent of the damage caused by the impact of a single ion of energy E can be determined from the degradation data of Fig. 3.8. The authors[120] derive an expression that relates the intensity ratio to bombardment dosage D, viz. $I/I_0 = \exp(-2AD)$, where A is defined as the mean surface area damaged per ion. Since LEED measurements are sensitive to relatively small disturbances, A includes not only the highly disordered region around the point of ion impact but also the surrounding region where minor lattice

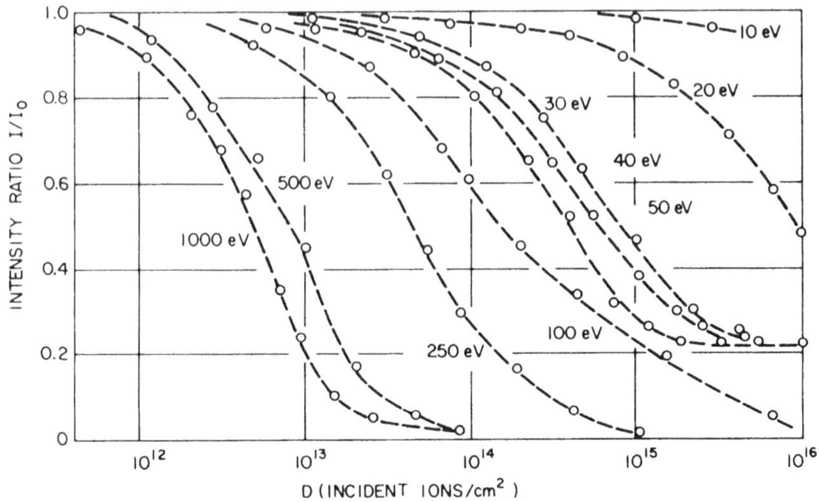

Fig. 3.8. LEED measurements of the degradation of the (10) beam at 36-eV electron energy versus the number of argon ions/cm² bombarding the germanium surface. Intensity ratios of 1.0 and 0 correspond to highly ordered and disordered surfaces, respectively. (From Jacobson and Wehner.[120])

distortion exists. They find that for a 1-keV argon ion, the area A is equivalent to about 50 surface atoms, or a circular damage area of approximately 15 Å radius. Jacobson and Wehner also studied the thermal recovery characteristics observed when the ion implanted target was annealed. They found that anneal temperatures greater than 300° C were required to return the intensity distribution to that of a well-ordered surface.

Although LEED studies of this nature are useful for shallow implants (i.e., for ion energies less than 1 keV), the technique does not provide information on the disorder produced at larger depths by the energies (typically 20–100 keV) used in most ion-implantation studies. Also LEED thermal-recovery data should be applied cautiously to implanted layers, since the proximity of the disorder to the surface may have a dominant effect in determining the recovery characteristics.

b. *Electron Diffraction.* Electron-diffraction techniques have been used to study the influence of ion implantation on silicon, germanium, and gallium arsenide surfaces.[97,113,124–127] The two basic techniques are reflection and transmission electron diffraction. In the latter case, samples thinned to less than 0.1 micron must be prepared because of the high absorption of electrons in solids. Hirsch *et al.*[128] and Large and Bicknell[126] have discussed typical procedures used to produce these thinned samples.

Figure 3.9 illustrates the transmission diffraction patterns obtained by Parsons[124] with crystalline germanium bombarded at room temperature with various doses of 100-keV oxygen ions. One can observe the appearance of diffuse halos superimposed on the characteristic spot patterns as the ion dose increases. Finally at a dose of 10^{15} ions/cm^2, a halo-diffraction pattern characteristic of an amorphous region is observed. Parsons found that the diffraction patterns were of a diffuse-ring type identical to that obtained by Parsons and Balluffi[129] for amorphous germanium. Similarly, the silicon investigations[113,126] indicated that the heavily implanted layers had characteristics very similar to those obtained by von Richter and Breitling[130] for amorphous silicon.

Electron-diffraction techniques have been used also to study the annealing behavior.[97,124–127] Figure 3.10 shows the reflection electron-diffraction patterns obtained by Itoh *et al.*[97] from silicon samples implanted at room temperature with doses of 10^{14}, 10^{15}, and 10^{16} aluminum ions/cm^2 at 10 keV. Spots and Kikuchi lines are not observed in the patterns of Fig. 3.10(b) and 3.10(c), but 3.10(e) and 3.10(f) show the reappearance of spots and lines after anneal at 800° C for 20 min.

The general conclusions from anneal studies with transmission methods are that the diffuse diffraction rings disappear, indicating that the material regains much of its crystalline character. There are, however, many residual

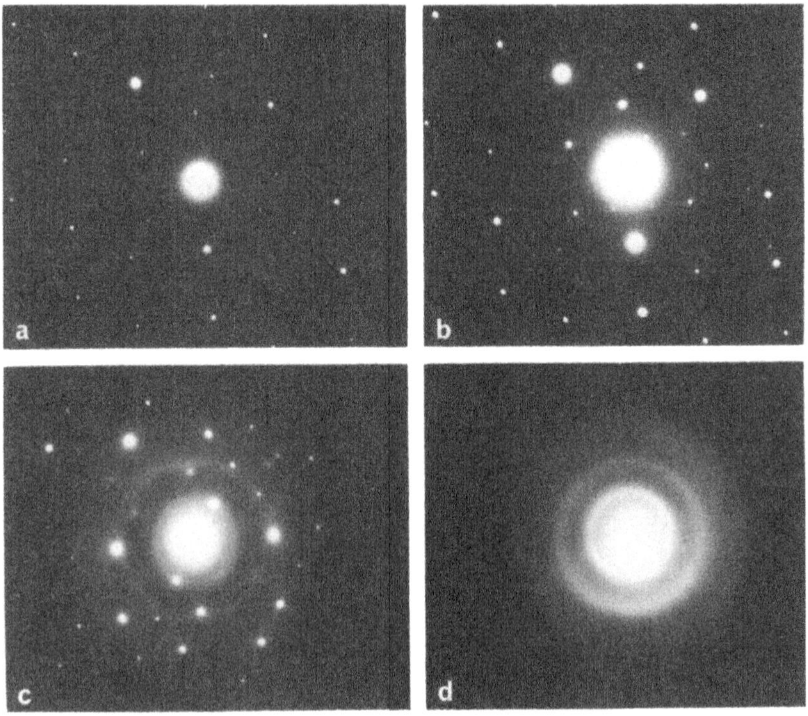

Fig. 3.9. Transmission electron-diffraction patterns from germanium bombarded at room temperature with 100-keV oxygen-ion doses as follows: (a) unirradiated; (b) 1.3×10^{12} ions/cm^2; (c) 4.3×10^{13} ions/cm^2; and (d) 1.0×10^{15} ions/cm^2. (From Parsons.[124])

defects in the layer. In some cases, twinning is observed,[113,126] and in other cases, polycrystalline areas are found.[131]

c. *Electron-Transmission Microscopy.* Some of the most conclusive evidence for the localized nature of the disorder created in ion implantation has been obtained in electron microscopy studies by Parsons,[124] Large and Bicknell,[126] and Mazey *et al.*[113] Qualitatively the results are the same for both silicon and germanium and are independent of the conductivity and type of crystal used. The investigations, to date, have been restricted to those cases in which the mass of the implanted ion (boron, oxygen, neon) was less than that of the substrate (germanium, silicon). An examination of the electron-microscope patterns indicates the presence of small, dark-contrast damage regions after implantation at room temperature, as shown in Fig. 3.11 for silicon. The number of these disordered regions increases with ion dose. Parsons found that the mean diameter of the damage region was approximately 70 Å for room-temperature bombardments of 100-keV oxygen

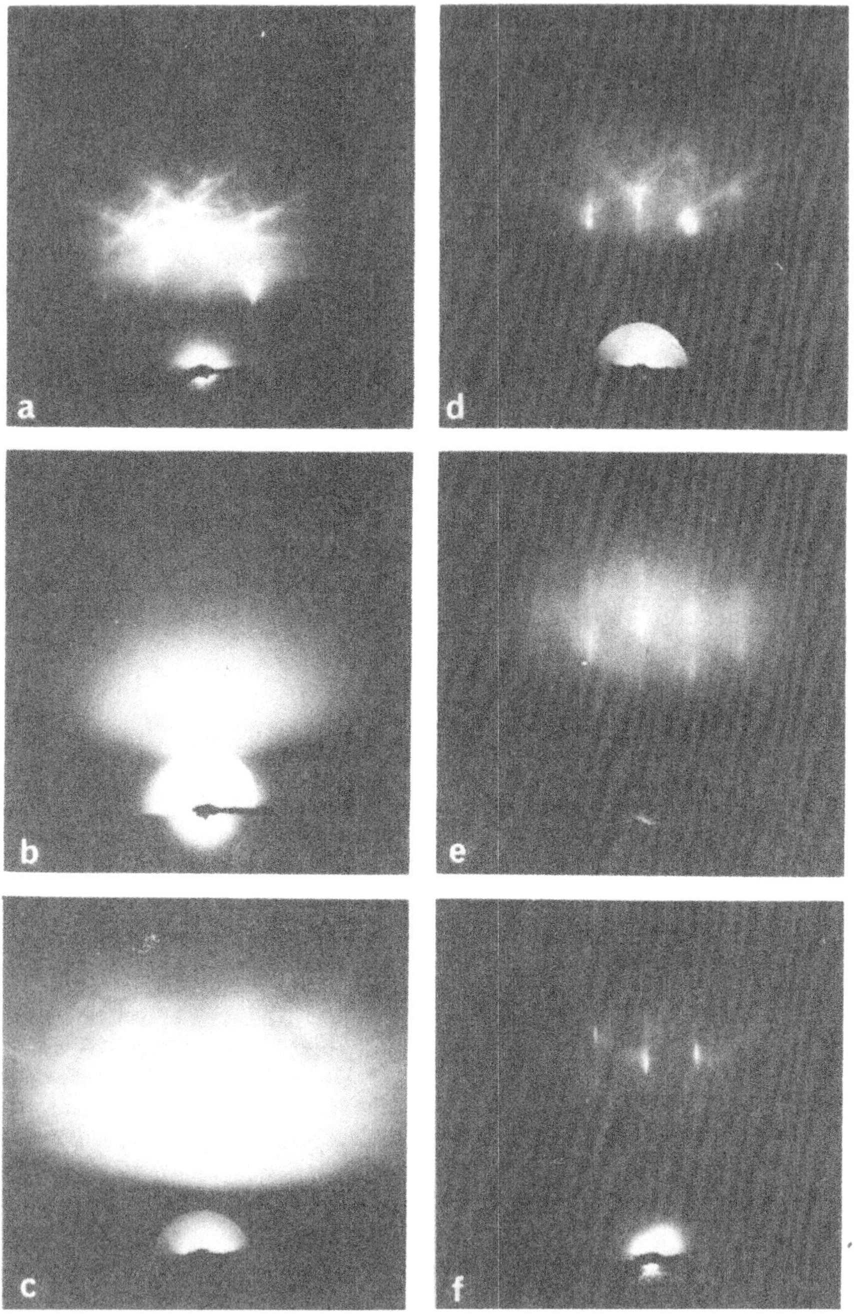

Fig. 3.10. Reflection electron-diffraction patterns obtained from silicon bombarded at room temperature with aluminum ion doses of: (a) and (d), 10^{14} ions/cm^2; (b) and (e), 10^{15} ions/cm^2; (c) and (f), 10^{16} ions/cm^2. Patterns (a), (b), and (c) were obtained before annealing and patterns (d), (e), and (f) after annealing at 800° C for 20 min. (From Itoh *et al.*[97])

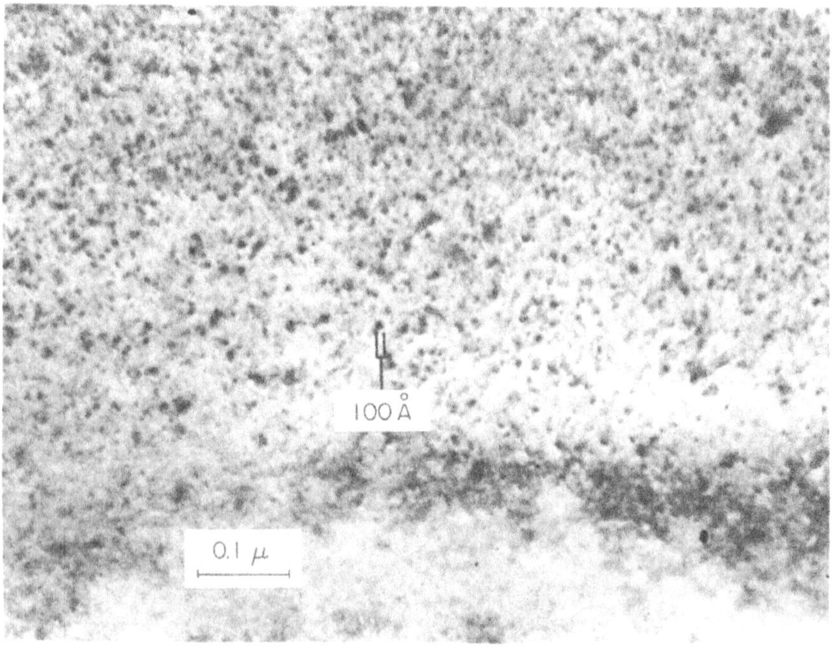

Fig. 3.11. Transmission electron micrograph illustrating the small regions of dark contrast near the edge of a thinned silicon sample. Implantation conditions: 1×10^{14} neon ions/cm^2 at 80 keV and 25° C. (From Mazey *et al.*[113])

ions on germanium[124] and approximately 40 Å for similar bombardments on silicon.[132] In ion-implanted germanium he found that the mean diameter of the damage region increased to 90 Å when the implantation was made at ~30° K.

Electron micrograph studies[124–126] of the anneal behavior of room-temperature implanted samples indicate the formation of long straight dislocations and dislocation loops. The loops, which may be either vacancy or interstitial type, increase in size with an increase in anneal temperature. For boron implants in silicon, visible loop defects start to grow at temperatures above 600° C and reach a maximum size of 1000 Å at 800° C.[126,131] Bicknell[131,133] has found dislocations in elevated-temperature implants (50 keV, 10^{15} boron ions/cm^2 at 300° C) that have been annealed to 600° C or higher (Fig. 3.12). Figure 3.12 shows the dislocations in a sample annealed at 700 and 800° C. Although there did not appear to be a significant difference in the *number* of defects per implanted ion (~5×10^{-6} defects/ion) between room-temperature and 300° C implants, the *type* of condensed defect is different. Annealed room-temperature implants contain interstitial loop defects lying on {111} planes, and long edge dislocations and dipoles lying

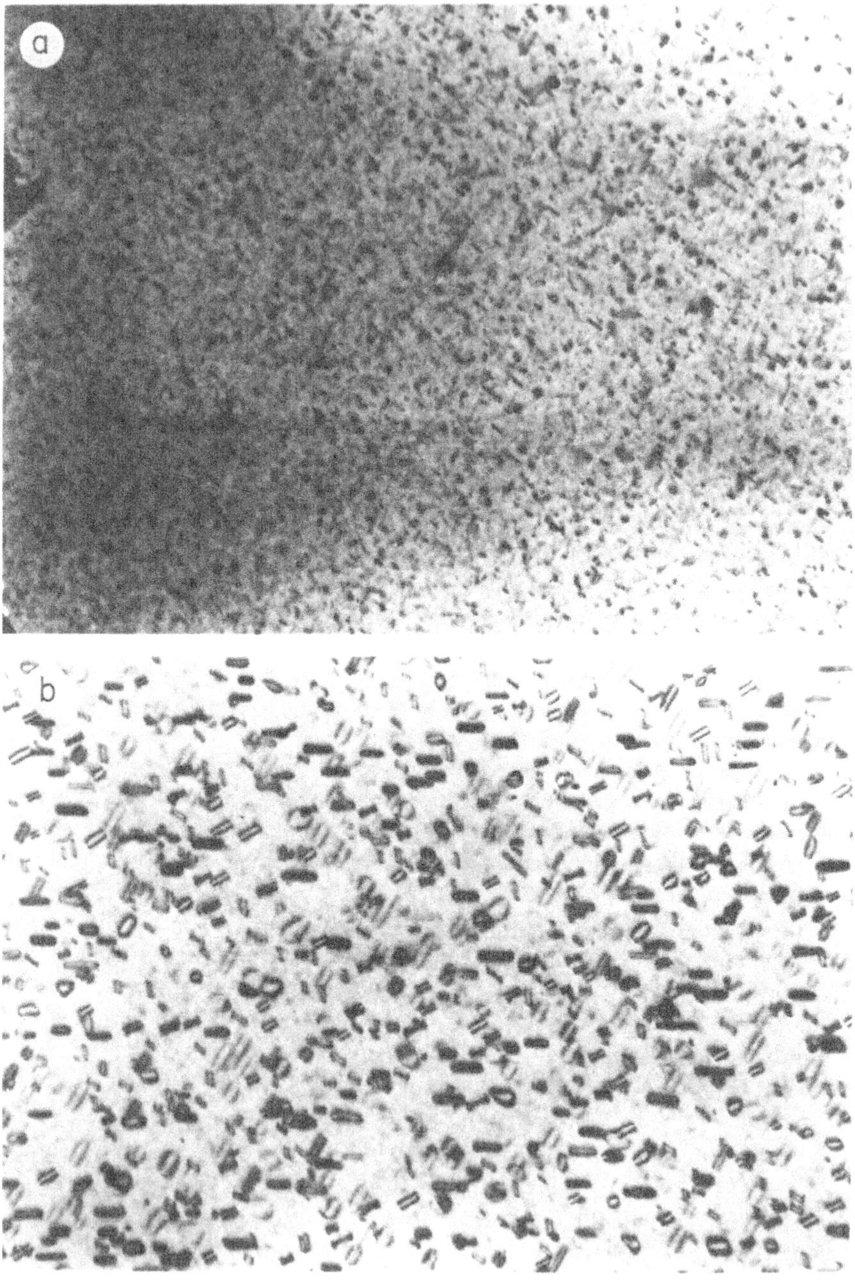

Fig. 3.12. Electron micrographs (40,000 ×) showing dislocation loops at anneal temperatures of (a) 700° C and (b) 800° C for a silicon sample implanted at 300° C with 2×10^{15} boron ions/cm^2 at 50 keV. (Figures supplied by Bicknell.[133])

parallel to ⟨110⟩ directions. The annealed 300° C boron implants also contain interstitial loop defects lying on {111} planes, but there is another set of loops and a marked *absence* of the long edge dislocations.[133]

3.3.3 X-Ray Transmission Techniques

Use of transmission X-ray diffraction provides an extremely sensitive tool for detecting strain or individual dislocations in single crystals. If the X-ray beam is incident at the Bragg angle, there is an enhanced transmission of X rays when a perfect crystal is set for Laue diffraction. This anomalous transmission was first discovered by Borrmann[134] and has been used extensively for the direct observation of imperfections in crystals.[135–139] It has also been used for the detection of defects produced by fast-neutron irradiation.[140] As described by Schwuttke,[138] two standing waves are set up if a plane monochromatic X-ray wavetrain enters a perfect crystal set for Laue diffraction. The net energy flow is mostly along the atomic plane. For a simple lattice, one of the standing waves may have its nodal planes coinciding with the position of a family of lattice planes (*hkl*). This wave passes through the crystal with full intensity, independent of crystal thickness (for real atoms and known structure factors, there is a finite absorption coefficient). The transmitted intensity depends critically on crystal perfection, with defects in the periodicity of the crystal reducing the intensity.

Anomalous transmission can be employed in X-ray topography to reveal local variations in crystal perfection and crystal orientation. In effect, the imperfections cast shadows in the field of the transmitted wave just as obstacles do in the path of light rays.[135] Schwuttke *et al.*[141] obtained large-area transmission topographs of silicon slices implanted with $\sim 10^{16}$ nitrogen ions/cm^2 at 2 MeV. These data along with transmission electron micrographs indicate that the silicon lattice stays almost perfect (some radiation defects are present) down to a depth of 1.7 microns. This undamaged layer is followed by a highly disturbed layer at approximately the depth of the nitrogen range distribution. In more recent studies,[141a] the position of the disordered region was also found to correspond closely to the ion range for 1–3-MeV boron, nitrogen, and phosphorus implantations.

Baldwin and Thomas[140] measured the intensity of the anomalous (Borrmann) transmission of X rays for the detection of defects produced in silicon and germanium by fast-neutron irradiation. They irradiated their samples with doses of fast neutrons at fluences of up to 4×10^{19} neutrons/cm^2. For the highest fluences the Borrmann intensities of silicon and germanium were decreased by nearly a factor of 2; the observed effects in silicon and germanium were almost identical. The decreases in intensity were attributed to the strains created in the neighborhood of disordered regions. From the

X-ray topographs, the authors concluded that no defects were created with strain fields greater than approximately 1 micron in size. In measurements of transmission intensity, they observed two stages in the anneal characteristics in both silicon and germanium, caused by the breakup of the disordered regions and the migration of point defects to sinks. These results are discussed in detail in Part 3.5.

Baldwin[142] has measured the effect of fast-neutron irradiation on the lattice parameters of silicon and germanium crystals for a dose of 4.2×10^{19} fast neutrons/cm^2. The results showed an increase in length $\Delta l/l \approx 1 \times 10^{-4}$ in both kinds of crystals. Isochronal annealing experiments on irradiated germanium again indicate two stages, which coincide with those in the anomalous X-ray transmission experiments discussed above.[140]

den Ouden[143] has made a study of neutron-irradiated germanium by measuring the small-angle scattering (double-Bragg) of X rays. He interprets his scattering data on the basis of localized damaged regions that have a disordered structure (possibly amorphous) with a density differing from that of the surrounding matrix by at most a few per cent. Annealing measurements reveal a distinct annealing stage (similar to the first stage observed in X-ray transmission measurements) in which approximately 60% of the small-angle scattering recovers.

The significance of these neutron-irradiation measurements to implantation analysis is that both germanium and silicon exhibit the same qualitative behavior and that the anneal stages in both materials show strong resemblance to those in ion-implanted samples. These comparisons are made explicitly in Part 3.5.

3.3.4 Ion Interactions and Electron Emission

Sputtering, elastic scattering, and secondary-electron emission during ion bombardment are sensitive to the presence of lattice disorder. Here, the same ion beam (usually one of the noble-gas ions) is used both to produce the disorder and to study the influence of disorder on the ion interactions. From such studies, the influence of substrate temperature on the disorder produced by the incident ion has been determined.

Anderson *et al.*[144–147] measured the substrate temperature at which ejection patterns could be observed in sputtering experiments in semiconductors. The existence of well-defined atom-ejection patterns requires that the target have single-crystal properties in the outermost layers. Hence, the absence of well-defined patterns or spots is characteristic of a polycrystalline or amorphous surface region. Anderson *et al.* found that ejection patterns occurred only above a certain implantation temperature. Figure 3.13 shows the influence of bombarding-ion energy E and target temperature T on

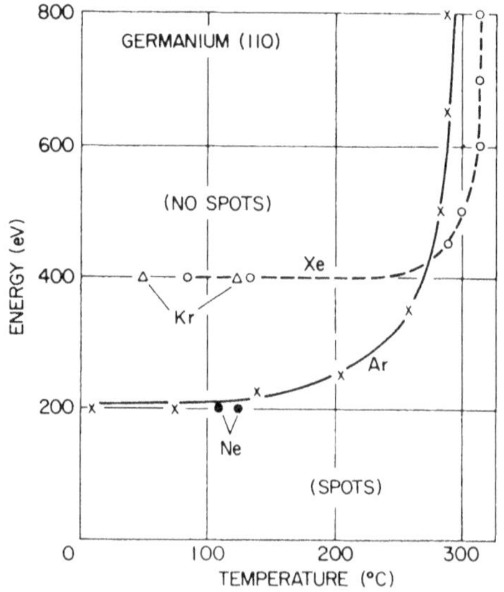

Fig. 3.13. The influence of bombarding-ion energy and target temperature on the formation of ejection patterns in sputtering experiments in germanium. The curves separate the regions where patterns are and are not observed. (From Anderson *et al.*[144])

the formation of ejection patterns in germanium for several ion species; the curves delineate the regions where deposit spots are and are not observed. The sharp rise of the E versus T curve for the higher temperatures is associated with an annealing process. At temperatures above the curves, the lattice disorder produced by the bombarding ions is rapidly annealed so that each bombarding ion impinges on a (nearly) perfect surface. At temperatures below $\sim 300°$ C, the damage is not annealed out rapidly enough, and the ejection patterns are smeared out as the target surface becomes amorphous. The annealing temperature depends primarily on the target materials and to a lesser extent on the ion species. The authors concluded that the transition temperature occurs where the annealing rate is equal to the rate of defect production. At higher temperatures the annealing rate predominates and the surface remains ordered; at lower temperatures the defect-production rate predominates and the surface becomes essentially amorphous.

This same temperature dependence was noted in ion scattering and in secondary-electron emission, both of which are sensitive to the degree of lattice arrangement near the surface. Mashkova and Molchanov[148,149] and Evdokimov *et al.*[150,151] measured the temperature dependence of the energy distribution of scattered ions and of the angular regularities of the secondary-electron emission coefficient. In the energy spectra of keV heavy ions scattered by single crystals, there are (in addition to the peaks caused by single scattering)

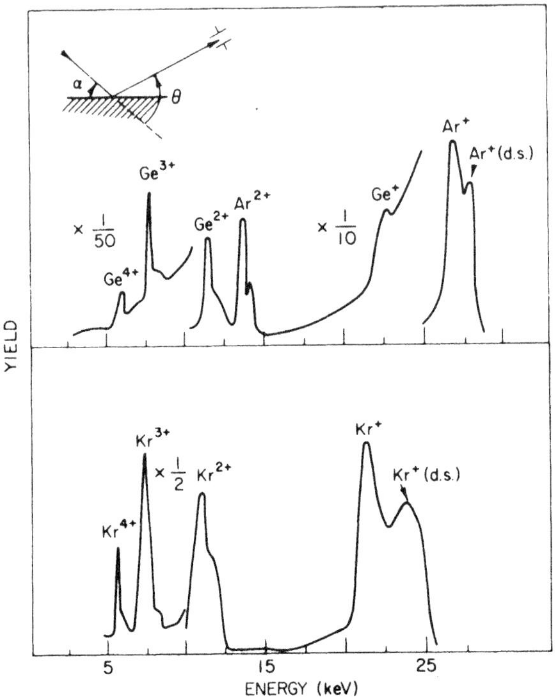

Fig. 3.14. Energy spectra for 30-keV argon ions (glancing angle $\alpha = 10°$) and for 27.5-keV krypton ions ($\alpha = 7°$) scattered from a germanium (111) surface. The scattering angle is $\theta = 25°$. The double-scattered (d.s.) peaks in the Kr^+ and Ar^+ spectra are indicated. (From Mashkova and Molchanov.[149])

also peaks caused by double scattering (Fig. 3.14). The intensity of the double-scattering peaks depends strongly on the location of neighboring scattering atoms. In the scattering measurements of Mashkova and Molchanov,[148] the energy distributions at low temperatures were similar to those observed for noncrystalline materials (i.e., without the double-scattering peaks). As the target temperature was increased, the spectral shape changed sharply and double-scattering peaks appeared. This was taken to indicate a transition from the disordered to the ordered state of the target surface.

Similar results were found in measurements of the secondary-electron emission coefficient.[150,151] At comparatively low temperatures the angular dependence was similar to that for polycrystals. However, at higher temperatures the angular dependence became strongly anisotropic and the secondary-electron emission coefficient showed a series of pronounced maxima and minima (Fig. 3.15) characteristic of a monocrystalline target.

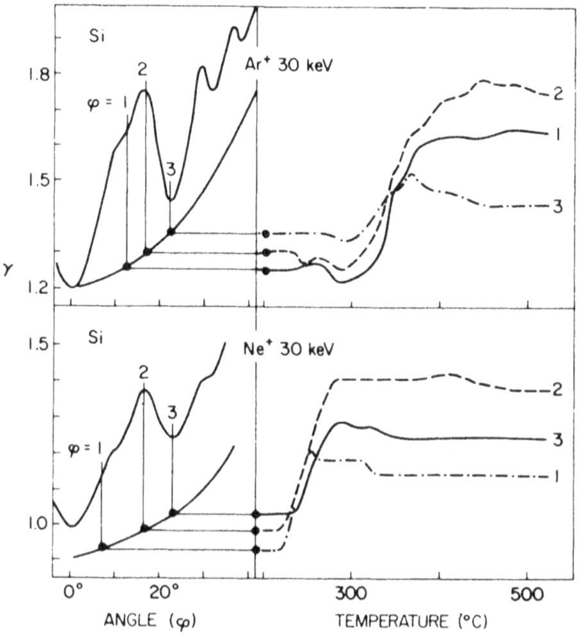

Fig. 3.15. Secondary-electron emission coefficient γ measured on (111) silicon targets as a function of rotation φ around $\langle 112 \rangle$ axes (left side) and target temperature for the indicated φ values (right side). (From Evdokimov *et al.*[151])

Table 3.2 shows a comparison of the anneal temperatures observed in these three types of ion-interaction studies. The comparison is made with argon ions as the bombarding species. It should be noted that the anneal temperatures in Table 3.2 depend on the ion-current density. Current densities in these experiments are of the order of mA/cm^2. These are 2–3 orders of magnitude higher than the typical currents used in ion-implantation doping of semiconductors. Vook and Stein[152] have considered anneal kinetics for

TABLE 3.2

ANNEAL TEMPERATURES OBSERVED IN ION-IMPLANTATION STUDIES[a]

	Sputtering	Ion scattering (30 keV)	Electron emission (30 keV)
Ge	300° C	~330° C	300–350° C
Si	430° C	~350° C	350° C

[a] Argon ions.

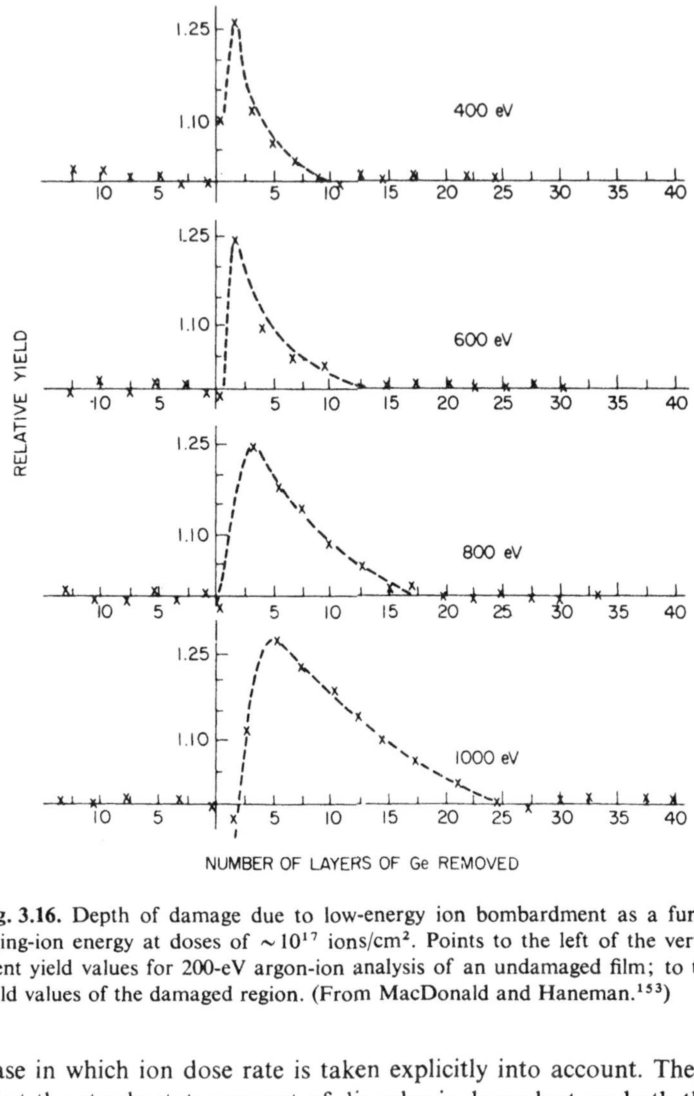

Fig. 3.16. Depth of damage due to low-energy ion bombardment as a function of damaging-ion energy at doses of $\sim 10^{17}$ ions/cm^2. Points to the left of the vertical line represent yield values for 200-eV argon-ion analysis of an undamaged film; to the right are yield values of the damaged region. (From MacDonald and Haneman.[153])

the case in which ion dose rate is taken explicitly into account. They point out that the steady-state amount of disorder is dependent on both the dose rate and the substrate temperature as well as on the activation energy of defect motion. Consequently, the anneal temperatures in Table 3.2 will be somewhat higher than those observed during ion implantation.

MacDonald and Haneman[153,154] have measured the depth of ion-bombardment disorder, using a method based on the differences in sputtering yield between damaged and undamaged germanium. The disorder is produced by ions at dose levels of 10^{16}–10^{17} ions/cm^2 with energy between 0.4 and

1 keV, and the sputtering yield is measured for lower energy ($\lesssim$0.2 keV) ions. The depth distribution is determined by monitoring the changes in the sputtering yield as the damaged layer is sputtered away. Figure 3.16 shows the relative yield in germanium as a function of the number of layers removed for initial bombardment doses of 10^{17} ions/cm^2. Their data[154] indicate that the disorder distributions at these low energies are markedly asymmetrical, with a tail extending to several times the mean depth. These distributions resemble closely the low-energy range distributions of argon in polycrystalline aluminum.[39]

3.3.5 Channeling-Effect Measurements

Analysis of the orientation dependence of the yield of backscattered MeV He^+ and H^+ particles has been used[155–157] to obtain a quantitative measure of the amount of disorder in ion-implanted silicon and germanium. First, the sample is *implanted* with heavy ions ($M \gtrsim 10$) at 10–100-keV energies; then the resultant disorder is *analyzed* with a channeled beam of lighter ions at higher energies. These channeling-effect measurements have also been used extensively to study the lattice location of implanted atoms, and the technique is discussed in detail in Chapter 4. However, it is appropriate to summarize here the basic concepts.

The amount and depth distribution of disorder can be determined from the energy spectrum of backscattered particles when the incident MeV beam is channeled—i.e., is aligned with a low-order crystallographic direction, typically the $\langle 111 \rangle$ or $\langle 110 \rangle$. A well-channeled particle is steered in such a fashion that it does not come closer to the lattice rows than 0.1–0.2 Å [the Thomas–Fermi screening distance a, Eq. (2.3)]. Thus, close-encounter collisions (such as backscattering) with the lattice atoms are not possible for these channeled particles. However, such interactions are possible with any substrate atoms located more than ~0.2 Å from a lattice site. Consequently, comparison of the yield of backscattered particles from both implanted and nonimplanted samples provides a measure of the lattice disorder.

This determination of lattice disorder counts *all* substrate atoms that lie outside the screening distance a of the rows. It includes not only the atoms displaced permanently from lattice sites by atomic collisions during implantation but also any local distortion ($\gtrsim$0.2 Å) in the lattice rows. Such distortions might occur in the neighborhood of defects such as divacancies created during implantation.[158] Hence, the measured amount of disorder may be somewhat greater than the actual number of displaced atoms. The experimental data in Part 3.4 suggest that the measured disorder is linearly related to the number of displaced atoms, but quantitatively may perhaps be as much as a factor of 2 to 3 greater, because of this "distortion" effect.

The energy spectra of 1-MeV helium ions backscattered from an implanted and an unimplanted silicon sample are indicated in Fig. 3.17. The random spectrum (i.e., the spectrum obtained when the helium-ion beam enters the crystal at a direction *not* coinciding with any of the major crystallographic axes, so that the crystal can be treated as a target of randomly distributed atoms) consists of a sharp edge at energy E^0 corresponding to scattering from the atoms in the surface region, followed by a smooth, slowly increasing yield at lower energy corresponding to scattering from atoms at greater depths in the crystal. In the aligned direction, the unimplanted crystal exhibits a 30-fold decrease in yield near the surface and a somewhat smaller decrease at greater depths. In considering the aligned spectrum it is convenient to treat the beam as divided into two components: a channeled portion consisting of particles that cannot interact with the lattice atoms, and a random portion that can interact equally with any atom. As the beam penetrates farther and farther into the crystal, the channeled component gradually decreases as

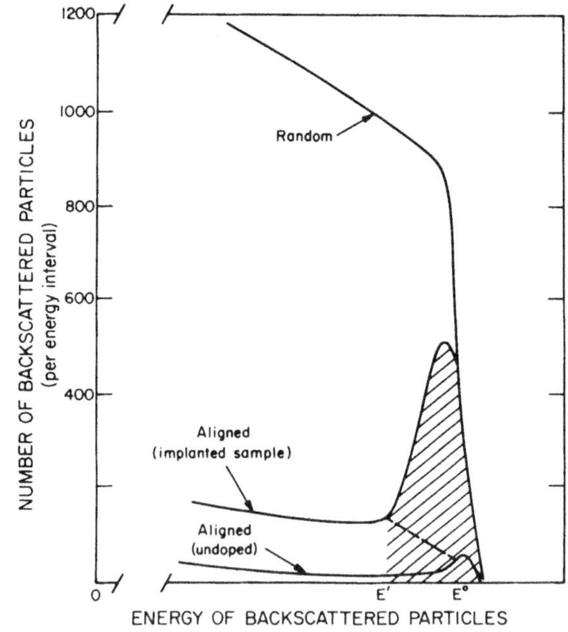

Fig. 3.17. Schematic representation of the random and aligned backscattering spectra (1.0-MeV helium beam) from a silicon crystal implanted at room temperature with $\sim 3 \times 10^{13}$ heavy ions/cm^2 at 40 keV. An aligned spectrum from an *un*doped crystal is included for comparison. The shaded area gives the total (uncorrected) amount of lattice disorder. The broken line illustrates approximately how the nonchanneled component of the aligned beam increases across the implanted region of the crystal.

particles become scattered out of the channeled component into the random component (dechanneling). This results in a gradual increase in the number of interactions in the aligned spectrum as the particles penetrate deeper into the crystal and hence an increase in the yield of backscattered particles.

In the implanted case, if a layer near the crystal surface contains a significant fraction of atoms located more than 0.2 Å from lattice sites, an increase in the yield of the aligned spectrum (as compared to the undoped spectrum) will occur in the energy region E' to E^0, Fig. 3.17. This energy region corresponds to the depth of the implanted region. The integrated yield in the shaded area is proportional to the amount of disorder after applying a small correction factor (as discussed later). The increase in yield for elastically scattered particles of energies less than E' (i.e., scattering events at depths below the disordered region) is caused by an increase in the amount of dechanneling as the particles penetrate the disordered layer, and hence an increase in the random component of the initially aligned beam.

Bøgh[155] has considered in detail the influence of displaced atoms on the yield of elastically scattered particles. Atoms located in off-lattice sites increase the measured backscattered yield in the aligned beam over that for a perfect crystal by two mechanisms: (1) Both the channeled and random components of the aligned beam can be backscattered directly from the displaced atoms; (2) the displaced atoms can deflect particles from the channeled beam into the random beam (dechanneling). Once in the random beam, these particles may then interact normally with *all* deeper atoms, and so contribute normally to the backscattering yield.

This leads to a yield y' as a function of depth given by

$$y'(t) = y_n(t)\left[\frac{N'(t)}{N} + \chi'(t)\frac{N - N'(t)}{N}\right] \tag{3.6}$$

where $y_n(t)$ is the yield for "random" orientation, N is the number of crystal atoms/cm^3, and χ' is the random fraction of the beam at a depth t for a crystal containing $N'(t)$ "off-lattice-site" atoms/cm^3.

The first term on the right-hand side of Eq. (3.6) is the contribution from mechanism (1) mentioned earlier; the second term is the contribution from mechanism (2), dechanneling. The influence of dechanneling can be seen directly in Fig. 3.17, where the backscattered yield at energies below E' is considerably larger than the yield in the perfect crystal, in spite of the fact that all atoms are on lattice sites in this depth region.

The random fraction χ' of the beam at a depth t is [155]

$$\chi'(t) = 1 - [1 - \chi(t)]\exp[-\gamma(t)] \tag{3.7}$$

where

$$\gamma(t) = \int_0^t P(t)\, N'(t)\, dt \tag{3.8}$$

$P(t)\ N'(t)\ dt$ is the probability of a particle in the aligned beam being scattered into the random beam by the atoms $N'(t)$, and $\chi(t)$ is the random fraction of the beam at t if all atoms are on lattice sites.

Equation (3.6) shows that the depth distribution of the defect concentration $N'(t)$ can be determined from measurements of the yield in the aligned direction, if P is known. Accurate analytical expressions for P are available for two limiting cases: (1) when the dechanneling is due entirely to single deflections greater than the critical angle for channeling; and (2) when dechanneling is due to multiple scattering in which a total deflection greater than the critical angle is a result of many deflections, each of which is much smaller than the critical angle. However, when the disordered region is less than 1000 Å thick, the contribution of dechanneling (although significant) can be small. The total amount of disorder in the layer can then be determined from the area of the peak in the aligned spectrum, as indicated by the shaded portion of the aligned spectrum in Fig. 3.17. Two corrections must first be subtracted from this shaded area: (1) the area of the surface peak in the aligned *undoped* spectrum, and (2) the contribution arising from interactions between the gradually increasing random component (shown approximately by a dotted line in Fig. 3.17) and the *nondisplaced* fraction of the lattice atoms. For shallow disorder, these corrections are both fairly small and can be estimated with sufficient accuracy by the dotted-line approximation in Fig. 3.17. However, whenever the aligned spectrum at energies $\lesssim E'$ is much greater than the undoped spectrum, (for example, for higher-energy implants where the disorder extends to depths greater than ~1000 Å), the analysis becomes complicated and a better procedure is required. Recently, Feldman[159] has developed an iterative computer program for handling such cases. This type of analysis has also been used by Eisen *et al.*[159a] to determine the depth distribution of disorder produced in silicon by boron-ion implantations at energies between 60 and 300 keV.

Figure 3.18 shows the energy spectrum that conceptually could be obtained from scattering measurements on a sample containing a lightly damaged and an amorphous layer. Although qualitative evaluation of the scattering spectra can be made concerning the disorder in the deeper layer (i.e., in the depth region t'' to t''' corresponding to energies between E'' and E'''), the determination of $N'(t)$ depends on a knowledge of the dechanneling mechanism. In constructing the graph, P was chosen as a constant, and only the linear term was used in the expansion of $\exp[-\gamma(t)]$.

Figure 3.19 shows the aligned and random spectra for germanium crystals implanted at room temperature with various doses of 40-keV indium ions.[157] A reference spectrum for an undoped crystal is included for comparison. For the lowest dose (6.3×10^{12} ions/cm^2), the number of atoms contributing to the surface peak is equivalent to ~15 atomic layers (~2×10^{16}

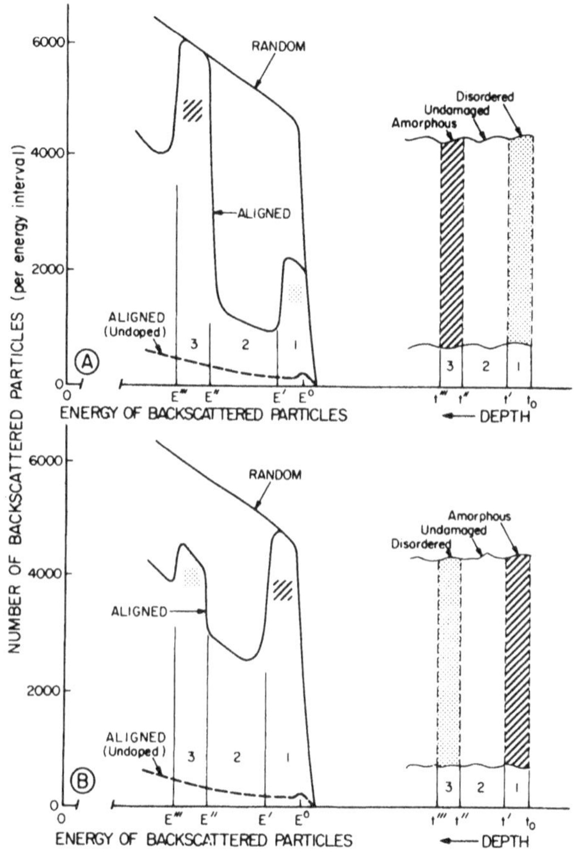

Fig. 3.18. Representation of the random and aligned spectra obtained from scattering measurements on crystals with disordered regions, represented on the right-hand side of the figure. The energy values E^0, E', E'', E''' in the aligned spectra correspond to crystal depths t_0 (=zero depth), t', t'', and t''', respectively. The broken lines represent the aligned spectra from a perfect crystal.

germanium atoms/cm^2). It can be seen that the height of the aligned spectrum increases with an increase in ion dose. The height of the surface peak depends, however, not only on the amount of disorder but also on the depth of the disordered region, compared to the depth resolution of the detector. The depth resolution is determined by the product of the energy resolution of the detector and the reciprocal of the stopping power dE/dx encountered by the analyzing beam (cf. Section 4.2.3). In the limiting case, when the dose of implanted ions is sufficient to create an amorphous layer whose thickness exceeds the depth resolution of the detector, the surface peak in the aligned spectrum

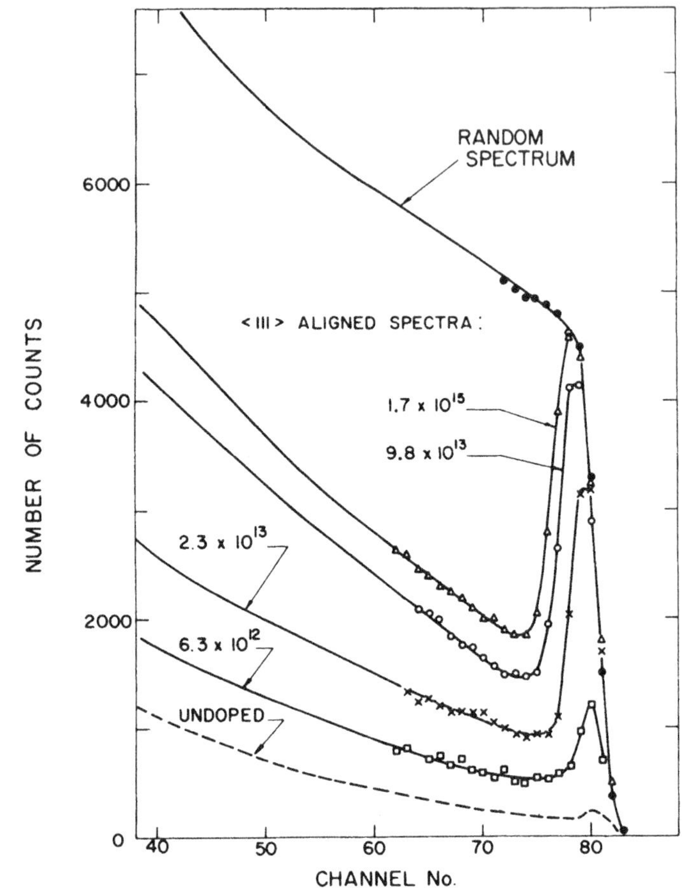

Fig. 3.19. Aligned (⟨111⟩) and random spectrum for backscattering from germanium crystals implanted at room temperature with various doses of 40-keV indium ions. The analyzing beam was 1.0-MeV helium, and all random spectra coincide within the statistical counting errors. (From Mayer *et al.*[157])

will coincide with the random yield. This is apparent in the high-dose (1.7×10^{15} ions/cm^2) spectrum of Fig. 3.19, where the depth (approximately 400 Å) of the amorphous region is larger than the detector depth resolution of approximately 300 Å.

The foregoing discussion covers only the simplest aspects of the channeling-effect method of studying lattice disorder. A more sensitive technique is to use double alignment where *both* the incoming and the elastically scattered (detected) particles are aligned with major crystallographic axes. Bøgh[155] has shown that this technique enhances the sensitivity from a total number of displaced atoms equivalent to $\sim 2 \times 10^{15}$ atoms/cm^2 with single alignment to $\sim 2 \times 10^{14}$ atoms/cm^2 with double alignment. Also, the optimum choice of beam parameters (species and energy) for such analyses will depend on the depth of the implanted layer and on the detector resolution[155,157] (see also Chapter 4.)

3.4 Results and Discussion

In this part and in Part 3.5, we will summarize the experimental data and try to reconcile them with theoretical predictions in hopes of constructing a unified picture. This attempt is somewhat premature, because there are many aspects of implantation disorder that have not yet been studied. In the first place, most experimental studies have been concerned with implantation processes at room temperature and above. At these temperatures diffusion and recombination of vacancies and interstitials are expected to occur. Further, in most cases, only one or two techniques have been used by a given group to study implantation processes. As a result it may be misleading for us to try to compare data obtained by different techniques, particularly since implantation conditions are often widely different.

3.4.1 Lattice Disorder Produced at Room Temperature

a. *Random Incidence*

In Part 3.3, we saw that isolated regions of disorder are formed in low-dose implantations. Parsons[124] concluded in his electron-microscopy investigation of 100-keV oxygen bombardment of germanium that the discrete, dark-contrast damaged regions have an amorphous structure. He first noted that their presence in the crystalline germanium lattice results in the superposition of their respective electron-diffraction patterns as shown in Fig. 3.9. Later, high-resolution dark-field techniques were used in which a segment of the second diffuse Debye ring in the electron-diffraction pattern was examined [Fig. 3.20(a)]. Parsons then observed that the damage regions exhibited white contrast and concluded that these regions were amorphous. Amorphous regions diffract more in this direction (i.e., in the diffuse Debye ring) than would be scattered by the crystalline germanium lattice. Thus they appear as small white spots on a dark-field positive print as shown in Fig. 3.20(b). Parsons also used the structure-factor contrast method of Ashby and Brown[160] to verify that the disordered regions have an amorphous character. It should be noted that Parsons[124] found only one disordered region for every five incident oxygen ions, although some of the disorder could have been below the detectability limit of the diffraction-contrast method. From anneal data

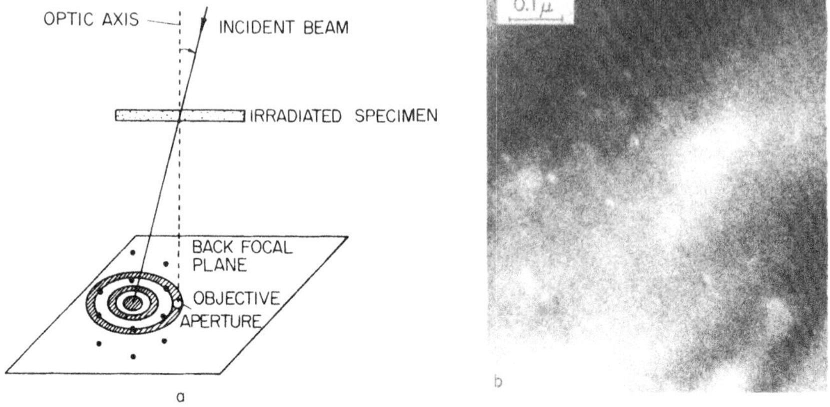

Fig. 3.20. Dark-field micrograph of the damaged regions in germanium implanted at room temperature with 100-keV oxygen ions. By placing the objective aperture as shown in (a), only those electrons diffracted by amorphous regions will contribute to the image. The white contrast of the damaged region in (b) suggests that the observed disordered regions are amorphous. (From Parsons[124].)

on implanted and neutron-irradiated samples, we believe that in most cases only a *small* fraction (<10%) of the disorder consists of amorphous regions. This conclusion (discussed in more detail in Part 3.5) has not been tested experimentally by direct comparison of electron-microscopy and channeling-effect data. Optical-absorption and electron-paramagnetic-resonance studies (cf. Part 3.5) have shown the existence of specific defects such as divacancies. The data indicate that these defects lie in crystalline regions. Hence, along the track of the particle (and energetic recoil atoms), there are heavily disordered regions surrounded by a crystalline zone containing defects. The extent of the amorphous region will depend on the ion mass and energy.

In a study[156,157] of the dose dependence of the amount of disorder, a series of room-temperature 40-keV gallium, arsenic, antimony, bismuth, and phosphorus implantations in silicon and 40-keV indium implantations in germanium have been analyzed by channeling-effect measurements. Figure 3.19 in the previous section showed some of the observed scattering spectra taken on germanium with the helium beam incident along the ⟨111⟩ direction. Under these implantation conditions (40 keV, $M_1 \gtrsim M_2$), the corrected area under the peak in the aligned spectrum is directly proportional to the number of atoms located more than ~0.2 Å from lattice sites. In Fig. 3.21 the lattice disorder is plotted versus the number of implanted ions. It can be seen that in all cases the amount of disorder increases linearly at first, but reaches a saturation value at doses $\gtrsim 10^{14}$ ions/cm^2. This indicates that at higher doses the individual disordered regions around each ion track have overlapped

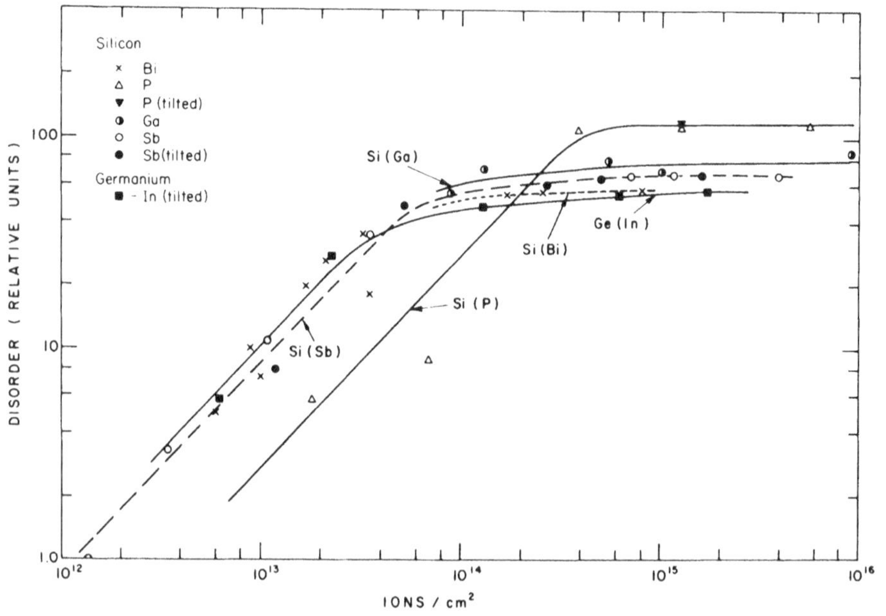

Fig. 3.21. Channeling-effect measurements of the amount of disorder produced in 40-keV room-temperature implantations in silicon and germanium. The implantations in silicon were at normal incidence (i.e., within 0.5° of the ⟨111⟩), except for a few cases where the crystals were tilted about 8° from the ⟨111⟩. The implantations in germanium were about 3° off the ⟨111⟩. (From Davies *et al.*[156] and Mayer *et al.*[157])

sufficiently to produce a completely disordered layer. In this "saturated" layer the height of the surface peak in the aligned spectrum agrees with the corresponding yield in the random spectrum. This fact, along with electron-diffraction studies performed on the same samples (and in the work of Parsons[124] and Large and Bicknell[126]), shows that the disordered layer does not possess long-range order.† We will define this condition as "amorphous." In channeling-effect measurements (cf. Chapter 4) there is a marked difference in the anneal characteristics between samples in which the dose is sufficient to form an amorphous layer and samples that received lower doses.

† Crowder *et al.*[160a] have found that the g-value, line-shape, and line-width of electron spin-resonance signals of silicon implanted with phosphorus and arsenic at room temperature are identical to those of the electron states observed in amorphous silicon films prepared by RF sputtering. The dose dependence of the number of spins per cm^2 exhibited the same linear relationship at low-dose levels and saturation at high-dose levels as the disorder determined by channeling-effect measurements. Optical-reflectivity measurements[116, 116a] also show a correlation between the characteristics of amorphous sputtered films and "saturated-dose" implanted layers.

Approximate values of the dose needed to form an amorphous layer are given in Table 3.3.

In the phosphorus implantations, the amount of disorder at saturation "doses" was significantly greater than in the corresponding gallium, arsenic, antimony, or bismuth implantations. This is a consequence of the fact that the mean range of a 40-keV phosphorus ion is significantly greater than that of the heavier ions. Therefore, the phosphorus-induced disorder extended to greater depths, thereby involving a larger number of lattice atoms; hence, the area of the peak in the aligned spectrum was larger.

In measurements[161–163] on samples that received ion doses $<10^{14}$ ions/cm^2, no significant difference was noted in the amount of disorder for aligned spectra along $\langle 111 \rangle$ and $\langle 110 \rangle$ directions. We therefore conclude that the displaced atoms are not located in the tetrahedral interstitial sites (cf. Chapter 4). This is not surprising in view of the high mobility of interstitial silicon atoms at room temperature.

From data similar to those in Fig. 3.21, it has been estimated that the disorder produced in germanium and silicon by each 40-keV antimony or indium ion is equivalent to approximately 3000 off-lattice atoms. The corresponding numbers for 40-keV phosphorus[157] and boron[161] implantations in silicon are 1200 and 10 respectively. Part (but not all) of this difference can be attributed to the increased contribution of electronic stopping processes relative to nuclear stopping for projectiles of low atomic number.

For room-temperature boron implantations, the dose required to form an amorphous layer is $>10^{16}$ ions/cm^2. In comparison with antimony implants, this dose is much larger than that calculated from the distribution between electronic and nuclear stopping, and suggests that for low-Z ions, a large amount of recombination of the disorder is occurring at room temperature. This is discussed in more detail in Section 3.4.3

The observed number of 3000 off-lattice atoms per 40-keV heavy ion is approximately a factor of 2 *greater* than the predicted *upper* limit of $N(E)$, based on a well-defined displacement energy [Eq. (3.4)]. Since some recombination of vacancies and interstitials is expected to occur during implantation,

TABLE 3.3

APPROXIMATE DOSE OF 40-keV IONS REQUIRED TO FORM AN AMORPHOUS LAYER IN ROOM-TEMPERATURE[a] IMPLANTATIONS IN SILICON

Ion	B	Al, P	Ga, As	In, Sb	Tl, Bi
Dose	$\gtrsim 2 \times 10^{16}$	$\gtrsim 5 \times 10^{14}$	2×10^{14}	10^{14}	5×10^{13}

[a] The amount of disorder depends on implantation temperature (cf. Section 3.4.3).

the measured number of off-lattice atoms should be *less* than this theoretical upper limit. There are at least two factors that could contribute to the observed higher value:

(1) The amount of disorder around the track is sufficiently great that in certain regions a large fraction of the atoms may be displaced. In such regions, as few as two bonds might have to be disrupted per displaced atom instead of the normal value of four. This would lower the average displacement energy. Marsden *et al.*[164] have estimated that the average concentration of displaced atoms around the path of a 40-keV antimony atom is only ~3% of the total number of lattice atoms contained within the volume of disorder. Hence it is unlikely that this factor will account for much of the above discrepancy.

(2) The amount of disorder determined by channeling-effect measurements may include a contribution from localized distortion of the surrounding lattice. For example, Watkins[165] has estimated that the six nearest neighbors of a divacancy are displaced by ~0.25 Å and would therefore be included in channeling-effect measurements of the disorder. If a similar distortion exists around other stable defects in the implanted region, this factor could account for the observed discrepancy. Further experimental and theoretical work is required to clarify the relative importance of these two factors.

b. *Influence of Channeling*

Some of the implantations (Fig. 3.21) were performed with crystals tilted 7–8° away from the ⟨111⟩ axial direction. In these cases the measured disorder agreed with that observed in samples implanted under perpendicular incidence (i.e., within 0.5° of the ⟨111⟩ direction), indicating that in these investigations channeling of the implanted ion was not an important factor in determining the disorder.† However, recent channeling-effect measurements[166] (Fig. 3.22) of the amount of disorder produced in silicon at much higher energies (1-MeV xenon ions) showed that the dose required to form an amorphous layer was ~5 times larger for an aligned beam compared to that for a random implantation. In this experiment, the samples were carefully aligned to within 0.05° of the ⟨110⟩ axis before implantation. In the subsequent backscattering spectra for a 1.8-MeV helium beam (Fig. 3.22), it can be seen that the peak in the disorder is located below the surface. This is an indication of the importance of electronic stopping at 1 MeV ($\varepsilon = 1.7$ for

† Bøgh[165a] finds that the amount of disorder produced in a well-aligned ($\leqslant 0.1°$ to the ⟨110⟩) implantation of 50-keV antimony in silicon is a factor of 2–3 smaller than that in a random implant. This difference decreases rapidly with increasing dose, presumably because of the enhanced dechanneling caused by the accumulated lattice disorder.

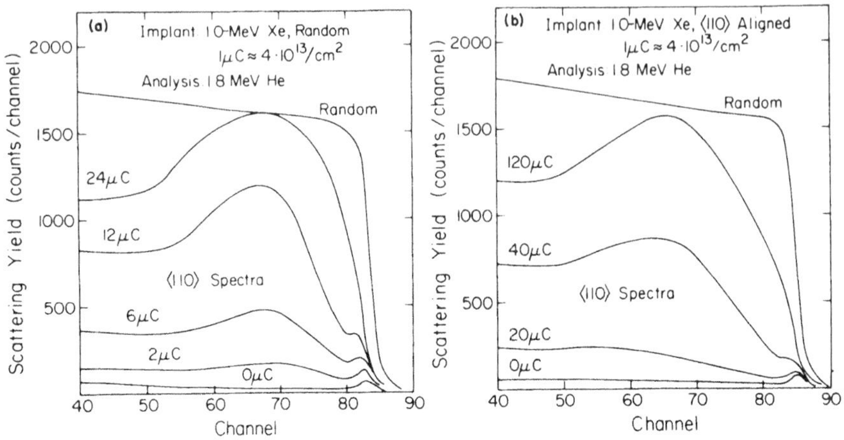

Fig. 3.22. Aligned (⟨110⟩) and random spectra for backscattering from silicon crystals implanted with 1.0-MeV xenon at (a) random beam incidence and (b) aligned (⟨110⟩) beam incidence. The analyzing beam was 1.8-MeV helium. (From Sigurd *et al.*[166])

xenon in silicon at 1 MeV). Note also that the initial introduction rate of disorder is more than 10 times greater for the random case.

In channeling-effect measurements on silicon samples implanted with boron ions at 200 keV, Eisen *et al.*[159a] found that the total amount of disorder observed in channeled implants relative to that in random implants depended on the implantation temperature. The ratio was about one-half for room-temperature implants of the same dose and appreciably lower for implants made at $-150°$ C.

In optical measurements of the disorder, Nelson *et al.*[114,115,167] found that channeling of the incident ion had a large effect on the radiation damage produced in silicon. In their work, a single crystal of silicon was mounted on a rotatable holder that allowed simultaneous rotation about two orthogonal axes. The face of the crystal was mounted in such a way that, as θ was increased in 1° steps (Fig. 3.23), the beam would describe a series of concentric circles on its surface as φ was rotated through 360°. The beam consisted of 80-keV neon ions, collimated to a divergence $\lesssim 0.01°$. After a dose just sufficient to form an amorphous layer under nonchanneling conditions (as determined from the onset of the milky appearance), the angle θ was changed by 1°, and the process repeated until the whole sample had been scanned with the ion beam. Figure 3.24(a) is a direct photographic reproduction of a (111) crystal surface after bombardment, and Fig. 3.24(b) gives the simple radial projections of the corresponding crystal directions for comparison. It is evident from the regular patterns mapped out on the crystal surface that those regions *not* exhibiting the milky appearance correspond to angles of incidence

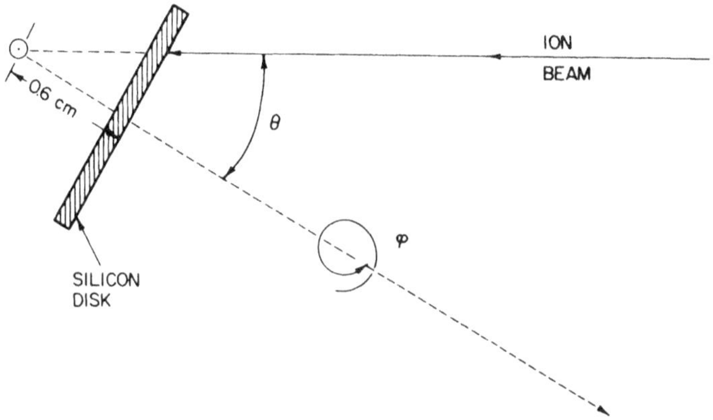

Fig. 3.23. Schematic of the experimental arrangement used by Nelson and Mazey[114] to determine the influence of channeling on lattice disorder. Indicated are the two orthogonal rotations (θ, φ) of the crystal relative to the beam.

close to major axial and planar channeling directions. The results indicate that a significant reduction in lattice disorder occurs whenever the incident neon ion is channeled. For the major channeling direction (i.e., $\langle 110 \rangle$), approximately eight times the standard dose was required to turn the surface milky, as compared to directions where the incident ions were not channeled. This implies a reduction in the radiation damage in the surface region of at least a factor of 8.

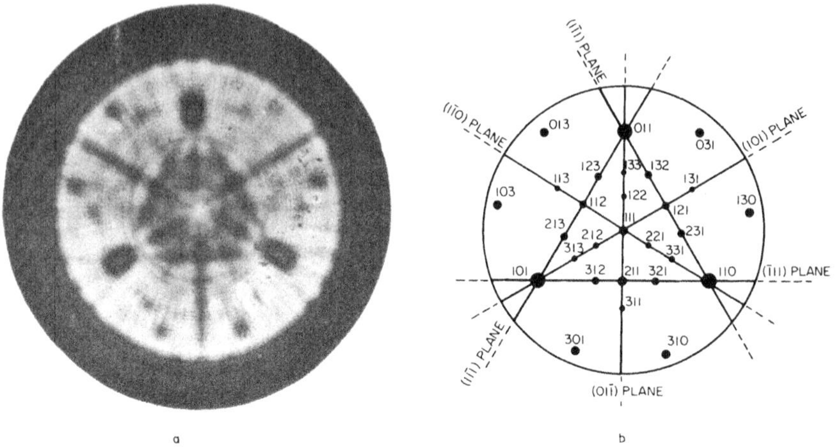

Fig. 3.24. (a) A photographic reproduction of a (111) crystal surface after bombardment with 80-keV neon ions (experimental arrangement as in Fig. 3.23). The dark regular patterns correspond to a reduction in disorder whenever the ion beam is channeled. (b) A radial projection of the {111} face of a fcc crystal showing the low-index planes and axes for comparison with (a). (From Nelson and Mazey[114].)

c. *Energy Dependence*

Marsden *et al.*[164] have measured the energy dependence of the amount of lattice disorder created in silicon by bismuth ions in the energy range 10–250 keV, using the channeling-effect technique. They find (Fig. 3.25) that the number of displaced atoms increases linearly with ion energy, as predicted from the theoretical considerations in Part 3.2. In this energy region the bismuth ions lose their energy primarily in nuclear collisions. Again, the amount of disorder per incident ion was equivalent to ~3000 off-lattice atoms per 40-keV heavy ion, similar to that found from the data of Fig. 3.21.

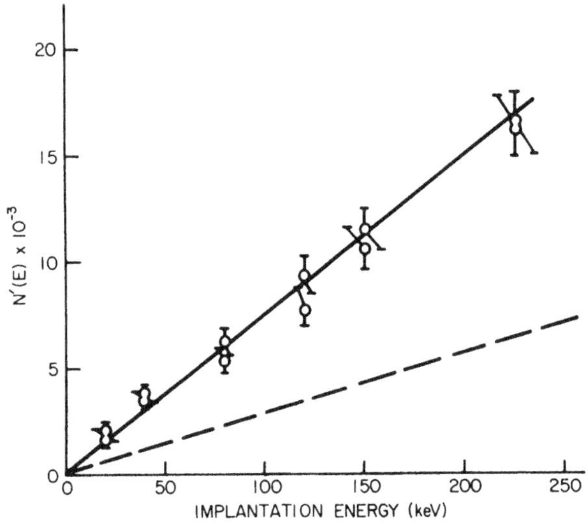

Fig. 3.25. Channeling-effect measurements of the amount of lattice disorder versus energy for bismuth implantations in silicon. The solid line is the number ($\times 10^{-3}$) of off-lattice-site silicon atoms per incident bismuth ion. The dashed line is the prediction of Eq. (3.1). (From Marsden *et al.*[164])

3.4.2 Depth Distribution of Lattice Disorder

There have been relatively few measurements of the depth distribution of disorder produced in ion implantation. For low-energy ions incident on germanium, MacDonald and Haneman[153,154] determined the depth distribution by sputtering (Section 3.3.4). They found that the depth of disorder increased with ion energy, as shown in Fig. 3.26 for argon ions. Sigmund and Sanders[15] compared their calculated range ($\langle x \rangle$) and damage ($\langle x \rangle_D$) distributions with the measured distributions of MacDonald and Haneman (Table 3.4). The calculations are in reasonable agreement with experiment, particularly in view of the very low ion energies, for which the calculations suffer from various uncertainties.

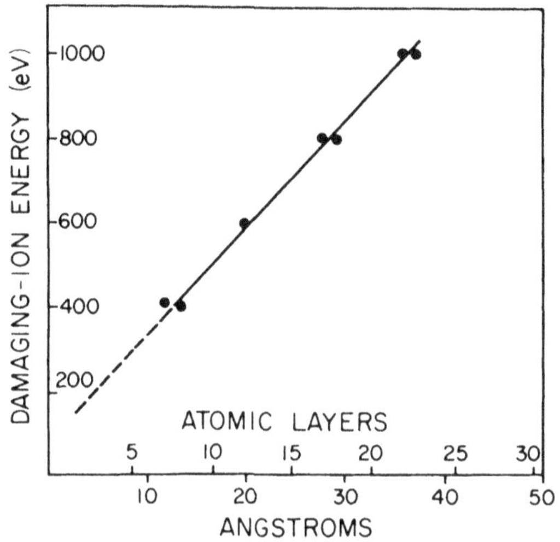

Fig. 3.26. The depth of damage produced by bombardment of germanium by low-energy argon ions. Sputtering techniques were used to determine the depth. (From MacDonald and Haneman.[154])

The disorder produced by high-energy (0.2–2-MeV) low-Z ions (where electronic stopping predominates) lies well below the surface. Schwuttke *et al.*[141] used transmission electron microscopy, X-ray topography, and microsectioning techniques to investigate silicon crystals bombarded with 2-MeV nitrogen ions. They found that the silicon lattice was relatively undamaged to a depth of 1.7 microns below the surface; however, dispersed in this region were precipitates and some radiation-damage defects. A sharp boundary line separated this relatively perfect region from a highly damaged

TABLE 3.4

Comparison of the Depth of Disorder (Å) with Calculated Ranges and Damage Depths[a]

	500 eV			1000 eV		
	Ne	Ar	Kr	Ne	Ar	Kr
$\langle x\rangle + \langle \Delta x^2\rangle^{1/2}$	32	21	14	51	34	23
$\langle x\rangle_D + \langle \Delta x^2\rangle_D{}^{1/2}$	27	18	13	42	29	20
Measured[b]	26	16	13	40	35	30

[a] Sigmund and Sanders.[15]
[b] MacDonald and Haneman.[153,154]

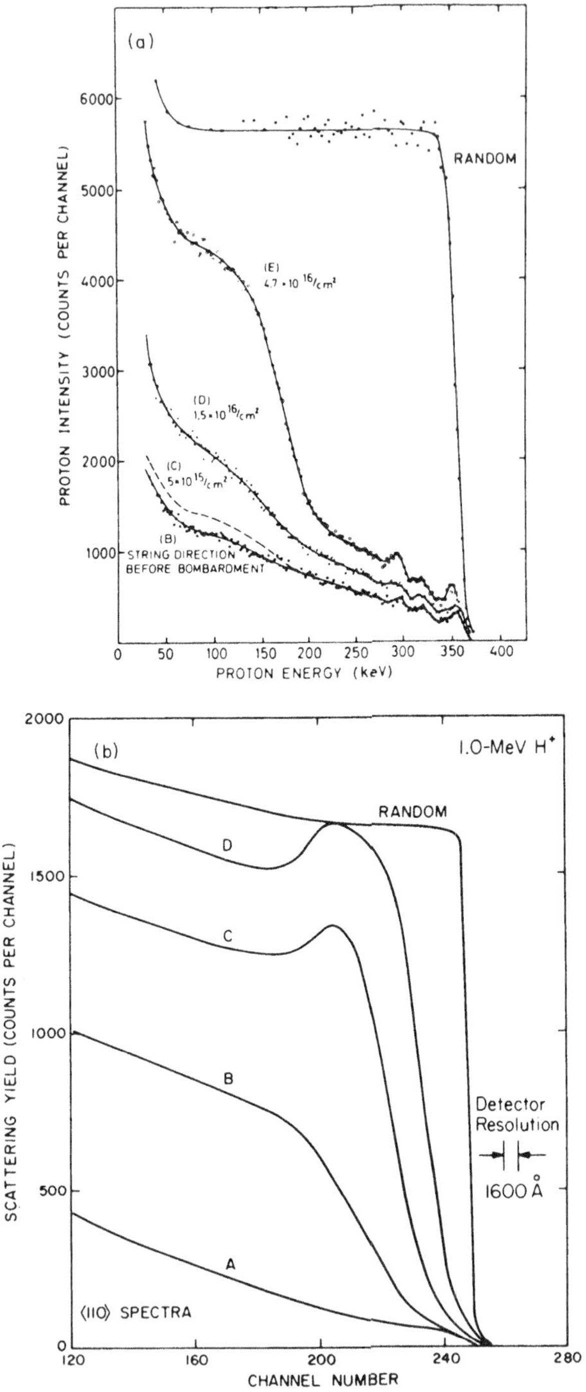
(a)
PROTON INTENSITY (COUNTS PER CHANNEL)
6000
5000
4000
3000
2000
1000
RANDOM
(E)
4.7 × 10 16/cm²
(D)
1.5 × 10 16/cm²
(C)
5 × 10 15/cm²
(B)
STRING DIRECTION
BEFORE BOMBARDMENT
0
50
100
150
200
250
300
350
400
PROTON ENERGY (keV)
2000
(b)
1.0-MeV H+
RANDOM
D
C
B
A
1500
1000
500
0
SCATTERING YIELD (COUNTS PER CHANNEL)
Detector
Resolution
1600 Å
⟨110⟩ SPECTRA
120
160
200
240
280
CHANNEL NUMBER

Fig. 3.27

layer extending to greater depths. Their results confirm the fact that a nitrogen ion of 2 MeV in silicon loses its energy primarily in electronic interactions over the major part of its path, and that the nuclear energy loss becomes significant only near the end of its range. In later studies[141a] on room-temperature, 1–3-MeV boron, nitrogen, and phosphorus implantations, it was found that only the region near the range of the ions was damaged severely and that the lattice parameters of the regions above and below the damaged layer remained unchanged.

There have been several channeling-effect measurements on high-energy low-Z implants that also illustrate the deep nature of the disorder for the predominantly electronic stopping regime. Gibson *et al.*[168] implanted 400-keV lithium ions in silicon and observed that the disorder peak occurred deep in the crystal and that the region adjacent to the surface was relatively free of damage [Fig. 3.27(a)]. A similar effect has been seen in germanium crystals implanted with 1.0-MeV carbon[169] [Fig. 3.27(b)]; in this case at the higher doses, a well-resolved disorder peak is observed at a depth of ~ 1.3 microns, a depth that corresponds roughly to R_p for the 1.0-MeV carbon beam. The same features are observed[170] also in Fig. 3.28 for 200-keV high-dose boron implantations in silicon; again, the depth of the disorder peak (~ 5400 Å) is comparable to R_p. This approximate agreement between the disorder peak and R_p is consistent with the predicted disorder distribution for the $M_1 \ll M_2$ case (Fig. 3.5). Figure 3.28 shows that the proper choice of analyzing-beam energy is important. This is discussed in more detail in Chapter 4. In subsequent work, Eisen *et al.*[159a] found that, for boron ions at energies between 60 and 300 keV, the peak in the disorder distribution occurred at depths comparable to or slightly less than the projected range R_p of the boron ions.

Because of the limited depth resolution of the detector system used in most channeling effect studies, the depth distributions of disorder in 2–50-keV implantations of high-mass ions, such as antimony or thallium, has not yet been studied in detail. Early investigations[156] combined with layer removal indicated that the disorder did *not* extend beyond the range distribution of the implanted ions. In fact, from the observation [171] [Part 4.3, Fig. 4.20(a)]

Fig. 3.27. (Left) (a) Energy spectra of 400-keV protons scattered from a silicon crystal. The curve marked *random* is for protons incident in a nonchanneling direction. All others are for protons incident parallel to the ⟨110⟩ axis after successive room-temperature implantations of 400-keV ^{7}Li ions incident parallel to the ⟨110⟩ direction. (From Gibson *et al.*[168]) (b) Random and aligned (⟨110⟩) energy spectra of 1.0-MeV protons backscattered from a germanium crystal implanted in a random direction with 1.0-MeV carbon ions. The upper curve (D) represents a dose of $\sim 2 \times 10^{16}$ carbon ions/cm^2. Curve A is the spectrum before carbon bombardment. The detector resolution was equivalent to 1600 Å. (From Björkqvist *et al.*[169])

that implanted bismuth and thallium atoms occupy substitutional sites in low-dose ($<2 \times 10^{13}$ ions/cm^2) implantations, it can be argued that the major amount of lattice disorder lies at shallower depths than the stopping point of the atoms.

The depth distribution of vacancy-associated defects produced in silicon by 400-keV oxygen ions was investigated by Stein *et al.*[170a] Divacancies were detected from the characteristic 1.8μ optical-absorption band, and Si–P3 centers (paramagnetic, divacancy-like structures that were first observed in neutron-irradiation studies) from electron-paramagnetic-resonance (EPR) measurements.[170b] The distributions were determined from repeated measurements following successive anodizations and strippings of the implanted layer. The depth distributions of divacancies and Si–P3 centers were in good agreement with calculations by Brice[112a] of the distribution of the energy spent in atomic collision processes.

It is somewhat surprising that more investigations have not been made of the depth distribution of lattice disorder. The techniques are at hand, and even if layer removal methods are required, it should not be an overly difficult task to resolve the depth distribution.

3.4.3 Influence of Substrate Temperature on Lattice Disorder

One of the most striking results of these investigations[156,172,173] was the marked reduction in the amount of lattice disorder in hot-substrate implantations in comparison with corresponding room-temperature implantations annealed to the same temperature. In implantations into substrates held at 450° C, it was found that antimony doses $\lesssim 10^{15}$ ions/cm^2 produced relatively little residual lattice disorder [Chapter 4, Fig. 4.19(b)], whereas in room-temperature implantations to the same dose, an amorphous layer was formed that required anneal temperatures of 600°C to restore lattice order. The fact that lattice disorder can be reduced by the use of heated substrates during implantation has also been shown in sputtering and in ion- and electron-emission experiments.[144–151]

Channeling-effect measurements have been used in several cases to determine the influence of substrate temperature on the residual lattice disorder after implantation. Eriksson *et al.*[174] analyzed 40-keV antimony implantations at a dose of 2×10^{14} ions/cm^2—a dose sufficient in room-temperature implantations to have formed a completely amorphous layer. As the implantation temperature was increased above 150° C, a marked decrease was observed in the lattice disorder (dose rate $\sim 10^{12}$ ions/cm^2 sec). In this transition region, the disorder should be dose-rate dependent; however, this has not yet been investigated. At substrate temperatures above 300° C, the lattice disorder was close to the detectability limit of the channeling-effect measurements (cf. Fig. 4.16 of Chapter 4).

Picraux *et al.*[162] have analyzed in more detail the dependence on substrate temperature for antimony implantations in silicon. Low-dose implantations were used so that the amount of disorder was insufficient to form an amorphous layer over the range of implant temperatures investigated. In this work, the ⟨111⟩ axis was tilted by approximately 5° with respect to the beam direction to reduce ion-channeling effects during implantation. Figure 3.29 shows the disorder per ion as a function of implant temperature as well as the anneal characteristics (dashed line in Fig. 3.29) of a low-dose room-temperature implant. Approximately 3300 displaced silicon atoms per incident antimony ion (6.8 on the ordinate scale of Fig. 3.29) remain after a room-temperature implantation; this agrees well with the data of Fig. 3.21. The temperature dependence of the disorder per incident ion is characterized by two regions. For implants in the region above 50° C, the amount of disorder per ion decreases rapidly with increasing implant temperature, whereas in the lower temperature region (less than 50° C) the disorder per incident ion is only

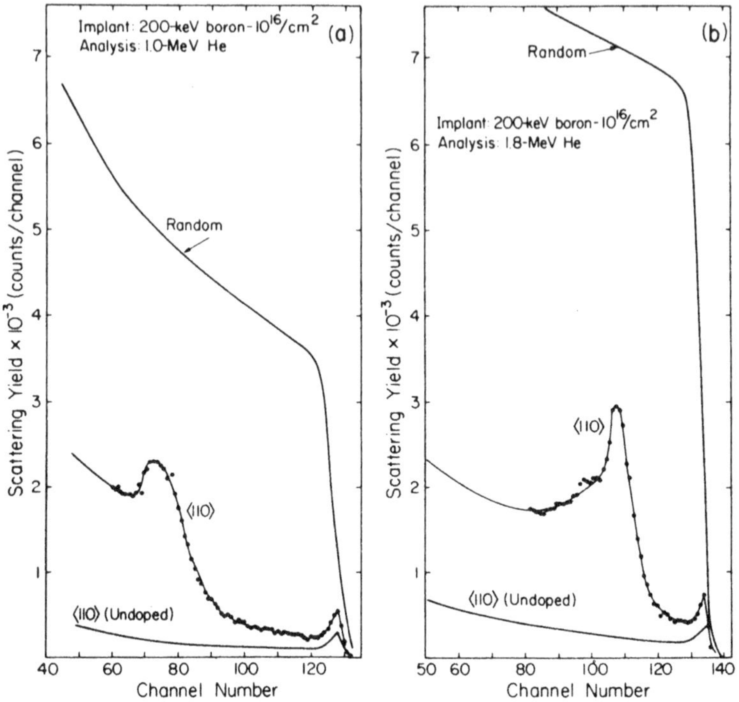

Fig. 3.28. Random and aligned (⟨110⟩) energy spectra for helium ions backscattered from a silicon crystal implanted at random incidence with 10^{16} boron ions/cm² at 200 keV. Analyzing beam was (a) 1.0 MeV and (b) 1.8 MeV. (From Eisen *et al.*[170])

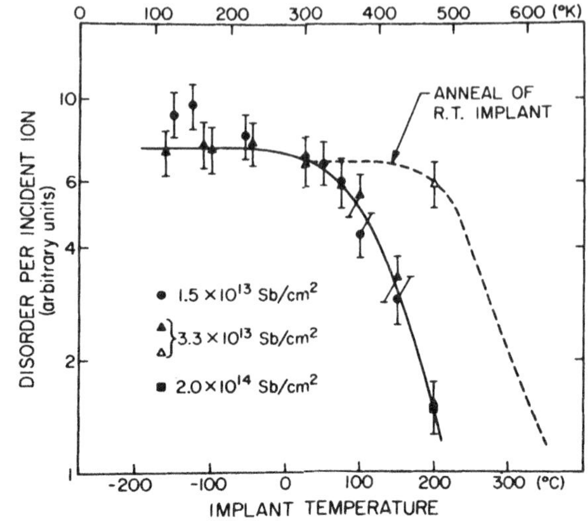

Fig. 3.29. Channeling-effect measurements of the amount of disorder per incident 40-keV antimony ion versus silicon-substrate temperature during implantation. The dashed line is the anneal of a room-temperature implantation given by Davies *et al.*[156] The open point represents the disorder in a sample implanted at room temperature and annealed at 200° C in the implantation target holder. Error bars represent the uncertainty in dose measurement. (From Picraux *et al.*[162])

mildly temperature dependent. One feature of this work was the observation that the temperature region in which the disorder per ion decreased strongly with temperature is approximately 100° C lower than the region of corresponding decrease in the anneal of a low-dose room-temperature implantation. This difference indicates that even at low doses the dynamic anneal occurring during implantation is not completely equivalent to anneal after the damage regions have been formed. Similar effects were also found[174a] in the analysis of the implantation-temperature dependence of disorder produced in silicon by 200-keV boron ions.

The temperature dependence of lattice disorder during implantation can also be established visibly from the onset of the characteristic milky appearance of the surface. Nelson and Mazey[115] and Mazey *et al.*[113] have used this as a measure of the ion dose required to produce an amorphous layer during implantation with a 60-keV beam of neon ions. The ion dose required to produce a visible coloration was determined for temperatures between 23 and 250° C. The results obtained are shown in a semilogarithmic plot in Fig. 3.30. These results show qualitatively the relation between substrate temperature and the amount of disorder produced in implantation. Channeling-effect measurements indicate that the observed dose of 10^{14} neon ions/cm^2

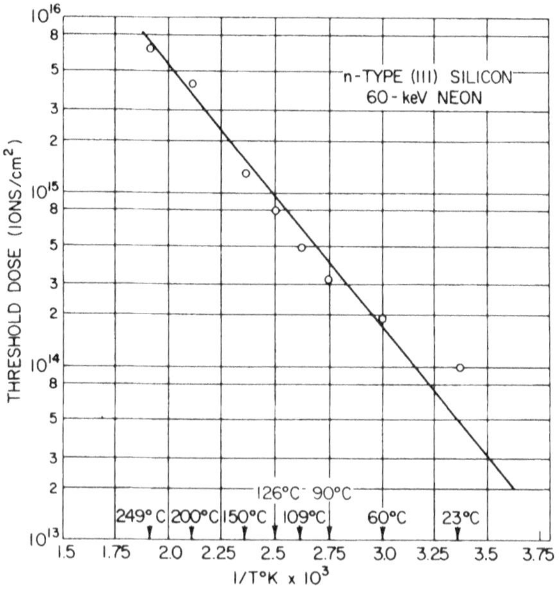

Fig. 3.30. Graph of the threshold ion dose (60-keV neon) required to produce (in silicon) a visible color change, plotted as a function of reciprocal temperature. (From Mazey *et al.*[113])

at room temperature (Fig. 3.30) is somwehat lower than that required to form an amorphous layer.†

In measurements of the amount of disorder as a function of substrate temperature, the dose *rate* should also be considered. Vook and Stein's analysis[152] indicates that this factor enters explicitly. If dose rate is not included in the analysis of the disorder versus implant temperature, the activation energy derived from such measurements may be incorrect.

For the case of light elements (i.e., boron, oxygen, fluorine) implanted in silicon or germanium at room temperature, annealing of disorder takes place during implantation. Measurements[174a, 176, 176a] on silicon implanted with boron at temperatures below $-100°$ C indicated that the amount of disorder is ten to twenty times higher than in a room-temperature implantation of comparable dose. Sigurd *et al.*[176b] observed a rather similar behavior in boron-implanted germanium. In low-temperature boron implantations in

† Nelson[175] recently pointed out that to form a completely amorphous layer requires a dose about eight times larger than that required to produce a visible change in appearance. Similar results (see also Fig. 3.6) are found for heavy ions such as gallium or antimony.

both silicon and germanium, the amount of disorder per unit energy lost in nuclear collisions (i.e., after correction for the significant amount of energy lost in electronic collisions) is comparable to that created by heavy ions.[176a, b] That is, for low-temperature implantations, the dependence of lattice disorder on ion mass reflects primarily the influence of electronic stopping.

The difference between the amount of disorder that anneals during room-temperature implantations and that at lower temperatures depends quite strongly on ion mass. Approximately six times less disorder was found in a silicon sample implanted with 200-keV fluorine at room temperature than in a similar sample implanted at $-150°$ C. A factor of 6 for fluorine and 10–20 for boron implants contrasts sharply with the results of antimony implantations in which there is not much temperature dependence of the disorder-production rate below 50° C.

It should be noted that at implantation temperatures below $-80°$ C, the dose of boron ions required to form an amorphous layer in silicon is approximately 10^{15} ions/cm^2, and a dose of 10^{16} ions/cm^2 is sufficient to extend this amorphous layer to the surface and to depths somewhat greater than the projected range R_p.[174a] It was also found[159a, 176a] that in low-dose, low-temperature boron implantations, the amount of disorder decreased by only 20 to 30% when the sample was warmed to room temperature. This indicates that the amount of annealing that takes place during a room-temperature implantation is an order of magnitude greater than that for a low-temperature boron implant annealed at room temperature.

3.4.4 Anneal Behavior of Lattice Disorder

Much of the gross lattice disorder produced by implantation in silicon and germanium can be annealed at temperatures of 600° C or lower.† Visual estimates and channeling-effect measurements both indicate that there are well-defined stages in the anneal process. Electron microscopy studies indicate that dislocation loops are formed on the annealing of the disordered regions, generally at temperatures above 600° C.

Channeling-effect measurements[157] of the isochronal annealing (10-min anneal periods) of lattice disorder are shown in Fig. 3.31 for silicon and germanium samples implanted at room temperature with 40-keV antimony and indium ions respectively. The uppermost curves represent amorphous implanted layers, while the lower curves represent low-dose implantations

† This does not imply that all *electrically active* defect centers disappear by 600° C. The data shown in Chapter 5 indicate that in implantations made at elevated substrate temperatures, electrical effects due to defect centers exist to 800–900° C.

characterized by isolated disordered regions. The data of Fig. 3.31 indicate that the anneal temperature required to reorder the lattice is considerably higher for the amorphous layer than for the isolated disordered regions. For heavy ($M_1 \gtrsim M_2$) ion implantations in both silicon and germanium, the anneal characteristics exhibit two well-defined regimes: (1) for low doses, where the particle tracks are well separated from each other, reordering occurs at 180° C in germanium and at 260° C in silicon; and (2) at higher doses, where the amorphous layer is formed, reordering occurs at 380° C in germanium and 570° C in silicon. These temperatures correspond to the midpoints in the 10-min isochronal anneal curves. As shown in Chapter 5 (Fig. 5.1), the anneal of the amorphous layer exerts a dominant effect on the electrical characteristics.

There should obviously be some sort of transition in the anneal behavior between the case where individual disordered regions are spatially well separated and that where a completely amorphous layer exists. In this transition-dose region, the anneal kinetics may depend on the relative amount of overlap of the disorder around each track. An example is seen in Fig. 3.31 for the intermediate dose of the three germanium implants. This transition

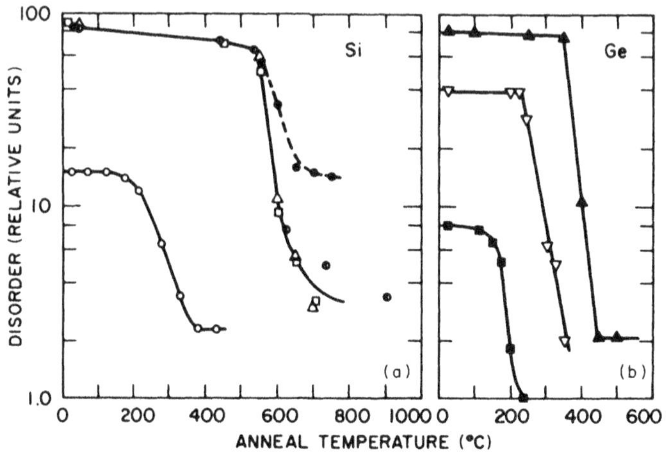

Fig. 3.31. (a) Lattice disorder versus annealing temperature for antimony, arsenic, and gallium implantations in silicon at room temperature. The data were determined from channeling-effect measurements. The doses were as follows: ○, 1.1×10^{13} Sb ions/cm²; ◐, 3×10^{15} Sb ions/cm²; ●, 2.6×10^{15} Sb ions/cm²; □, 4×10^{14} Ga ions/cm²; △, 2.5×10^{15} As ions/cm². (Identical anneal behavior was also obtained for a 40-keV phosphorous implantation of 2.0×10^{13} ions/cm².) (b) Lattice disorder versus anneal temperature for indium implantations into germanium at room temperature: ■, 6.3×10^{12} ions/cm²; ▽, 2.3×10^{13} ions/cm²; and ▲, 6.2×10^{14} ions/cm². (From Mayer *et al.*[157])

region has not been explored in detail, and hence both the anneal behavior and the ion doses cannot be specified exactly.

A significant difference has recently been found[174a] between the anneal of low-dose, low-temperature boron implants and comparable heavier-mass implants. In the boron case, the disorder began to anneal at temperatures above $-80°$ C and had decreased to half its initial value after annealing to approximately 100° C. However, following an anneal at 200° C, there was still appreciable disorder remaining. Considerably more work is required before a complete picture can be given of the dependence of anneal behavior on ion-mass and implant temperature.

It should be pointed out that the amorphous layer formed in silicon does not always fully anneal at temperatures around 600° C. In 40-keV implantations, doses greater than approximately 10^{15} ions/cm^2 sometimes require much higher annealing temperatures, as indicated in Fig. 3.31. At these high doses the data suggested that the anneal temperature was species dependent and that a correlation might exist between the anneal behavior of high-dose implantations and the solubility of the implanted species. At such high-dose levels the concentration of implanted ions can amount to several atomic percent of the substrate; hence, the anneal kinetics may also be influenced by the formation of alloy phases, gas bubbles, large numbers of dislocation loops, or polycrystalline regions.

In both germanium and silicon, the anneal characteristics of the lattice disorder produced by ion implantation exhibit strong correlations with fast-neutron-irradiated samples (cf. Section 3.5). For the isolated-disorder regions in ion-implanted crystals, the anneal behavior of divacancies[177] and of the total amount of disorder[156,157] follows closely that of divacancies[178] and of X-ray transmission intensities[140,142] in fast-neutron-irradiated samples. This suggests not only that the type of disorder is similar in the two cases but also that the anneal characteristics may be governed by the migration and dissociation of the same type of defects.

There is also a correlation between the annealing of the amorphous layer in silicon and germanium and the formation of crystalline films of these materials during vacuum evaporation.[179,180] Hass and Scott[179] have studied the formation of these films during evaporation on heated substrates. With silicon they found that when the evaporation was performed on substrates at temperatures over 600° C, crystalline coatings were formed that showed sharp diffraction patterns. At lower temperatures amorphous films were formed; these required a subsequent anneal at $\sim$700° C to convert them into crystalline form. With germanium, crystalline films were formed at temperatures between 350° and 400° C.

There is fairly convincing evidence that the amorphous layer reorders by epitaxial growth on the underlying silicon substrate. This can be seen both in the channeling-effect measurements and in the electron diffraction and microscopy work. Figure 3.32 shows channeling spectra obtained from a silicon sample implanted with 1.3×10^{15} phosphorus ions/cm^2 at room temperature and subjected to various annealing temperatures. Up to 500° C the spectra are essentially identical with the unannealed case. After a 10-min anneal at 550° C, there is a significant reduction in the width of the surface peak, but no appreciable change in its height. This decrease in the peak width indicates that the reordering of the amorphous region is initiated at the interface between the undamaged and disordered regions. The annealing is nearly completed by 655° C. Electron-diffraction patterns have been taken on thinned silicon specimens bombarded with ions to a dose such that a region to the depth t (shown in Fig. 3.33) has become completely noncrystalline. Upon anneal the region marked A (which is disordered throughout its thickness)

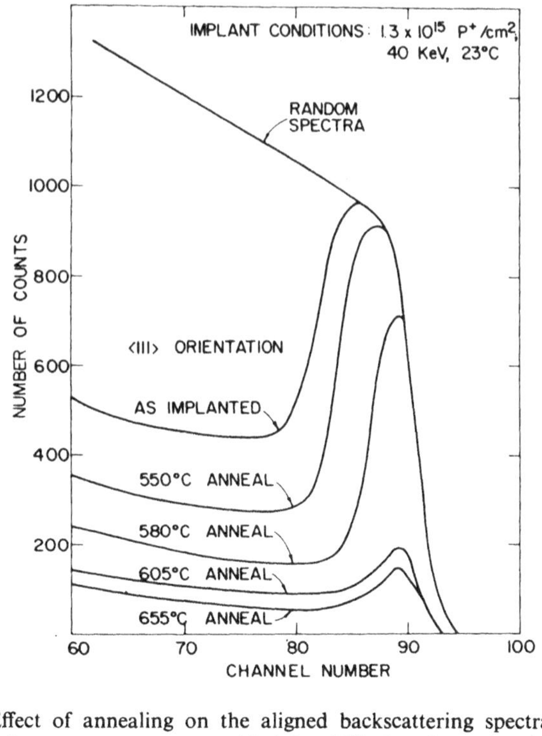

Fig. 3.32. Effect of annealing on the aligned backscattering spectra from a silicon sample implanted at room temperature with 1.3×10^{15} phosphorus ions/cm^2. All random spectra coincide within the statistical counting errors. The analyzing beam was 1.0-MeV helium ions. (From Mayer *et al.*[157])

became polycrystalline (Large and Bicknell[126]). Neighboring areas (region B) that were completely amorphous but which were over crystalline silicon recrystallized epitaxially with the layer underneath. These regions, however, sometimes contained twin planes.[113,126]

It should be emphasized that although well-defined annealing stages occur, there can still be an appreciable amount of lattice disorder left in the crystal. The electron-microscopy studies of Parsons,[124] Large and Bicknell,[126] and Mazey *et al.*[113,126] indicate that dislocations develop during anneal. Dislocation loops lying on (111) planes, edge dislocations, and also dipoles lying parallel to the ⟨110⟩ directions appear after anneal. The dislocation loops (Bicknell[131,133]) increase in size with anneal temperature from a diameter of approximately 100 Å at 700° C to a maximum of about 1000 Å at 800° C. At this point the loops appear to stabilize and do not increase in size with further increase of anneal temperature up to 900° C.

One of the interesting areas in the study of ion-implantation may lie in the use of electron microscopy to investigate the dislocations formed during anneal. Because of the complexity of the results that have been obtained so far, it is somewhat difficult to give a comprehensive summary. The dislocation loops found in electron-microscopy studies represent an amount of disorder that is in general below the detectability level for channeling-effect measurements (even if the more sensitive technique of double alignment, as suggested by Bøgh,[155] is used). For example, Bicknell estimates that for an implantation of 2×10^{15} boron ions/cm^2 at 50 keV and 300° C, the number of atoms in the dislocation loops is $\sim 10^{15}$/cm^2 after anneal at 800° C.

One anneal observation that may have a large bearing on the electrical properties of these layers is the study of the out-diffusion[180a] of radioactive gases during the anneal process. In this work radioactive gases such as ^{85}Kr are implanted at room temperature (40–100 keV energies) and the anneal temperature at which the gas diffuses out of the crystal is determined (see, for example, Matzke,[181] Kelly and Matzke,[182] Kelly,[183] and Jech and Kelly[54]).

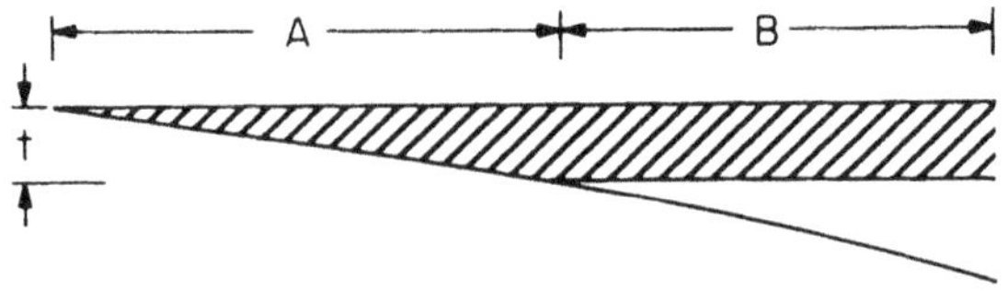

Fig. 3.33. Schematic cross section of a thinned silicon sample that has been implanted such that a noncrystalline layer is formed to a depth *t*.

It has been found that at low doses such that isolated disorder regions are formed, most of the gas diffusion does not occur until approximately the temperature for self-diffusion of the substrate.† At higher doses in semiconductor materials, however, there is a low-temperature anneal stage in the gas-release measurements, corresponding approximately to the temperature at which the amorphous layer reorders. Although the exact diffusion mechanism that leads to the release of gas at these low temperatures is not known, it is clearly associated with the reordering of the lattice.

Jech and Kelly[54,184] have recently investigated the release of ^{85}Kr from germanium and silicon that had been bombarded with doses of $\sim 10^{14}$ ions/cm^2. During annealing, they used a constant rate of heating of 23° C/min, and observed a well-defined release of ^{85}Kr in germanium at 480° C and in silicon at 735° C, as shown in Fig. 3.34. These temperatures are well below the normal self-diffusion temperatures, and correspond to the end points in the lattice-reordering data (see Fig. 3.31). It should be emphasized that these out-diffusion phenomena are not restricted to inert gases alone, but can also

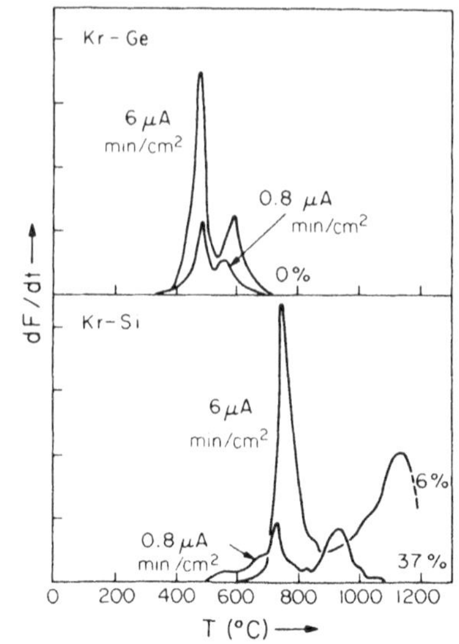

Fig. 3.34. Differential gas-release measurements for germanium and silicon implanted at room temperature with krypton at 10 keV. The quantity *dF/dt* represents fractional gas release and the percentages refer to the amount of gas remaining at the highest temperature reached. The doses are given in units of μA min/cm^2 and correspond to 2.2×10^{15} and 3×10^{14} ions/cm^2. (From Jech and Kelly.[54])

† In some cases the diffusion rates are higher than those for self-diffusion.[185]

occur for dopants which have a very low solubility (or low surface-sticking probability) at the anneal temperature. For example, Picraux *et al.*[55] have found that in silicon samples implanted with cadmium to doses $>10^{14}$ ions/cm^2, a marked decrease (over an order of magnitude in some cases) in the amount of cadmium contained within the crystal occurs during the annealing. At present, however, it is not possible to establish general guidelines as to which dopant species will exhibit these strong out-diffusion effects when the lattice reorders. For example, tellurium and bismuth (both of which have relatively low solubilities in silicon) do not exhibit such an effect. In fact, bismuth can be retained in the lattice during the annealing of the amorphous phase, at concentration levels two orders of magnitude above its equilibrium solid solubility.[174] At sufficiently high anneal temperatures, however, those dopants whose concentrations are above the solubility limits usually exhibit out-diffusion or precipitation effects.

At present there are only data on the *isochronal* anneal behavior of implanted layers. It would be very desirable to study also the *isothermal* anneal kinetics in the hope of obtaining a physical description of the defect dissociation. As will be pointed out in Part 3.5, a strong correlation has been seen between the anneal of fast-neutron-irradiated and ion-implanted germanium and silicon. Consequently, it may be possible to obtain some physical insight from an analysis of both types of irradiation disorder.

3.5 Comparison with Fast-Neutron Irradiation

Over the past 20 years there have been extensive investigations of the influence of particle irradiation on semiconductors, metals, and insulators. Of these studies, fast-neutron irradiation (i.e., neutron energies greater than 0.2 MeV) has the most features in common with those experienced in ion-implantation studies. When a fast neutron collides with a lattice atom, the average energy transmitted in the collision can be of the order of 20–100 keV. The lattice atom recoiling from such a collision will produce disorder around its track similar to that produced by, for example, a 20–100-keV phosphorus ion. However, there are some important differences between fast-neutron irradiation and ion implantation. These are:

1. In neutron-irradiated samples, the localized regions of lattice disorder are distributed uniformly throughout the sample volume. In implanted layers, the disorder is confined to a region close to the surface; this proximity to the surface may influence the anneal characteristics, particularly if the surface acts as a sink for the vacancies, interstitials, and other defects as they escape from the disordered regions.
2. Even for low-dose implantations, the average concentration of defects

in a region within 1000 Å of the surface is some orders of magnitude higher than that experienced in most neutron-irradiation studies. On the basis of a neutron-encounter probability of 0.16 cm^{-1}, a fast-neutron fluence of $\sim 10^{18}$–$10^{19}/cm^2$ is required if the concentration of displaced atoms is to approach that seen in implantations of $\sim 10^{13}$ antimony ions/cm^2. Furthermore, even if comparable fluences are used, the time scales are vastly different, viz. weeks for the neutron case, versus seconds for the ion implantation.

3. In neutron irradiation the disordered regions are spatially well separated, and there is a low probability that the dopant atoms (in the substrate) are enclosed within the disordered regions. In ion implantation, on the other hand, the implanted dopant atom comes to rest at the end of a region of disorder. These differences could be reflected in the anneal characteristics of electrical behavior.

To provide a reference to the neutron fluences commonly used, it should be noted that investigations of lattice parameter changes, X-ray transmission, or small-angle X-ray scattering studies require irradiations to dose levels of the order of 10^{19} fast neutrons/cm^2. On the other hand, *electrical* evaluation of neutron-irradiated samples are usually concerned with neutron irradiations of much lower dose (i.e., several orders of magnitude lower). Furthermore, most of the electrical measurements of neutron-irradiated samples have been carried out in the anneal-temperature range *below* 400° C, whereas the electrical characteristics of ion-implanted layers are generally studied in the anneal-temperature region *above* 400° C. Consequently, it is often difficult to make direct comparisons between neutron-irradiated and ion-implanted samples.

3.5.1 Lattice Disorder

Comparison of the anneal data on implanted and fast-neutron-irradiated samples suggests that there is a strong correlation between the types of disorder produced in the two processes. Figure 3.35 shows the fraction of disorder remaining in a germanium sample implanted at room temperature to a low dose (6×10^{13} indium ions/cm^2 at 40 keV) as compared to the data obtained in the anneal of fast-neutron-irradiated samples (fluence $\sim 5 \times 10^{19}/cm^2$). Disorder was measured in the neutron-irradiated sample by changes in small-angle scattering of X rays,[143] lattice parameter,[142] and anomalous (Borrmann) X-ray transmission.[140] In the latter two measurements, the results were virtually identical,[142] and the same curve was used for both in Fig. 3.35. There are two stages in the anneal of neutron-irradiated samples. The low-temperature stage (extending to 200° C), in which 60% of the lattice disorder and strain recovers, coincides with that observed by channeling-effect measurements on low-dose ion-implanted samples. Baldwin[142] and den Ouden[143] attribute this stage to the breakup of small disorder clusters or the migration

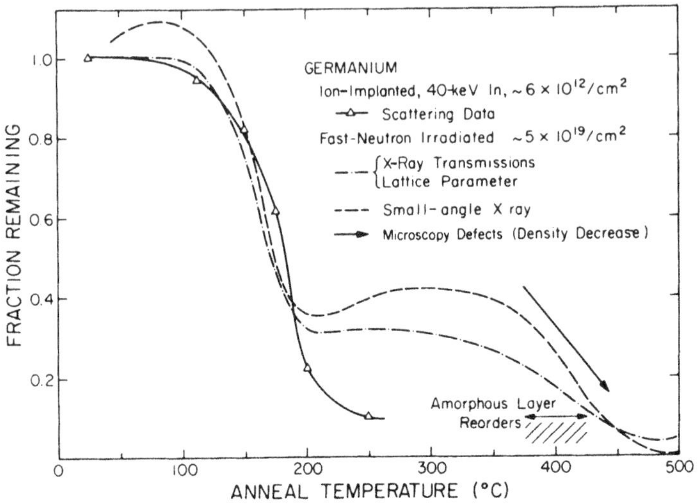

Fig. 3.35. Comparison of anneal of disorder in fast-neutron-irradiated (fluence $\sim 5 \times 10^{19}/cm^2$) and ion-implanted (dose $\sim 6 \times 10^{12}$ ions/cm^2) germanium by anomalous (Borrmann) X-ray transmission (Baldwin and Thomas[140]), lattice-parameter changes (Baldwin[142]), small-angle X-ray scattering (den Ouden[143]), electron micrographs of dark-contrast defects (den Ouden[186]), and channeling and scattering data (Mayer *et al.*[157]). The shaded region indicates the temperature region over which the amorphous layer reorders in high-dose implanted samples.

of point defects such as vacancies or divacancies. The higher-temperature stage is in the same region as that at which the amorphous layer reorders (shaded region in Fig. 3.35) in high-dose implantations. This stage has been attributed to the recovery of heavily disordered regions.

A very significant correlation can be made with the discrete, dark-contrast disordered regions observed in electron micrographs of implanted germanium by Parsons[124] and with the similar regions observed by den Ouden[186] in fast-neutron-irradiated germanium. den Ouden found dark-contrast regions with a mean diameter of ~45 Å (irradiation ambient of ~40° C). During anneal treatment up to 350° C, *no* significant change in density or contrast of the damaged regions was observed. Annealing in the temperature region above 350° C resulted in a decrease in the density of the damaged regions (as indicated by the arrow in Fig. 3.35). Parsons[124] also found that the regions were stable up to 300° C and that samples heated to 210° C (above the first anneal stage) also exhibited dark-contrast damaged regions. On the basis of these observations plus the channeling data, we can speculate that (at least when $M_1 \gtrsim M_2$) the *majority* of the disorder produced in ion-implanted samples is *not* associated with these discrete (dark-contrast) disordered regions. If this speculation is correct, it implies that the major

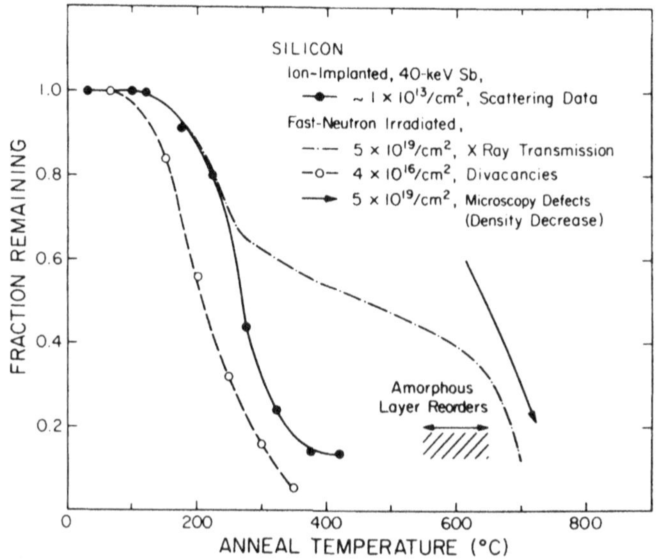

Fig. 3.36. Comparison of anneal of disorder in fast-neutron-irradiated and ion-implanted silicon by anomalous (Borrmann) X-ray transmission (Baldwin and Thomas[140]), divacancies (Cheng and Lori[178]), transmission electron micrographs of visible defects (Pankratz *et al.*[187]), and channeling and scattering data (Mayer *et al.*[157]). The shaded region indicates the temperature region over which the amorphous layer reorders in high-dose implanted samples.

part of the disorder around the track of the incident ion is not in the form of an amorphous zone.

A similar, but not quite as well-defined situation occurs in silicon, as indicated in Fig. 3.36. The lower anneal stage[140] in neutron-irradiated material again follows the anneal behavior (channeling data) of low-dose implanted samples, and the higher-temperature stage also occurs in approximately the same temperature region as the reordering of the amorphous layer in high-dose implantations. In the isolated-disorder dose range, there are heavily damaged (in some cases, amorphous) regions, and surrounding these is a crystalline zone containing defects. The existence of this zone is clearly evident from optical-absorption studies of divacancies[177] and paramagnetic-resonance investigations.[186a]

Pankratz *et al.*[187] have observed defects by bright- and dark-field transmission electron microscopy in silicon samples irradiated with fast neutrons at 125° C. The density was approximately proportional to the neutron dose. They noted a rapid decrease in the density of defects for anneal treatments above 700° C (as indicated by the arrow in Fig. 3.36). As noted in ion-implanted samples,[113,126,131] dislocation loops grow for anneal temperatures greater than 600° C.

The data in Fig. 3.36 indicate also that the anneal behavior of the low-temperature stage is similar to the anneal of divacancies,[178]† and hence may be associated with the migration of point defects.[142,188] The divacancies were determined from infrared absorption[189] and, since they retained their individual properties in terms of absorption frequencies, it was concluded[178] that they were in crystalline regions. The number of observed divacancies is small compared to the amount of disorder: 38 per scattered neutron[178] and 5 per incident 400-keV oxygen ion.[177] Of course, many divacancies could be present within disordered regions and hence not observable with optical techniques.

Investigations of specific defects in ion-implanted silicon[170a,b,177,186a] have shown that there is a correlation between the types of defects produced in ion-implantation and in fast-neutron irradiation. In some cases, even the number of specific defects produced is comparable in the two processes.[177,186a]

It can be anticipated that there will also be differences. In fact, Daly and Pikar,[186a] in their paramagnetic-resonance investigations of implanted silicon, have identified a new defect that has not been seen in neutron-irradiated material.

In summary, it is clear that the anneal of the disorder produced in both fast-neutron-irradiated and ion-implanted silicon and germanium show similar features. The physical processes occurring in the various anneal stages have not been identified. The low-temperature anneal stage may involve the dissociation of defect clusters and the migration of defects such as divacancies. Except for the relatively small number of observed divacancies, one could speculate that the anneal of the isolated disordered regions is associated with the same mechanism that leads to the anneal of divacancies. At high ion doses, the anneal is governed by the epitaxial reordering of the amorphous layer.

3.5.2 Electrical and Optical Characteristics

The study of the effect of irradiation (neutron, electron, etc.) on the electrical characteristics of semiconductors is an exceedingly complex field (see, for example, Corbett[190] or Vook[191]). It is beyond the scope of this monograph to provide a complete summary; however, it is necessary to point out some of the aspects of irradiation damage effects which may have a bearing on the electrical properties of ion-implanted silicon.

† Stein *et al.*[177] have recently found that the anneal behavior of divacancies in 400-keV oxygen-implanted silicon follows closely that observed in fast-neutron-irradiated silicon. Brower *et al.*[170b] also conclude from electron-paramagnetic-resonance studies that the anneal of damage associated with a room-temperature, 400-keV, oxygen-ion implantation, is similar to that in a room-temperature, high-fluence neutron irradiation.

In electron or gamma irradiation of semiconductors in which isolated vacancies and interstitials are created, the major changes in electrical characteristics at room temperature are not due to these vacancies or interstitials, but to their interaction with dopant atoms, oxygen atoms, or other vacancies. In silicon, a vacancy adjacent to a donor dopant atom can form a stable pair which introduces an acceptor level. The change from a positive donor to a negative acceptor requires removal of two conduction electrons A vacancy trapped by a neutral oxygen atom also forms a negative center which removes one conduction electron. Isolated defects such as these have been identified for the most part by EPR measurements and have well-defined association temperatures.[190] A major point to be made is that the dissociation of these defects usually occurs below 300° C in silicon.

It is now generally accepted that in neutron-irradiated semiconductors, clusters of defects also influence the electrical behavior. The cluster model was first used to explain the observed changes in *n*- and *p*-type germanium following fast-neutron bombardment.[192] Gossick[193] developed a detailed description for the cluster in which he proposed that the cluster be treated as a sphere of *p*-type material forming a *p-n* junction with the surrounding host lattice. More recently it has been pointed out by Stein and coworkers[194–198] that the anneal characteristics of neutron-irradiated silicon differ significantly from those found in electron- and gamma-irradiated samples. There are two major factors that must be considered in evaluating the role of these defect clusters. The first is the presence of the clusters themselves. They have been found to cause large changes in carrier mobility, to introduce trapping centers for majority and minority carriers, and to produce light-sensitive defects.[198,199] The second effect is that these clusters serve as either a source or sink of vacancies and interstitials during the annealing of the disordered region. An example of this latter effect is shown in Fig. 3.37, where the anneal characteristics of vacancy–oxygen pairs[200] are determined from optical absorption studies. In electron-irradiated material there is a decrease in the intensity of the absorption line, indicating the dissociation of the vacancy–oxygen pairs. In neutron-irradiated material, on the other hand, the absorption first *increases*, then decreases. This initial increase is attributed[200] to the escape of vacancies from the disordered regions in the defect cluster and then the subsequent capture of these vacancies by oxygen atoms to form the vacancy–oxygen pairs. In another example, Stein[198] attributes some of the reverse-anneal characteristics noted around 300° C in neutron-irradiated silicon to the liberation of vacancies from disordered regions and their subsequent recapture by the dopant species. Again, it should be noted that most of the reported anneal behavior is for the temperature region below 400° C.

One can anticipate some obvious similarities and differences between

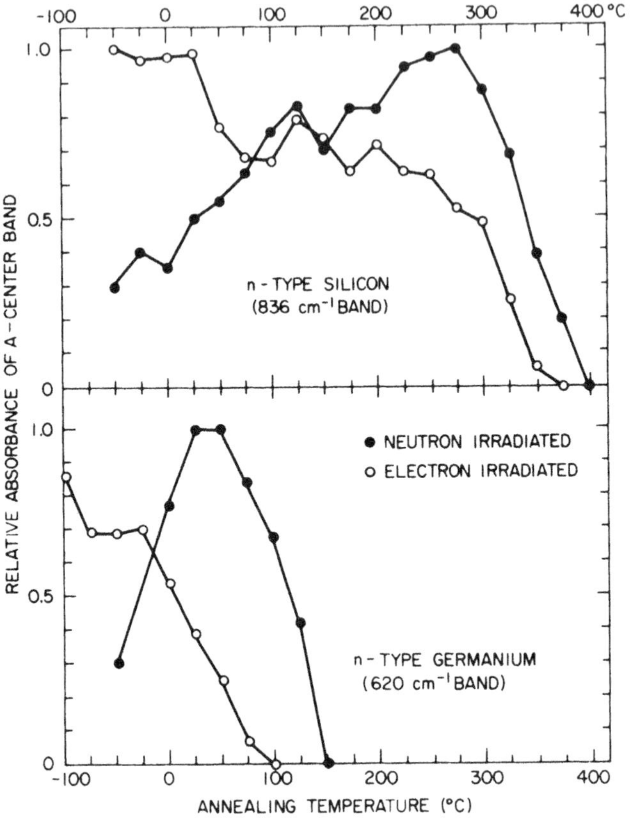

Fig. 3.37. Annealing behavior of the infrared absorption band associated with the oxygen–vacancy complex in silicon and germanium. Upper half, annealing behavior in silicon after neutron and electron irradiation at −50° C. Lower half, annealing behavior in germanium after neutron irradiation at −50° C and electron irradiation at −196° C. (From Whan.[200])

neutron-irradiated and ion-implanted samples. In implanted samples the dopant atom comes to rest in a region of disorder. This disorder should produce a significant decrease in carrier mobility; hence, complete electrical activity of the dopant atom will be obtained only when the defects in the disordered region dissociate and migrate away from the dopant atom. This is perhaps the reason why higher anneal temperatures are required in implanted samples than in neutron-irradiated ones.

Further discussion of electrical effects in irradiated semiconductors is deferred until Chapter 5, where the electrical behavior of implanted samples is treated.

4

The Lattice Location of Implanted Atoms

4.1 Introduction

The objective of most ion-implantation work is to implant specific dopant atoms into well-defined (usually substitutional) locations in the crystal lattice. Since the dynamics of the slowing-down process is complex and is accompanied by a large amount of atomic displacement (Chapter 3), one might expect this objective to be hard to attain. However, the behavior of the Group V dopants is rather straightforward in that the implanted atoms, both in silicon and in germanium, tend to occupy substitutional sites. The Group III behavior is more complex, but even here the majority of the implanted atoms usually end up in well-defined locations in the lattice.

Until recently, most lattice-location studies have been based on either the optical or the electrical behavior (Chapter 5). Such measurements are usually indirect and so do not provide a unique correlation with the fraction of implanted atoms on substitutional sites. Using the recently developed channeling-effect technique, however, the fraction of the dopant atoms on lattice sites can be *directly* measured. Hopefully, this technique, in combination with electrical and optical studies, will eventually enable the contribution of the various types of defects to the conductivity of an implanted layer to be resolved. The technique has so far been applied largely to silicon and germanium; in fact, the majority of measurements are for the heavier Group III and Group V dopants in silicon.

Part 4.2 describes this rather new technique in some detail. Part 4.3 contains a general summary of the results thus far plus a brief discussion, with the object of providing a framework for the subsequent sections. The experimental work is then reviewed in detail in Part 4.4, and its significance is discussed in Part 4.5.

4.2 The Channeling Effect Technique

Channeling phenomena have two important consequences to ion implantation. As already discussed in Chapter 2, they significantly influence the implanted dopant profile, and thus the junction depth. Also, as we shall discuss here, they provide a technique for evaluating directly both the lattice location of implanted dopant atoms and the number of displaced lattice atoms. The latter application was discussed in detail in Chapter 3.

In Chapter 2 we saw how the motion of energetic atomic particles can be strongly influenced (i.e., channeled) by the regular arrangement of atoms composing a crystal lattice (cf. Fig. 2.12). For the relatively low velocities used in most ion implantations, this channeling effect depends strongly on implantation conditions, and so the heavy-ion range distribution cannot be accurately predicted. On the other hand, for high-velocity particles (such as MeV protons, helium ions, etc.), the situation is quite different: here, the channeling behavior is both reproducible and predictable, and the observed effects can be interpreted quantitatively in the theoretical framework given by Lindhard.[57] We shall not attempt to give a detailed discussion of the basic theoretical and experimental studies of channeling. Excellent surveys of the more recent work are contained in the proceedings of the conferences held at Aarhus[10] and Chalk River,[11] and in a review article by Datz *et al.*[201] Here we will consider only the basic framework needed for the present application. Some details of the channeling behavior of MeV low-Z atoms in several semiconductor lattices are contained in the appendix.

4.2.1 General Principles

An energetic charged particle (such as a 1-MeV proton or helium ion) moving through a crystal lattice will be steered by correlated collisions with the lattice atoms whenever its angle of approach to a close-packed atomic row (or plane) is smaller than a certain critical angle. According to Lindhard,[57] such a channeled particle is deflected without penetrating closer to the center of an atomic row than approximately a, the Thomas–Fermi screening distance given by Eq. (2.3).† Some typical values of a are given in Table 4.1. Further-

† Strictly speaking, this rough estimate of the "steering distance" for channeled particles is applicable only to a static lattice. In a real (vibrating) lattice, a better estimate is usually the mean vibrational amplitude ρ. In silicon and germanium, however, both a and ρ are comparable in magnitude, as can be seen from Table 4.1.

TABLE 4.1

THE THOMAS–FERMI SCREENING DISTANCE[a] IN Å FOR VARIOUS Z_1, Z_2 COMBINATIONS

	Si[b]	Ge[b]	Sn
H	0.18	0.14	0.12
He	0.17	0.14	0.12
C	0.16	0.13	0.11
Ar	0.13	0.11	0.10
Xe	0.10	0.09	0.09

[a] As obtained from Eq. (2.3).
[b] At 20° C, the mean vibrational amplitude is 0.14 Å in silicon and 0.15 Å in germanium.

more, in the axial case the critical angle of approach $\psi_{1/2}$ for such a steered motion is given by the formula†

$$\psi_{1/2} = \alpha(2Z_1Z_2e^2/Ed)^{1/2} \tag{4.1}$$

Z_1e and Z_2e are the nuclear charges of the moving particle and the lattice atoms respectively, E is the energy of the projectile, and d the atomic spacing along the lattice row. Here, α may be considered a proportionality constant, whose value depends weakly on the vibrational amplitude of the lattice atoms, and hence on the temperature. Usually, the value of α is between 0.8 and 1.2 (cf. the appendix). The value of $\psi_{1/2}$ is typically of the order of 1°.

A particle is thus said to be channeled if its angle of approach ψ to an atomic row (or plane) is within the critical angle. If, on the other hand, ψ is larger than the critical angle, no correlated motion will occur. In this case, the particle will not "feel" the lattice structure, but instead will move through the lattice as if it were composed of randomly distributed atoms. Since there are many low-index directions in a crystal, it is sometimes difficult to select an incident direction that does not lie within the critical angle of some plane or axial row.

Assume that a beam of particles enters a *perfect* single crystal parallel to one of the major axes (Fig. 4.1). Even so, not *all* of the particles in the beam become channeled. A small fraction of the particles (as a result of their point of entry on the crystal surface) undergo a sufficiently violent collision to be deflected immediately through an angle larger than the critical angle [cf.

† For the lower velocities typical of implantation energies, a different critical-angle relationship is applicable, as was discussed in Section 2.4.1.

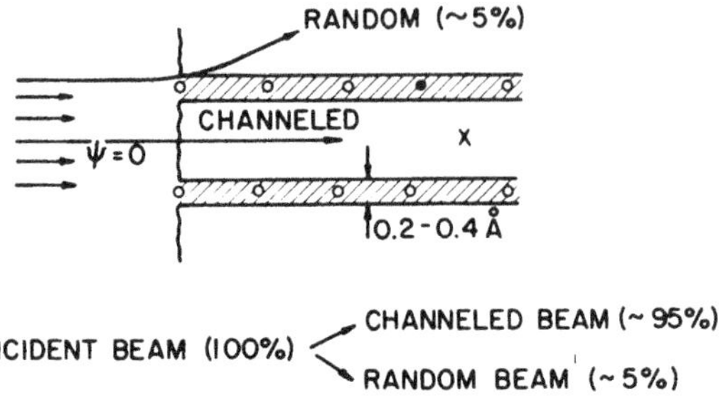

	FRACTION OF BEAM THAT CAN INTERACT WITH	
	•	X
ALIGNED CASE	~5%	100%
RANDOM CASE	100%	100%

Fig. 4.1. The splitting of a perfectly aligned beam into channeled and random components in passing through the crystal surface. The table indicates how the division of the crystal into allowed and forbidden regions for the channeled component can in principle be used to locate atoms.

Fig. 2.13(b)]; this fraction of the beam is referred to as the *random component*. We can thus define two special conditions:

1. The *aligned case* ($\psi = 0$), in which the incident beam is split into channeled and random components. Typically, the channeled component contains 95–99% of the incident beam, depending on the magnitude of the closest distance of approach between a channeled particle and a lattice atom (cf. Table A.1 in the appendix).
2. The *random case* ($\psi > \psi_{1/2}$), in which the incident beam is equal to the random beam, i.e., there are no channeled particles.

One aspect of the channeling phenomenon is of particular importance for the present discussion, namely, the fact that a channeled particle cannot approach closer to a lattice site than ~0.1–0.2 Å (Table 4.1). The crystal is therefore divided into allowed and forbidden regions (Fig. 4.1). Hence, all physical processes requiring a closer impact than 0.1–0.2 Å (for example, wide-angle elastic scattering, the inducement of nuclear reactions, or the production of inner-shell X rays) are not possible for a channeled particle, as was demonstrated by Bøgh and Uggerhøj,[202] Andersen *et al.*,[65] and Khan *et al.*[203] In consequence, a strong attenuation (tenfold to one-hundred-fold) in the yield of such processes is observed for the aligned case in comparison with the random case. The yield cannot go completely to zero, since even for $\psi = 0$ there is always a small random fraction (1–5%). These pronounced orientation effects have been observed in numerous experiments. One example is shown in Fig. 4.2.

Large attenuations in the interaction yield occur not only with lattice atoms but also with any foreign atoms located in the forbidden region for the channeled beam—for example, on lattice sites (position ● in Fig. 4.1). If, on the other hand, the atom being studied is located somewhere in the allowed region for the channeled beam (e.g., position × in Fig. 4.1), then no attenuation in its interaction yield will be observed, since now both the channeled and the random parts of the beam can interact normally with the atom in question.† Two examples are illustrated in Fig. 4.3. The antimony atoms (●)

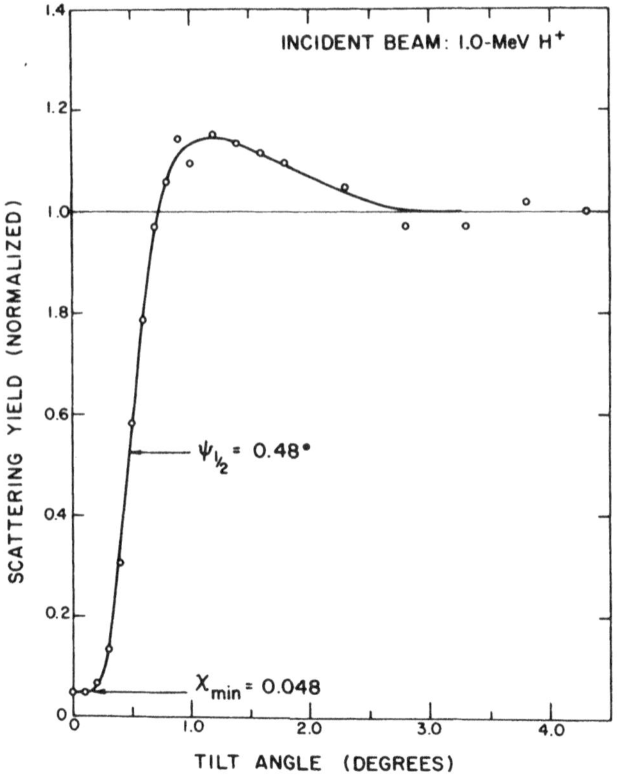

Fig. 4.2. Backscattering yield of 1.0-MeV protons from a diamond crystal as the angle of incidence is changed relative to the ⟨111⟩ direction. The yield for random incidence is normalized to 1.0. The critical angle $\psi_{1/2}$ and the minimum yield χ_{min} are indicated in the figure.

† This statement is not strictly true for the *well*-channeled part of the beam. In principle, a foreign atom located right at the center of a channel would show a considerably higher interaction yield for a well-channeled beam than for a random one. In practice, however, the channeled beam ordinarily acquires enough transverse momentum because of beam spread, surface penetration, lattice disorder, and imprecise alignment (i.e., not better than ~0.1°) that this effect should probably not exceed 20%. With low implantation dose and careful alignment, considerably larger effects have been observed.[203a]

are located on lattice sites and hence show a large attenuation along the ⟨110⟩ direction; the gold atoms (×), on the other hand, are not on regular sites and hence interact normally with the channeled beam.

So far, we have considered the channeling behavior of particles approaching the crystal from the outside (the external-beam case). The same separation into allowed and forbidden particle trajectories occurs, however, if the charged particle originates from a position *inside* the lattice—for example, as a consequence of a radioactive decay[204] or an elastic-scattering event.[205,206] If it originates from a position contained in a row of atoms (such as a lattice site), it cannot emerge from the crystal within the critical angle characteristic of that row—i.e., it cannot become channeled, since it started out within one of the *forbidden* regions for channeled particles (Fig. 4.1). Hence, a reduction in the number of emitted particles is observed in the direction of the row (Fig. 4.4). This phenomenon is usually termed *blocking*. Conversely,

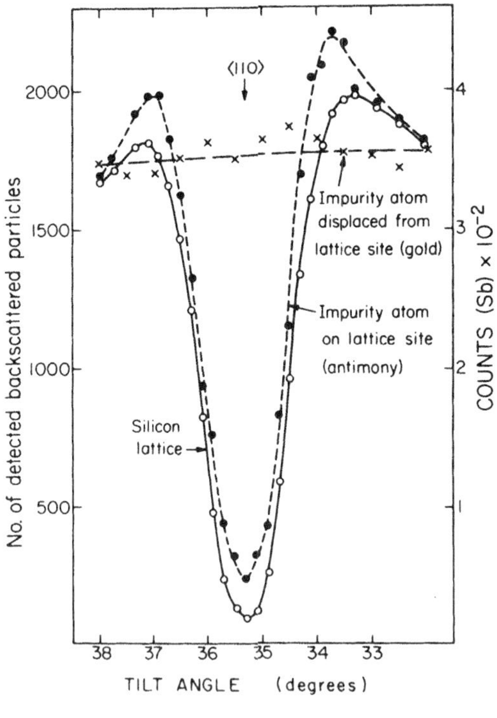

Fig. 4.3. Backscattering yield of 1.0-MeV helium ions from substitutionally located impurity atoms (antimony) and from ones that are displaced from lattice sites (gold) as a function of the angle of incidence relative to the ⟨110⟩ direction. The corresponding yield from the lattice atoms (solid curve) is included for comparison.

if the charged particle does not originate from a position lying within ~0.1–0.2 Å of a close-packed row, then the emission rate will not exhibit any significant anisotropy. The existence or absence of such a "blocking" effect is therefore determined by the location of the emitting nucleus.

Lindhard[57] has shown from a reversibility argument that channeling and blocking experiments are completely equivalent, provided depth effects (cf. the appendix) are negligible. This equivalence has been quantitatively verified in the experimental work of Bøgh and Whitton[208] and Andersen and Uggerhøj.[209]

A typical arrangement for an external-beam experiment is shown in Fig. 4.5(a). The incident beam (e.g., 1-MeV helium ions, collimated to better than 0.1°) strikes a crystal mounted on a suitable goniometer, which permits simultaneous rotation about two orthogonal axes. In this case, particles that have undergone wide-angle (~150°) scattering are detected by means of a solid-state detector (X rays or nuclear-reaction products may be detected by substituting the appropriate type of detector). In the measurements, the detector subtends a large enough solid angle that possible channeling or blocking effects on the *outgoing* trajectories are minimized. Figure 4.5(b) shows a similar arrangement for an emission-type experiment; in this case, of course, the solid angle to the detector is deliberately made sufficiently small (i.e., $\ll\psi_{1/2}$) that angular effects (Fig. 4.4) for the outgoing trajectories can be resolved.

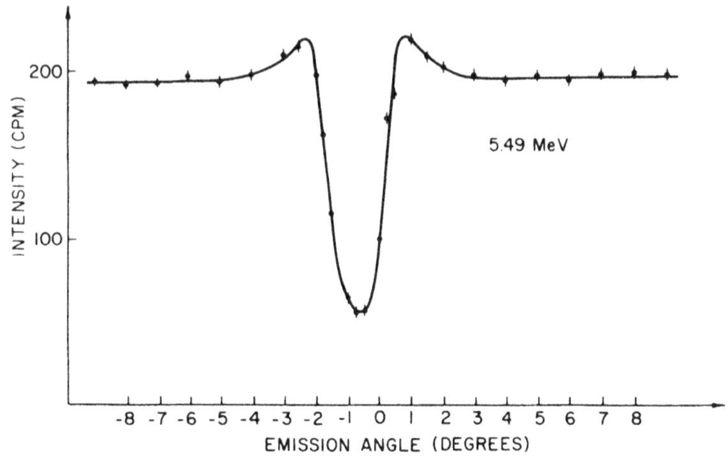

Fig. 4.4. The yield, as a function of emission angle, of 5.49-MeV α-particles from radioactive ^{222}Rn embedded in a ⟨111⟩ tungsten single crystal. Implantation conditions: 60 keV at 25° C. The zero in the abscissa scale was chosen arbitrarily, and does not correspond exactly to the ⟨111⟩ direction. (From Domeij and Björkqvist.[207])

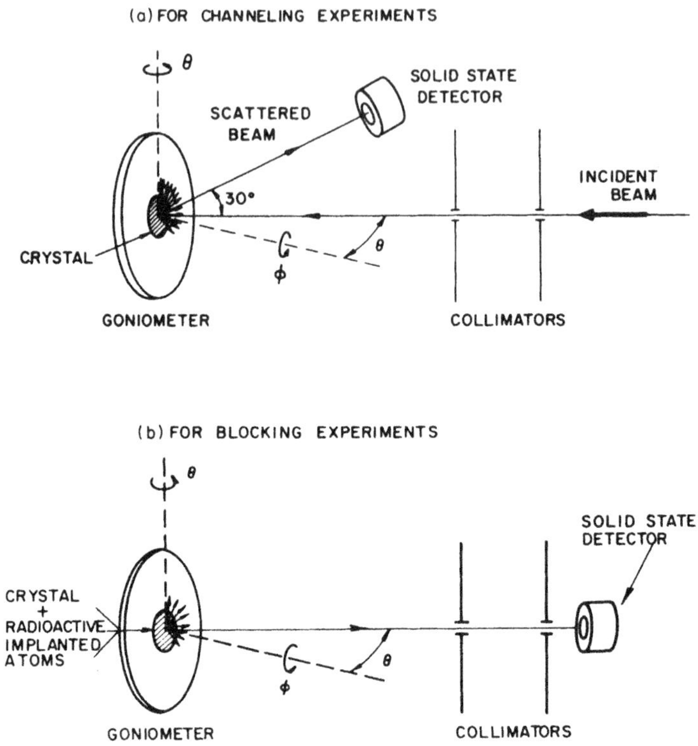

Fig. 4.5. Schematic diagram of a typical experimental assembly used (a) for channeling experiments, and (b) for blocking experiments.

In certain cases, particularly in the study of lattice disorder, it is advantageous to utilize both channeling and blocking effects simultaneously. This *double-alignment* technique is discussed briefly later in this chapter (Section 4.2.3).

We shall now consider how the channeling-effect technique can be applied to locate foreign atoms on various types of lattice sites. Since most analyses are performed with an external probing beam, channeling (rather than blocking) is used for the purpose of illustration. Figures 4.1 and 4.3 indicate that simultaneous observation of the interaction of a channeled beam with lattice atoms and with specific embedded foreign atoms yields a *quantitative* measure of the fraction of those foreign atoms located within 0.1–0.2 Å of a particular row or plane.

If the interaction yields between the incident beam and the dopant atoms for an aligned and random incidence are denoted by N_A and N_R respectively, the percentage of dopant atoms located along the particular axial row is given by

$$100 \times (N_R' - N_A')/N_R' \quad (4.2)$$

where the prime indicates that N_R and N_A have been corrected for the fact that in the aligned case a small fraction of the incident beam in the region of the implanted atoms is *not* channeled. In silicon, typically 95% of the beam is channeled, and the correction to be subtracted from both N_R and N_A is therefore $\sim 0.05\ N_R$.

From such measurements, one can often deduce the exact location(s) of the foreign atoms: (1) by repeating the measurement along different axial directions, as suggested in Fig. 4.6; (2) by considering the symmetry properties of the lattice.

Let us consider the situation in the diamond-type lattice, as this is the lattice structure of most semiconductor crystals. Figure 4.6 depicts the configuration of atoms in the {110} plane. Note that the ⟨111⟩ and ⟨100⟩ rows intersect at two different types of site: (1) at the substitutional sites (○, ●), and (2) at a set of unoccupied sites, the so-called tetrahedral interstitial sites (▲). Both of these positions (and possibly others) are potential sites for dopant atoms. Obviously, channeling-effect measurements along the ⟨111⟩ or ⟨100⟩ direction do not distinguish between the substitutional and the tetrahedral interstitial site, since along these sets of atomic rows both positions lie within the forbidden (shaded) regions for the channeled beam. However, in the ⟨110⟩ direction, the interstitial sites are not contained along the atomic rows, but lie within the ⟨110⟩ channels; thus, atoms in these sites will be exposed to the ⟨110⟩ channeled beam. Consequently, by making measurements along a ⟨111⟩ and a ⟨110⟩ direction, together with a measurement along a properly chosen "random" direction (i.e., ψ significantly greater than $\psi_{1/2}$), one can in principle determine the fraction(s) of dopant atoms located (1) within 0.1–0.2 Å of the *substitutional* lattice sites; (2) within 0.1–0.2 Å of the *tetrahedral interstitial* sites (from here on, we will often refer to these simply as *interstitial* atoms); (3) in some *random* locations other than (1) or (2): for example, on a dislocation, at the surface, or in some "irregular"† interstitial site (for example, at the position marked (×) in Fig. 4.6).

In any one measurement, the beam probes only one of several equivalent axes; hence, additional information may often be obtained by considering lattice symmetries. For instance, suppose we observe that 80% of the dopant atoms are located along a particular ⟨110⟩ row. One can then conclude that a similar fraction is located along *each* of the other five sets of ⟨110⟩ rows.

† By "irregular," we mean any site that does *not* lie along one of the low-index atomic rows.

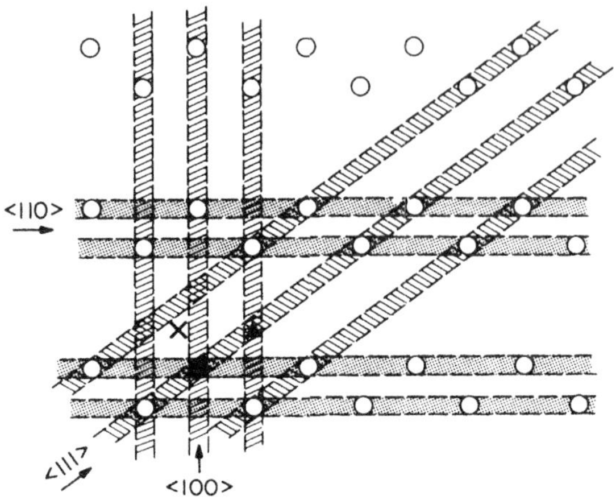

○ - silicon lattice atoms

	Directional Effect <111>	<110>	<100>	Impurity Location
●	yes	yes	yes	substitutional
▲	yes	no	yes	tetrahedral interstitial
×	no	no	no	off-lattice site

Fig. 4.6. The atomic configuration in a {110} plane, showing how impurity-atom locations (●, ▲ and ×) can be resolved by channeling-effect measurements along different crystallographic directions. The shaded areas represent the respective forbidden regions into which a channeled beam (along the ⟨100⟩, ⟨110⟩, or ⟨111⟩) cannot penetrate. To establish the directional effect, in each case a measurement is also made along a properly chosen random direction. Open symbols refer to silicon lattice atoms.

Furthermore, since in the diamond structure the intersection of two or more ⟨110⟩ rows occurs only at the substitutional sites, we deduce that no more than 4% (i.e., $\frac{1}{5}$ of the missing 20%) of the dopant atoms can be located in a *non*substitutional position along any particular set of ⟨110⟩ rows. Hence, at least 76% of the dopant atoms must be on the substitutional sites. In consequence, whenever a *large* attenuation in yield is observed along a major direction, the location of the dopant atoms in the crystal can be specified quite accurately. It also follows from this discussion that whenever a *small* attenuation in yield occurs, considerable uncertainty arises in the interpretation. For example, a 15% attenuation along the ⟨110⟩ direction might suggest that 15% of the atoms were on substitutional sites; but it could equally well mean that up to 90% (i.e., 6 × 15%) were located elsewhere along the ⟨110⟩ rows. Furthermore, in this situation, the statistical accuracy of Eq. (4.2) will be rather poor.

A critical question is how the foreign atoms can be detected at low enough concentrations for the technique to be useful. In the case of an external probing beam, the interaction with the foreign atoms must give rise to a signal distinguishable from the much more numerous interactions with the lattice atoms. Several different processes have been utilized. For the lighter atoms (such as lithium or boron), a specific nuclear reaction may be available (Section 4.2.4). In certain other cases, the yields of characteristic inner-shell X rays may be used, or the emission behavior (blocking) of charged particles from radioactive implants. So far, however, the most widely used process for analyzing implanted layers has been the yield of back-scattered particles (Section 4.2.3), since it can be applied to almost *all* cases in which the implanted foreign atom is heavier than the substrate. It also has the advantage of measuring simultaneously the amount of lattice disorder introduced by the implantation, as was discussed in Chapter 3. In very thin crystals, the scattering technique can even be used to study foreign atoms that are *lighter* than the substrate. For lattice-location studies, however, this method has not yet been extensively used.

The sensitivity of these external-beam measurements is typically 10^{-3}–10^{-4} atomic fraction. In emission-type measurements, where the implanted atoms are radioactive and generate their own signal, the sensitivity can be enhanced somewhat, since there are no background signals from the lattice.

4.2.2 Alignment of Single Crystals

Accurate orientation of the crystal is of course a necessary requisite for the channeling-effect type of analysis that is being discussed in this chapter. It is also of importance in determining the range profile during the implantation, as was discussed in Chapter 2. In both cases, one wants to orient the crystal at least to 0.1°. Fortunately, the channeling effect itself (Andersen *et al.*[65], Dearnaley and Wilkins[64]) provides a convenient tool for performing such precise alignments *in situ*. Accurate orientation can also be made with quite low accelerator voltages. It is feasible, for example, to align silicon crystals to within 0.1°, using a 50-keV proton beam (Mayer[210]); this may be important in cases where a controlled implantation into an aligned crystal is desired, since many ion-implantation systems do not operate above 100 keV. Higher-energy beams, however, offer certain advantages: more precise alignment, and less lattice disorder introduced into the implanted region by the analyzing beam.

4.2.3 Backscattering Measurements

We shall now discuss briefly the various processes that can be used in the channeling-effect technique of detecting and locating foreign atoms in crystals (Andersen *et al.*,[211] Bøgh,[212] Davies *et al.*,[156] Matzke and Davies [213]).

The main features of the backscattering technique [using the geometry of Fig. 4.5(a)] are schematically illustrated in Fig. 4.7, in which an amorphous silicon target having antimony atoms embedded close to the surface is bombarded with helium ions of energy E_0. The energy spectrum of backscattered particles consists of a sharp edge at $0.59E_0$ (corresponding to elastic scattering from silicon atoms in the surface region) followed by a smooth, slowly changing yield at lower energies (corresponding to scattering from silicon atoms at progressively larger depths). Backscattering of helium ions from the embedded antimony atoms involves a smaller elastic energy loss than from silicon, and so the edge of the antimony spectrum occurs at a

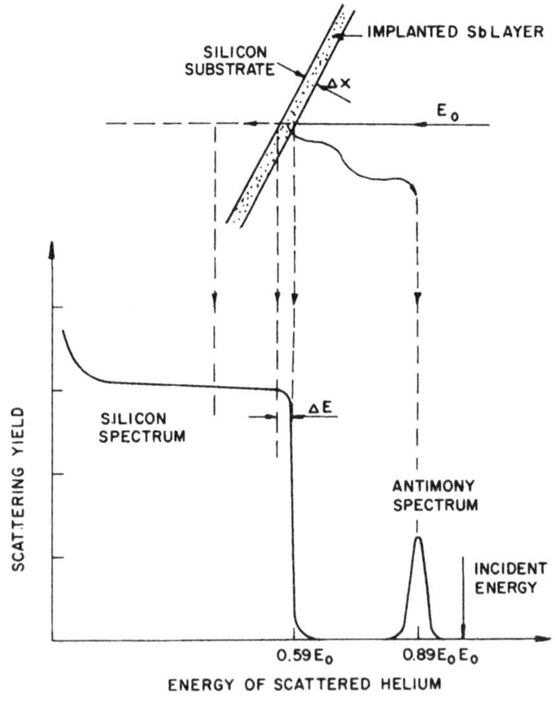

Fig. 4.7. Principles of dopant-atom detection by means of backscattering. An amorphous silicon substrate, containing a small fraction of heavier atoms (e.g., antimony) in the surface region, is bombarded with helium ions of energy E_0. The lower part of the figure shows the resulting energy spectrum of scattered particles.

higher energy ($0.89E_0$). Furthermore, as the implanted atoms are confined to a shallow surface region (for most implantations, $\lesssim 0.1$ micron), the antimony spectrum consists of a fairly sharp peak. It may be noted at this point that the use of the backscattering technique for the chemical analysis of surfaces is well established (see, for example, Patterson *et al.*[213a]). It has recently been used by Thompson *et al.*[213b] to study surface contamination on silicon. The technique has also been used to determine range profiles, as was discussed in Section 2.2.2.

The detailed shape of the energy spectrum from a crystal lattice is determined by (1) the kinematics of the elastic scattering process, (2) the rate of inelastic energy loss which a particle undergoes as it penetrates the crystal, (3) the energy dependence of the scattering cross section, and (4) channeling effects.

The cross section as a function of scattering angle is given by the Rutherford scattering formula

$$\frac{d\sigma}{d\Omega} = 1.3 \times 10^{-27} \left(\frac{Z_1 Z_2}{E}\right)^2 \left(\frac{M_1 + M_2}{M_2}\right)^2 \frac{1}{\sin^4 \theta/2} \quad \text{cm}^2/\text{sr} \tag{4.3}$$

where E is the energy (in MeV) of the particle at the moment of scattering and θ is the scattering angle in center-of-mass coordinates. Because of the strong forward-peaking of the Rutherford scattering cross section, only a small fraction of the incident particles are backscattered from the target.

With the geometry of Fig. 4.8, the detected energy of a particle backscattered from a depth x is given by

$$E_{\text{obs}} = \left[E_0 - \int_0^{x/\cos\theta_1} S^*(E)\, dl_1\right] k_s - \int_{x/\cos\theta_2}^{0} S(E)\, dl_2 \tag{4.4}$$

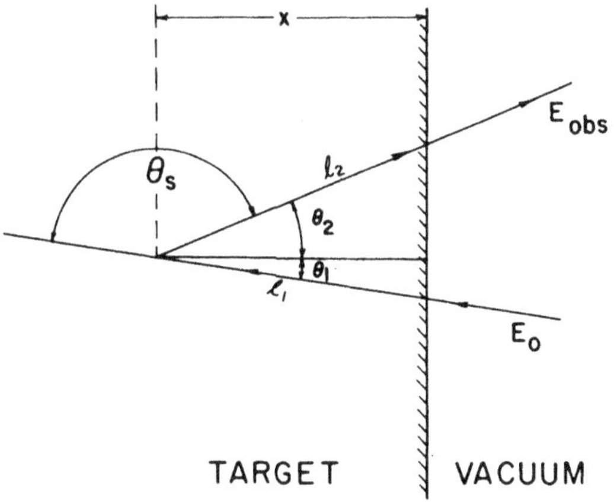

Fig. 4.8. Schematic diagram of the various angles involved in converting energy scales into depth scales by means of Eqs. (4.4) and (4.5).

where

$$k_s = \left[\frac{M_1 \cos \theta_s + (M_2^2 - M_1^2 \sin^2 \theta_s)^{1/2}}{M_1 + M_2}\right]^2 \tag{4.5}$$

k_s is the ratio of the energy before and after scattering, and the first and second integrals are the energy loss along the incoming and outgoing trajectories l_1 and l_2. The laboratory scattering angle θ_s is typically $\sim 150°$. $S^*(E)$ and $S(E)$ are the stopping powers[214] along the incident and outgoing trajectories, respectively.

From a knowledge of the stopping powers $S^*(E)$ and $S(E)$, the energy scale from the substrate can be converted into a *depth scale*. In the aligned case, $S^*(E)$ is the stopping power for a channeled beam and is usually *not* well known; however, a reasonable estimate can be $S^*(E) \approx 0.7$ $S(E)$.† Since for most geometries the major part of the energy loss occurs along the random outward trajectory, for which $S(E)$ is accurately known, this conversion from keV to Å is not too sensitive to uncertainties in the value of $S^*(E)$.[215] As an example, even a 30% uncertainty in the value of $S^*(E)$ introduces only a 12% uncertainty in the depth scale. Note that a detector resolution of 15 keV (FWHM) is equivalent to a depth resolution of about 300 Å for 1-MeV helium in silicon.

Using Eqs. (4.4) and (4.5), the energy scale can also be converted into an atomic-mass scale for identifying foreign-atom peaks, such as the antimony peak in Fig. 4.7—provided the foreign atoms are located close to the surface (as shallow dopants or as surface contaminants).

Figure 4.9 compares the backscattered helium spectra from an unimplanted silicon crystal with the beam incident along an aligned (⟨111⟩) and a random direction. Depth and mass scales are indicated according to the previous discussion. In the aligned case, a 35-fold attenuation in scattering yield is observed close to the spectrum edge due to the channeling effect; this agrees reasonably well with the theoretically predicted attenuation for a perfect crystal (cf. Table A.1 in the appendix). A small increase in the aligned yield relative to the random yield can be seen with increasing depth; this is attributed to scattering effects (dechanneling) that gradually transfer helium ions from the aligned to the random beam (cf. Part 2.4 and the appendix). When obtaining the curves shown in Figs. 4.2 and 4.3, a narrow portion of the energy spectrum close to the surface is selected in which to measure the scattering yield as a function of incident angle.

† For the well-channeled part of the beam, $S^*(E) \approx 0.5$ $S(E)$ is probably a better relationship. However, recent measurements by Bøgh[165a] suggest that the *average* stopping power encountered by that part of the channeled beam contributing to the backscattering yield is not significantly less than the random value. For most purposes, the relationship $S^*(E) \approx 0.7$ $S(E)$ is therefore a reasonable compromise.

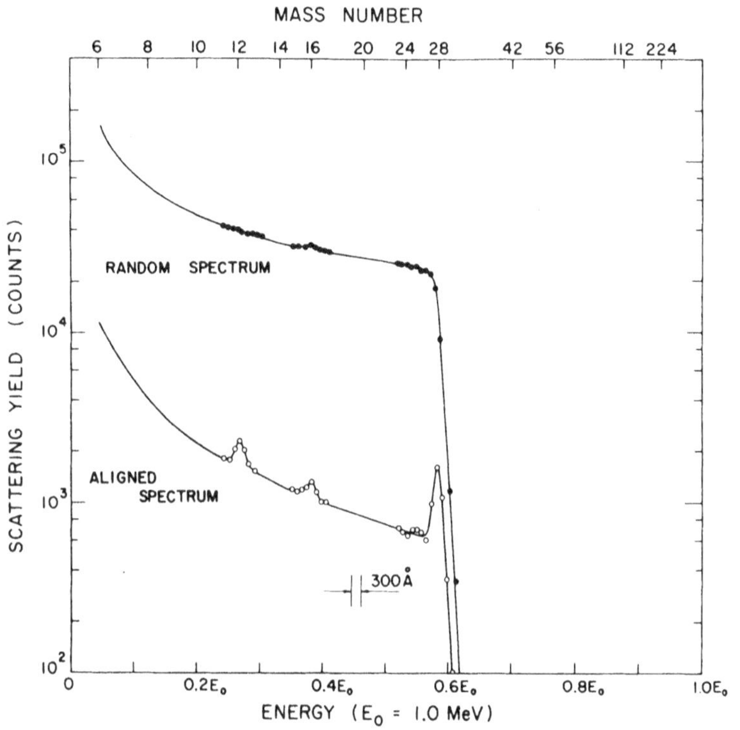

Fig. 4.9. Aligned (⟨111⟩) and random backscattering spectra for an undoped silicon crystal, recorded for the same dose of 1.0-MeV helium ions. The depth resolution for the aligned case (viz. 300 Å) is indicated. A mass-number scale (for impurities located near the crystal surface) is also given. The three peaks in the aligned spectrum—at masses 12, 16, and 28—indicate the amounts of carbon, oxygen, and silicon in the surface region that are not aligned with the underlying substrate. In order to simplify the figure, data points have been omitted except in the more important regions of the spectra. (From Davies *et al.*[156])

The peak in the ⟨111⟩ spectrum at the spectrum edge is caused by randomly distributed silicon atoms in the surface oxide layer; the next peak (as seen by the mass scale) is due to the oxygen atoms in this surface oxide. The third peak is attributed to a thin layer of carbon contamination that gradually accumulated on the surface during analysis. Neither the carbon film nor the surface oxide is thick enough to reduce significantly the channeled fraction of the transmitted beam. Note that in the random spectrum, the carbon and oxygen peaks are barely detectable above the large background from the silicon atoms. Therefore, by aligning a crystal, the sensitivity for detecting low-Z atoms on the surface is increased considerably. This sensitivity can be increased still further (and also the sensitivity for detecting lattice disorder) by the double-alignment technique referred to in Section 3.3.5, as has been demonstrated by Bøgh,[155] and more recently by Feldman and Appleton.[215a]

As was discussed in Section 3.3.5, the area of the silicon peak in the ⟨111⟩ spectrum of Fig. 4.9 is directly proportional to the amount of disordered silicon in the surface region (i.e., in the surface oxide plus any rearrangement in the underlying silicon lattice). Similarly, the area of each contaminant peak (carbon, oxygen in Fig. 4.9) and of the heavy-foreign-atom peak (antimony in Fig. 4.7) is directly proportional to the amount of that specific foreign atom present, according to the relationship in Eq. (4.3). In ion-implantation studies, this proportionality can be conveniently used to obtain an accurate measure of the number of implanted atoms per cm^2.

In Fig. 4.9, the thickness of the oxide layer was only ~30 Å and thus was considerably less than the available detector resolution. Recently, Meyer *et al.*[215b, 215c] have demonstrated how the above principles can be conveniently applied to analyze the composition of thick (1000–3000 Å) surface layers of silicon oxide or nitride as a function of depth. They were able to study variations in the depth distribution of the composition of these layers with a resolution of ~100 Å.

One particular advantage of Rutherford scattering for lattice-location analysis over most other close-impact processes is its relatively large cross section, which varies smoothly with atomic number and with energy (as opposed to resonance reactions), thereby making a wide range of combinations accessible to analysis. Typical beam currents used in these experiments are of the order of 10^{-8} Å.

The main limitation of the scattering technique is that the interaction yield with the dopant atoms must appear at higher energy than that with the substrate atoms; and this occurs only if $M_{\mathrm{dopant}} > M_{\mathrm{substrate}}$. For the opposite case, the dopant spectrum will be superimposed on the large substrate background, (e.g., the oxygen and carbon peaks of Fig. 4.9), and would not be resolved at the dopant concentrations of interest (i.e., $\lesssim 10^{-3}$ atomic fraction). Hence, many of the common dopants (Li, B, Al, and P) have to be studied by some other close-impact process (cf. Section 4.2.4). As remarked earlier, this limitation can be removed by using a very thin substrate (i.e., 1–10 microns depending on the energy of the analyzing beam). In such a case, the substrate spectrum would extend down only to some lower energy E^*, and the low-mass dopant spectrum would then appear in the energy range below E^* (cf. Fig. 4.7). To date, this method has not been applied to the analysis of implanted layers.

Figure 4.10 shows an example of the scattering technique in which a 1.0-MeV helium beam has been used to analyze a thallium-implanted silicon crystal. Spectra of backscattered particles have been recorded by means of a solid-state detector for random, ⟨111⟩, and ⟨110⟩ directions of incidence. As discussed above, the thallium spectra consist of well-defined peaks, since

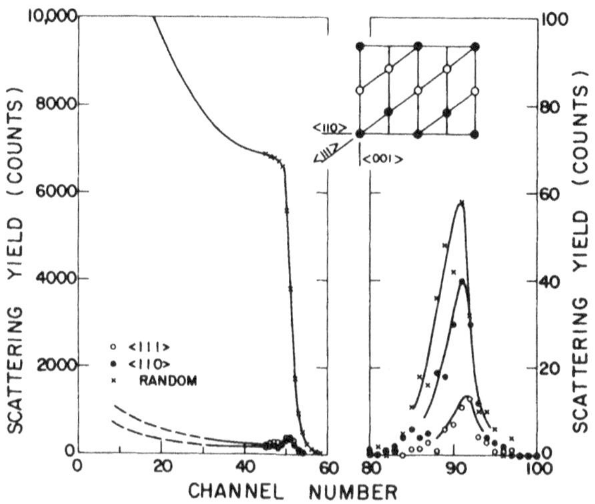

Fig. 4.10. Energy spectra of helium ions backscattered from a thallium-implanted crystal, with the 1.0-MeV helium beam incident along a "random" direction (×), a ⟨111⟩ direction (O), and a ⟨110⟩ direction (●), respectively. The sharp peak around channel 90 corresponds to scattering from thallium atoms in the implanted region, and the continuum at lower energies (channels 0–50) to scattering from the thick silicon crystal. Implantation conditions: 1.1×10^{14} thallium ions/cm^2 at 40 keV, with the substrate at 350° C. The insert shows the atomic configuration in a {110} plane in silicon. ●, substitutional sites; O, tetrahedral interstitial sites. (From Eriksson *et al.*[174])

the implanted atoms are confined to a shallow depth ($\lesssim$0.05 microns) below the surface. When the crystal is aligned with either the ⟨111⟩ or the ⟨110⟩ direction, equally large attenuations are observed in the backscattering yield from the substrate silicon atoms. For the thallium region of the spectra, however, the ⟨111⟩ attenuation is much greater than that along the ⟨110⟩. By comparing the areas under the thallium peaks, one deduces that approximately 40% of the thallium atoms are located on substitutional sites and another 40% on the tetrahedral interstitial sites. The "missing" 20% of the thallium atoms are in some irregular location, such as a dislocation or at the surface.

The clear separation between the dopant peak and the edge of the silicon spectrum in Fig. 4.10 makes the measurement relatively easy to interpret. The optimum choice of analyzing beam and its energy depends, however, upon the particular combination of substrate and impurity atom. The following factors must be considered: the *mass separation* (i.e., the energy difference between particles backscattered from dopant and from substrate atoms at the same depth), the *detector resolution*, and the *backscattering yield* per incident particle. It is also important to ensure that significant damage is not introduced by the analyzing beam, and that radiation-induced defects are

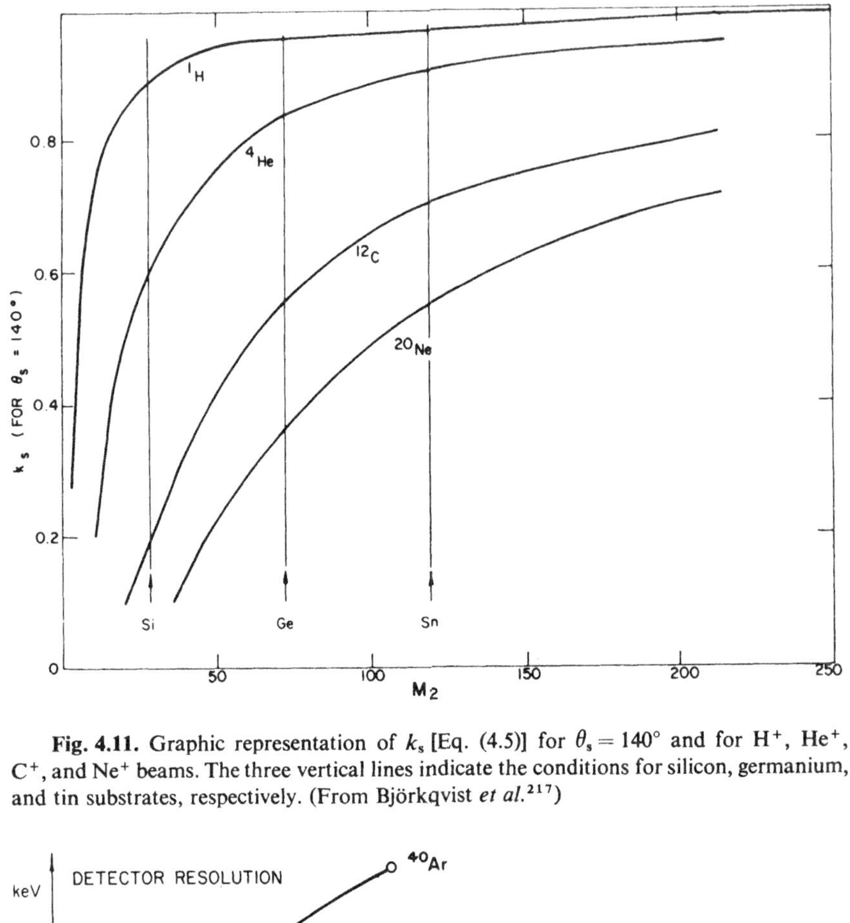

Fig. 4.11. Graphic representation of k_s [Eq. (4.5)] for $\theta_s = 140°$ and for H^+, He^+, C^+, and Ne^+ beams. The three vertical lines indicate the conditions for silicon, germanium, and tin substrates, respectively. (From Björkqvist *et al.*[217])

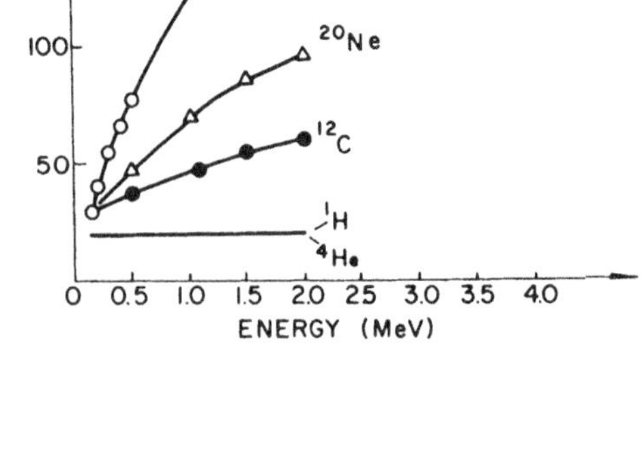

Fig. 4.12. The resolution (keV) of a standard surface-barrier detector as a function of energy for H^+, He^+, C^+, Ne^+, and Ar^+ beams. (From Bergström *et al.*[218])

not interacting with the dopants.[156,216] Each of these factors contributes to the overall *sensitivity* of the method (i.e., the minimum concentration of impurity atoms that can be studied). The detector resolution in combination with the stopping power of the incident beam also determines the *depth resolution*, as we have already discussed.

Sufficient mass resolution is by far the most stringent requirement for this type of analysis. The energy separation between the dopant peak and the edge of the substrate spectrum is given by $E_0 \Delta k_s$, where Δk_s is the difference between the k_s values [Eq. (4.5)] for scattering from the dopant and substrate atoms. In Fig. 4.11, k_s values are plotted[217] as a function of M_2 for H^+, He^+, C^+, and Ne^+ beams. For optimum mass resolution, $E_0 \Delta k_s$ should be as large as possible, preferably at least ten times the detector resolution. It is therefore necessary that an increased energy separation *not* be offset by a worsening of the detector resolution. However, as Fig. 4.12 shows,[218] the resolution of a surface-barrier detector changes relatively slowly with Z_1 or E, so this is not usually the limiting factor.

There are therefore two ways to improve the mass separation when analyzing implanted semiconductors: (1) use a heavier projectile, and (2) increase the projectile energy. Point (1) is illustrated in Fig. 4.13 for the case of a 40-keV antimony implant in silicon. Using a 1.0-MeV beam of protons, there is no clear separation of the antimony peak from the substrate spectrum, whereas a 1.0-MeV helium beam provides good separation. Similarly, in germanium, a helium beam does not provide sufficient mass separation, but a 1.8-MeV carbon beam [Fig. 4.14(a)] is able to resolve antimony and thallium peaks not only from the germanium spectrum but also from each other. Even better separation can be achieved by increasing the energy of the beam [Fig. 4.14(b)]. For most laboratories, however, (2) presents a difficult and costly alternative. There are, on the other hand, cases in which a very-high-energy beam would be a definite advantage.

In selecting the optimum projectile (disregarding for the moment the practical restriction of having only certain beams available), there are other aspects to consider besides good mass resolution:

1. Since the back-scattering yield varies as $(Z_1/E)^2$, one would in principle like to keep the incident energy as low as possible for a given Z_1.
2. Since it is usually desirable to have good *depth* resolution, one would like the stopping cross section to be as large as possible: i.e., $E_0 \approx \varepsilon_3$ (cf. Fig. 2.5 and Table 2.3).
3. Since the depth sampled by the analyzing beam should cover the entire implanted distribution, E_0 must be large enough for the impurity spectrum to be separated completely from the substrate spectrum. For example, the 1.8-MeV carbon beam cited above could not be used to analyze deep ($\gtrsim$1-micron) implantations.

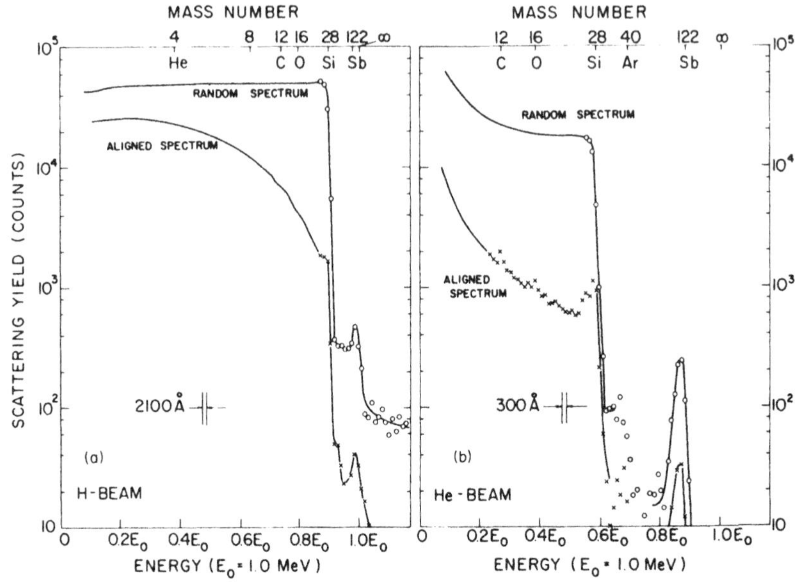

Fig. 4.13. Comparison of backscattering spectra obtained with (a) 1.0-MeV H^+ and (b) 1.0-MeV He^+ beams. The target in both cases was a silicon crystal implanted at 450° C with 3×10^{14} antimony ions/cm^2. The implantation energy was 30 keV. Both sets of spectra were obtained with the same electronic gain, the same total analyzing dose, and comparable beam currents. Mass and depth resolution scales are also included for comparison. In order to simplify the figure, data points have been omitted except in the more important regions of the spectra. (From Davies *et al.*[156])

Assuming a detector resolution of ~15 keV FWHM, a 1-MeV proton beam gives a depth resolution in silicon of only 2000 Å, whereas a 1-MeV helium beam improves the resolution to 300 Å (Fig. 4.13); a 1.8-MeV carbon beam in germanium has a depth resolution of ~250 Å. Since the mean implantation depth is usually only a few hundred angstroms (Table 2.4), none of these beams has sufficient depth resolution to resolve details in the dopant distribution.

One can further improve the energy resolution and hence the depth resolution by using either magnetic or electrostatic spectrometers for the analysis[212] or by cooling a semiconductor detector to reduce the leakage current. In the latter case, 6-keV energy resolution (for 1-MeV helium ions) can be obtained.

Practical limits to the backscattering technique are normally set by background impurities (such as surface contaminants), pulse "pileup,"† and radia-

† The problem of pulse "pileup" can be eliminated almost entirely with a well-designed electronic pulse-handling system.

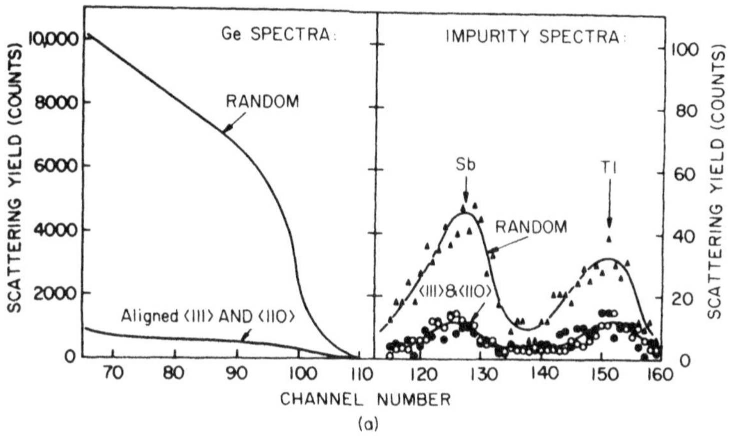

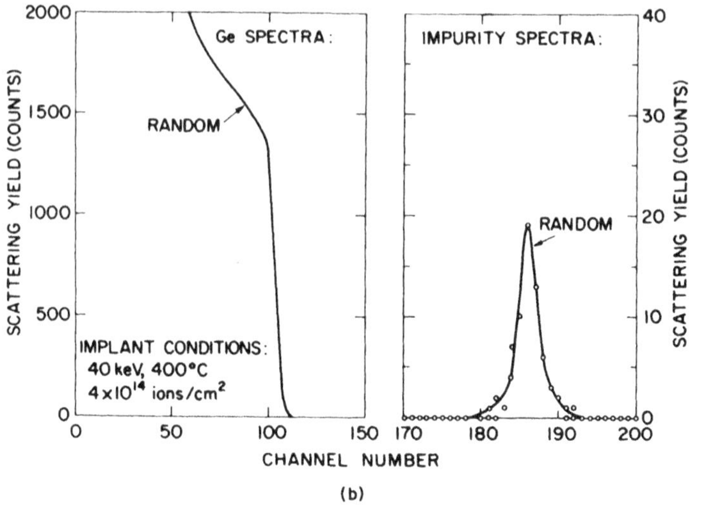

Fig. 4.14. (a) Backscattering energy spectra (▲, random; ○, ⟨111⟩; and ●, ⟨110⟩) of 1.8-MeV ^{12}C ions from a germanium crystal implanted at 30 keV and 350° C with $\sim 4 \times 10^{14}$ antimony ions/cm^2 followed by $\sim 2 \times 10^{14}$ thallium ions/cm^2. In the germanium spectra, the data points have been omitted for clarity. (From Eriksson *et al.*[219]) (b) Energy spectra of 20-MeV ^{16}O ions backscattered from a germanium crystal, implanted with 3×10^{14} thallium ions/cm^2 at 40 keV, illustrating the enhanced mass resolution obtainable with higher-energy beams. (From Björkqvist *et al.*[217])

tion damage [cf. Fig. 3.27(b)] by the analyzing beam. Since the scattering cross section is proportional to Z_2^2, the sensitivity is greatly enhanced for heavier dopants; in addition, the mass separation is also better. In favorable cases, atom-location experiments can be conducted with 10^{12}–10^{13} ions/cm^2, but 10^{14} ions/cm^2 is a more typical dopant level for the measurements to date. Since implanted regions normally extend ~1000 Å into the crystal, dopant concentrations that have been studied range from typically 10^{19} atoms/cm^3 down to 10^{17}–10^{18} atoms/cm^3.

4.2.4 Nuclear-Reaction Techniques

Successful application of the backscattering technique (Section 4.2.3) requires the dopant atoms to be heavier than the lattice atoms. In order to study lighter dopant atoms (for example, Li, B, N, or P), some other close-impact process can be used. We will discuss in this section the possibility of using specific nuclear reactions between a probing beam of charged particles and the dopant atoms. Experimentally, the procedure is identical to that of the scattering technique [Fig. 4.5(a)], except that one or more nuclear reaction products are now monitored instead of particles backscattered from the foreign atoms. An example is shown in Fig. 4.15, in which lithium atoms, implanted into a silicon substrate, are detected by means of the ^{7}Li (p, α) ^{4}He

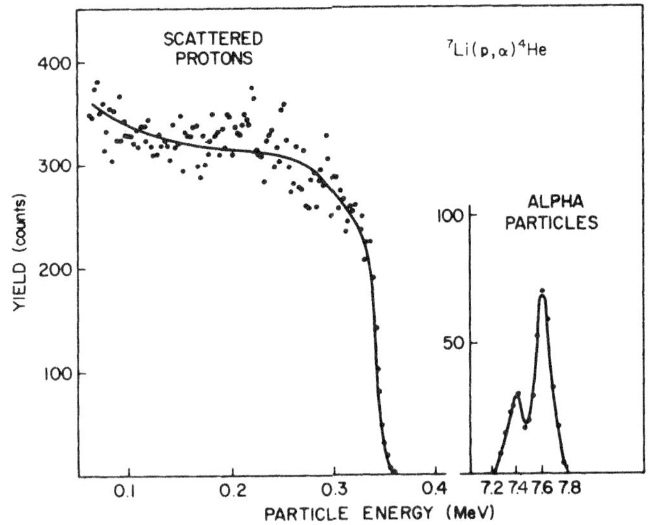

Fig. 4.15. Energy spectrum of particles measured at an angle of 135° to the direction of a 450-keV proton beam striking a silicon crystal that had previously been implanted with 6×10^{16} lithium ions/cm^2. Detected at high energy are the α-particles from the ^{7}Li(p, α) ^{4}He reaction. (From Gibson *et al.*[220])

reaction, using a beam of 450-keV protons.[220] As in the scattering technique, one observes on the left the typical backscattered spectrum (protons) from the silicon substrate, and at much higher energies (7.5 MeV) the α-particles from the (p, α) reaction.

The major restriction on the use of this technique is that most nuclear-reaction cross sections are not large enough for low dopant concentrations to be studied. At moderate projectile energies (i.e., <3 MeV) a sufficiently high cross section is available only for the lighter elements ($Z_2 < 10$); however, this includes several important dopant elements. Some of the more suitable reactions are listed in Table 4.2. For elements heavier than fluorine, the cross sections generally drop by at least an order of magnitude; for example, the (p, α) reactions on ^{27}Al and ^{31}P require 10^{15}–10^{16} atoms/cm^2 in order to obtain reasonable statistics.

In selecting a suitable reaction, competing reactions from the substrate atoms and also from other low-Z impurities with larger cross sections must be considered.

The limitations of the nuclear-reaction technique can be summarized as follows. If the reaction products to be detected are charged particles, then the detector–analyzer system also has to handle the much larger counting rate due to backscattered particles from the substrate. This restriction is in principle the same as for the scattering technique. However, sometimes the reaction product has a significantly longer range than the backscattered

TABLE 4.2

NUCLEAR REACTIONS FOR THE STUDY OF LIGHT IMPURITIES IN CRYSTALS BY CHANNELING-EFFECT MEASUREMENTS

Impurity	Suitable reaction	Detect	Incident energy (MeV)
Li	^{7}Li(p, α) ^{4}He	α	Nonresonant reactions (~1 MeV)
Be	^{9}Be(d, p)^{10} Be	p	
B [a]	^{11}B(p, α) ^{8}Be	α	0.67
C [a]	^{12}C(d, p) ^{13}C	p	1.0
N	^{15}N(p, $\alpha\gamma$) ^{12}C	γ	1.21
O [a]	^{18}O(p, α) ^{15}N	α	0.63, 0.83
	^{16}O(d, p) ^{17}O	p	0.97
F	^{19}F(p, $\alpha\gamma$) ^{16}O	γ	0.87, 1.38
Al	^{27}Al(p, α) ^{24}Mg	α	1.18, 1.36
P	^{31}P(p, α) ^{28}Si	α	1.51, 1.89

[a] Cases in which the sensitivity is better than 10^{14} atoms/cm^2.

primary beam (for example, $^{11}B(p, \alpha)\,^{8}Be$, $^{12}C(d, p)\,^{13}C$ and $^{16}O(d, p)\,^{17}O$ reactions); in such cases, the scattered beam can be eliminated completely by use of an appropriate absorber foil. If the reaction products are γ-rays, the major limitation is the relatively high background of the γ-detector.

Most of the reactions in Table 4.2 have an additional disadvantage compared to the scattering technique: the depth resolution is not nearly as good, because of the broad character of these nuclear resonances (some of the reactions are even nonresonant). Whereas depth resolutions of the order of a few hundred angstroms can be obtained with the scattering technique (Section 4.2.3), the resolution attainable for most nuclear reactions is of the order of 1 micron. For most implanted layers, since the thickness is small ($\lesssim 0.1$ micron), this is not a severe problem. For thicker layers (deep implantations), however, two complications arise:

1. Significant dechanneling of the probing beam (see the appendix) can occur over the depth of the implanted dopant atoms. This complicates considerably the interpretation of the angular dependence of the reaction yield in terms of lattice-location information.
2. Since the stopping power of the incident beam depends upon the orientation of the beam with respect to the lattice, serious complications arise when the nuclear cross section has a significant energy dependence over the interesting depth region. An example of this effect is given in Gibson *et al.*[168]

Since only a relatively small number of elements are accessible to the nuclear reaction technique, alternate methods will have to be developed in many cases where the scattering technique fails. There are at least two other processes that may be of advantage in such cases: X-ray production[221,221a] and internal emission.[204,222,223] Neither of these techniques has been extensively applied as yet.

4.3 Survey of Lattice-Location Experiments

4.3.1 General Observations

In order to make the experimental results presented in Part 4.4 somewhat easier to digest, we shall discuss here some general observations and conclusions that have been made from channeling-effect measurements. As yet, there is no adequate theoretical basis for predicting the lattice location of implanted atoms. However, the experimental results display many systematic regularities, and we shall therefore suggest possible mechanisms and attempt to formulate a reasonable model. Since the results indicate that the lattice-location behavior of implanted dopants is far from obvious, a valid model would certainly be important for the full realization of the potentialities of ion implantation.

Table 4.3 summarizes the various implanted species studied and the techniques used. With only a few exceptions, the detailed channeling-effect measurements to date have been confined to the Group III and Group V dopants; even here, very few such measurements have been made on nitrogen,

TABLE 4.3

LATTICE-LOCATION EXPERIMENTS IN SILICON AND GERMANIUM

Element studied	Technique used	References
Silicon experiments		
Li	(p, α) reaction	Gibson *et al.*[168]
B	(p, α) reaction	Davies *et al.*,[156] Gibson *et al.*,[168] Fladda *et al.*,[231b] Eriksson *et al*,[231a] North and Gibson[231c]
P	(p, α) reaction	Mitchell[225]
P	β^- emission	Gibson *et al.*[168]
Zn	Backscattering	Meyer *et al.*[231d]
Ga	Backscattering	Mayer *et al.*,[157,173] Eriksson *et al.*,[174,219,225a]
Ga	X-ray emission	van der Weg and Saris[221a]
As	Backscattering	Davies *et al.*,[156] Eriksson *et al.*[174,219]
Se	Backscattering	Meyer *et al.*[231d]
Cd	Backscattering	Picraux *et al.*,[55] Eriksson *et al.*[174]
In	Backscattering	Davies *et al.*,[94,156] Eriksson *et al.*,[174]
Sn	Backscattering	Eriksson *et al.*[174]
Sb	Backscattering	Davies *et al.*,[156] Mayer *et al.*,[157] Eriksson *et al.*[172,174,219]
Sb	X-ray emission	Cairns and Nelson[221]
Te	Backscattering	Picraux *et al.*,[55] Eriksson *et al.*[174]
Xe	Backscattering	Davies *et al.*[156]
Cs	Backscattering	Eriksson *et al.*[174]
Tm	Backscattering	Bøgh[212]
Yb	β^- emission	Uggerhøj *et al.*[203a, 222]
Au	Backscattering	Eriksson *et al.*,[174]
Hg	Backscattering	Meyer *et al.*[231d]
Tl	Backscattering	Davies *et al.*,[94] Eriksson *et al.*,[171,174,219] Fladda *et al.*[216]
Bi	Backscattering	Eriksson *et al.*,[171,174] Mayer *et al.*[173]
Germanium experiments		
In	Backscattering	Mayer *et al.*,[157,173] Eriksson *et al.*,[219] Björkqvist *et al.*[226]
Sb	Backscattering	Mayer *et al.*,[173] Eriksson *et al.*,[219] Björkqvist *et al.*[226]
Tl	Backscattering	Eriksson *et al.*,[219] Björkqvist *et al.*[226]
Bi	Backscattering	Björkqvist *et al.*[226]
Pb	Backscattering	Björkqvist *et al.*[226]
Hg	Backscattering	Fladda *et al.*[224]

aluminum, or phosphorus implants. It therefore remains to be established whether or not the picture that can be outlined for the heavier dopants can be extrapolated to the lighter ones. Also, no channeling-effect measurements have been reported on the lattice-location behavior in semiconducting substrates other than silicon and germanium.

Implantation is a violent process that creates a large amount of disorder per incident ion. As was seen in Chapter 3, this disorder can accumulate, and at higher doses leads to the formation of an amorphous phase—except when the substrate temperature during implantation is maintained above a critical value (i.e. ~300° C in silicon and ~200° C in germanium).† It is therefore natural to treat room-temperature implantations separate from those performed at higher temperatures.

The importance of substrate temperature during implantation is illustrated in Fig. 4.16. At 25° C and 2×10^{14} ions/cm^2, the individual disordered regions around each antimony ion track overlap sufficiently to produce a completely

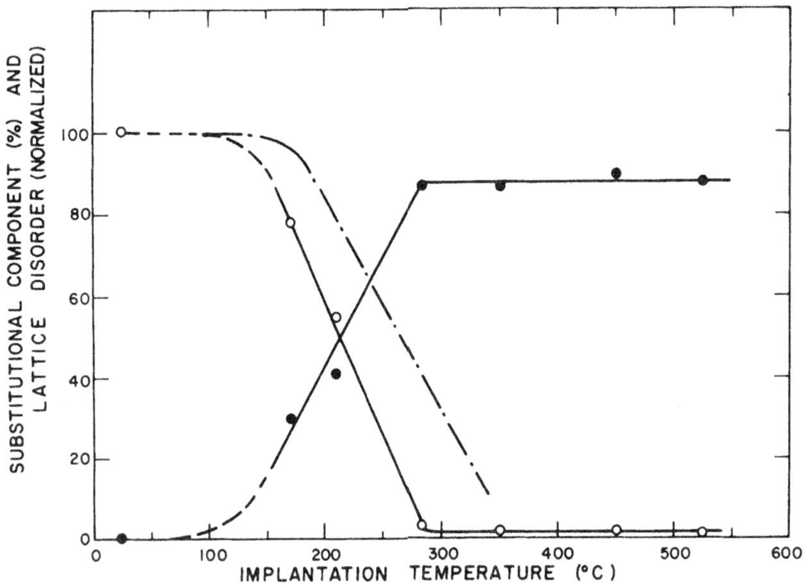

Fig. 4.16. Influence of implantation temperature on the substitutional antimony component (●) and on the lattice disorder (○) in silicon. Implant conditions: 2×10^{14} antimony ions/cm^2 at 40 keV. The anneal behavior of the lattice disorder for a low-dose (~10^{13} antimony ions/cm^2) room-temperature implant is also shown (dot-dash curve) for comparison. (From Eriksson *et al.*[174])

† These temperatures depend somewhat on the dose rates used; they may also differ between light and heavy ion species.

disordered (amorphous) region. In the channeling-effect measurements, *no* substitutional antimony component is found, since most of the antimony atoms lie within this amorphous region. As the implantation temperature (silicon) is increased above ~150° C, the lattice disorder (○) decreases and the observed substitutional antimony fraction (●) increases markedly. Above 300° C, there is essentially no further change in these parameters, i.e., the annealing of the disordered region around each ion track is now sufficiently rapid to prevent significant overlap from occurring at the dose *rates* used (10^{12}–10^{13} ions/cm^2 sec). This correlation between the increase in substitutional antimony and the decrease in lattice disorder indicates that even below 300° C, those antimony atoms outside the heavily disordered regions have a strong preference for substitutional sites. Similar evidence is found in low-dose (~10^{13} ions/cm^2) room-temperature implants of bismuth or thallium, where high substitutional levels are observed even before annealing [cf. Fig. 4.20(a)]. It can be anticipated from Fig. 4.16 that the implantation behavior (both regarding lattice location and disorder) is particularly critical in the temperature region 150–250° C, where the total implanted dose and the dose rate may both be important.

One might ask whether a *channeled* implantation introduces the implanted species into a different lattice location than a *non*channeled implantation. Almost no information on this problem exists as yet. However, such an orientation effect would presumably be restricted to those well-channeled particles that remain channeled throughout their path. Since in semiconductors the number of such well-channeled atoms appears to be relatively small (Part 2.4), we would not expect this to be a large effect. Nevertheless, the problem should be investigated; in particular, it might determine the electrical behavior at the junction depth.

In all the measurements reviewed here, isotopically pure ion beams from electromagnetic isotope separators have been used. Implantation energies have been typically 30–40 keV, and in a few cases higher. As long as one does not consider channeled implantations, a change in energy represents mainly a change in the average dopant concentration (Chapter 2); this can easily be compensated by a change in implantation dose. The silicon and germanium crystals used have usually been 1–100 Ω-cm *n*- or *p*-type, etch-polished, float-zone material, cut almost perpendicular to a ⟨111⟩ (or occasionally a ⟨110⟩) direction. Generally, *p*-type dopants have been implanted into *n*-type material and vice versa, to allow also for electrical measurements. Occasionally, base material of the same type as the implanted atoms has been used, with no observable difference in the lattice-location behavior. This is not surprising, since the doping properties are normally dominated by the large concentration of implanted atoms. However, if the substrate conductivity is similar in magnitude and opposite in type to that of the implanted dopant, then it can sometimes affect the lattice location of the implanted species (Section 4.4.5). Doses have ranged from 10^{13} to 10^{16} ions/cm^2 and implantation temperatures from 25–600° C, the majority being either at 25° C or in the interval 350–450° C. Subsequent anneal treatments usually extended up to 800–900° C.

4.3.2 Summary of Observations

As was discussed in Section 4.2.1, channeling-effect measurements along ⟨111⟩, ⟨110⟩, and "random" directions of an implanted crystal can be used to determine (with respect to the underlying substrate) the fraction of dopant atoms (1) within 0.1–0.2 Å of lattice sites (S), (2) within 0.1–0.2 Å of the tetrahedral interstitial sites (I), and (3) in positions other than (1) or (2), i.e., in "random" locations (R). From such measurements on implants in which disorder is not dominant, one can distinguish the following four types of behavior:

1. $S \neq 0$, $I \approx 0$; i.e., a significant fraction of the dopant atoms are substitutional, with no interstitial fraction. Group IV and Group V dopants implanted in silicon and germanium typically are 50–90% substitutional. An example is shown in Fig. 4.3.
2. $S \approx 0$, $I \neq 0$; i.e., a significant fraction of the dopant atoms are interstitial, with no substitutional fraction. This behavior is not often observed; however, it has been seen for thallium and the Group II elements in silicon under special conditions. Typically $I \lesssim 0.6$. An example is seen in Fig. 4.18(b) around 600° C.
3. $S \neq 0$, $I \neq 0$; i.e., both substitutional and interstitial fractions exist. This is a common observation in silicon and germanium samples implanted with Group III dopants. Very often $S \approx I \approx 0.3$. An example is shown in Fig. 4.10. This behavior has *not* been observed for any Group V dopant.
4. $S \approx I \approx 0$; $R \approx 1$; i.e., *no* significant substitutional or interstitial fractions are observed. This is often seen with Group III dopants at high implantation temperatures, or after anneal to high temperatures. The crosses in Fig. 4.3 represent such a case, where gold had been implanted into silicon at 450° C.

The fraction R is usually interpreted as representing dopants pinned on dislocations, situated in locally disordered regions, or contained in bulk or surface precipitates. However, in cases where S and I are small but not zero (i.e., where $R \approx 0.7$–0.9), other interpretations are possible. For example, the dopant atoms may occupy some interstitial site other than the tetrahedral—such as the hexagonal interstitial. Alternatively, they may all be on substitutional sites, but (due to differences in ion size, bond length, etc.) their equilibrium position may be displaced from the normal lattice site by $\gtrsim 0.2$ Å. In either case, the *observed* S or I component would be reduced by an appreciable factor, depending on the lattice symmetry involved—as was discussed in Section 4.2.1.

In the following discussion, it is important to bear in mind that most ion implantations involve doses well in excess of solid-solubility concentrations, and this may affect significantly the anneal properties, particularly at high enough temperatures for ordinary diffusion to occur. From the results presented in the following section, it will be evident that the implantation behavior of dopants in semiconductors is *not* simple. Nevertheless, certain general observations and correlations are found.

The following discussion is based mainly on silicon data, since the dependence on various implantation parameters has not yet been studied nearly as thoroughly in germanium. However, with a few exceptions, similar behavior is noted in the two materials. The behavior in silicon can be summarized as follows:

1. Strong correlations exist among all the Group V elements studied. As shown in Fig. 4.19(a), they exhibit a large substitutional component under those implantation conditions where the lattice structure is maintained (i.e., at implant temperatures above $\sim 300°$ C and at dose levels less than $\sim 10^{15}$ ions/cm^2). They do *not* show any interstitial component. Both for implants at elevated temperatures and for annealed room-temperature implants, there is *no* decrease in the substitutional level during anneal (Fig. 4.17), except at temperatures where normal diffusion processes become significant (i.e., > 800–$900°$ C). Significant dose effects on the substitutional fraction are not observed below 10^{15} ions/cm^2.

Large substitutional components are also observed for the Group IV elements (tin and lead) and for the Group VI elements (selenium and tellurium).

2. Correlations exist also for the Group III elements studied, although they exhibit a markedly different behavior than that of the Group V elements. At comparable dose levels ($\sim 10^{14}$ ions/cm^2), and at implant temperatures $> 300°$ C, they show a considerably smaller substitutional component than the Group V elements. An appreciable fraction is usually observed in the tetrahedral interstitial sites.† For implantations at 350–450° C, this interstitial component in most cases‡ is roughly equal in magnitude to the substitutional one, as shown in Fig. 4.18. The interstitial component is *not* observed if a sufficiently large concentration of n-type dopant is present in the substrate. The implantation and subsequent annealing behavior varies

† No interstitial component has yet been found for indium implantations in germanium.

‡ For thallium implantations in silicon, the relation between substitutional and interstitial fractions varies strongly over a wider range of implantation temperatures (cf. Fig. 4.21).

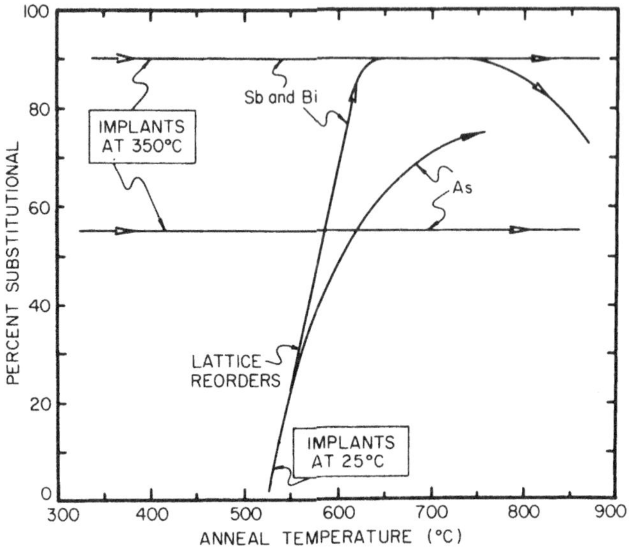

Fig. 4.17. Anneal behavior (schematic) of arsenic, antimony, and bismuth implantations into silicon at 350° and 25° C. The doses are typically 1–5 × 10^{14} ions/cm^2. Above ~700° C a dose dependence is observed for antimony and bismuth; arsenic data are incomplete. (Modified from Mayer *et al.*[157] and Eriksson *et al.*[174])

from one element to another, as can be seen by comparing the behavior of gallium and indium in Fig. 4.18(a) with that of thallium in Fig. 4.18(b). However, for all three elements the interstitial component decreases strongly on annealing to 650–750° C; this is considerably below the temperatures at which normal diffusion becomes significant.

Large interstitial components are found also for 350° C implantations of Group II elements (zinc, cadmium, and mercury) in silicon. In contrast to the behavior of Group III elements, however, there is virtually no substitutional component present in 350° C implants.

3. There are also some similarities in behavior between those Group III and Group V elements that have been studied in detail: For most of the dopants studied, the lattice-location behavior in germanium and in silicon is almost identical. The amount and subsequent annealing of the lattice disorder introduced by the implantation is largely independent of ion species, except for low-Z implants such as boron (Chapter 3). The concentration of substitutionally located ions is not limited by thermal-equilibrium considerations. High-temperature annealing tends to reduce the substitutional component (and the interstitial). In low-dose ($\lesssim 10^{13}$ ions/cm^2) room-temperature implants (into silicon), thallium and bismuth are both highly substitutional before anneal, despite the presence of individual disordered regions [Fig. 4.20(a)].

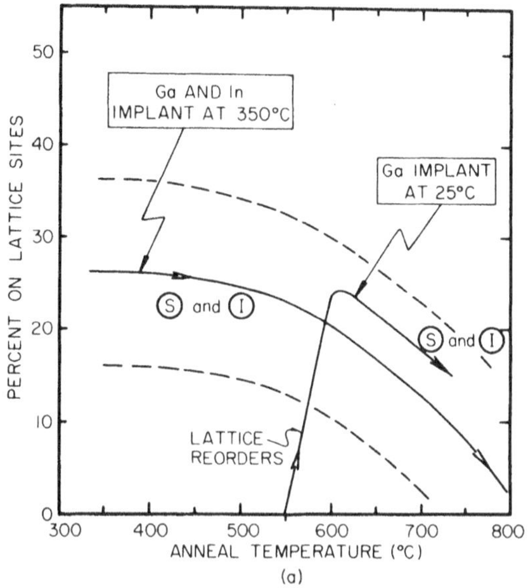

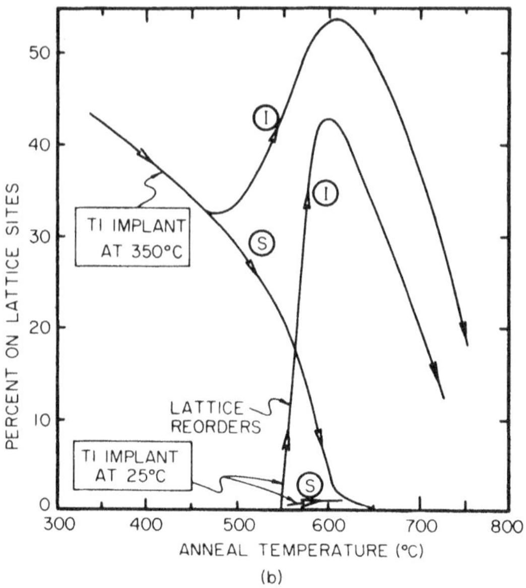

Fig. 4.18. Anneal behavior (10 min at each temperature) of Group III implants in silicon. S denotes the substitutional component and I the tetrahedral interstitial component. Typical doses: 2×10^{14} ions/cm^2. (Derived from Eriksson *et al.*[174]) (a) Gallium and indium. Note that the substitutional and interstitial components are approximately equal and decrease in the same manner with increasing anneal temperature. The dashed lines indicate the maximum spread in the data obtained from different samples. (b) Thallium. Here, the substitutional and interstitial components exhibit different anneal behavior.

4.3.3 Simple Picture of the Behavior of Implanted Atoms

As a framework for the following sections, we list here some of the factors that we believe are important in determining the lattice location of implanted atoms. This simple picture appears to be reasonably consistent with present experimental information.

a. *Lattice Defects.* One of the major differences between implantation doping and doping by thermal means is the creation of a large number of defects (interstitial substrate atoms as well as vacancies) in the former case. Various interactions may occur between these defects and the implanted atoms. It might be appropriate to distinguish two quite different situations:

1. The implanted atom is contained in a completely disordered volume (an amorphous layer, for example). Until the crystal structure has been restored by annealing, concepts such as lattice location, vacancies, silicon interstitials, etc. are rather meaningless.
2. The implanted atom acquires a definite lattice location, which may subsequently be changed by interaction with various defects. These defects may originate from other implanted atoms striking nearby, or they may be released during a subsequent anneal. For a typical implantation (involving $\gtrsim 10^{14}$ ions/cm^2), each implanted atom lies within the range of disorder created by several neighboring implanted atoms. Consequently, even in an implantation at elevated temperatures, the final fate of each dopant atom may not be determined until near the end of the bombardment.

Under ordinary implantation conditions, a significant number of dislocations are also formed, and these persist to temperatures of 900° C or even higher. They involve considerably more displaced atoms than the total implantation dose, and hence may provide trapping sites for an appreciable fraction of the implanted dopants if the latter become mobile either through normal or enhanced diffusion processes. Such a trapping mechanism may account at least partly for the fraction of implanted atoms that are neither substitutional nor interstitial.

b. *Diffusion Properties.* The close similarity in the lattice-location behavior of a given dopant in silicon and in germanium indicates that this behavior can hardly be correlated to the diffusion constant. In germanium, the Group III elements are the slower diffusants and the Group V elements the faster diffusants; in silicon, this behavior is reversed. However, the diffusion constant does probably determine the rate at which an implanted nonequilibrium system can relax toward equilibrium during anneal.

c. *The Replacement Reaction.* Experimental evidence for the existence of a replacement reaction between silicon interstitials and Group III substitutional dopants was first obtained by Watkins[165,227] in a series of EPR studies on aluminum-doped silicon during low-temperature (4° K) electron irradiation, and recently also by Davies *et al.*[156] in proton-channeling studies at 25° C on boron-diffused silicon. Watkins proposed that silicon interstitials interact with the substitutional dopants and displace the latter from their lattice sites. This problem has been investigated by Fladda *et al.*[216] and will be treated in more detail in Part 4.4.

d. *Charge-State Effects.* One major difference between the Group III and Group V dopants is that on a substitutional site the Group III ion is an acceptor and the Group V ion a donor, whereas when interstitially located, both presumably behave as donors. In addition, the charge states of simple defects, such as vacancies, are likely to depend on whether the implanted region is *n*- or *p*-type.[227] Consequently, effects of this type may be important in some of the implantation reactions. It is also possible that differences in mobilities between the various species involved may be a contributing factor.

e. *Overall Picture.* When an energetic ion comes to rest, a large number of vacancies are created in its vicinity. The implanted atom is thus readily able to find a substitutional site. From this point on, the behavior becomes species- and temperature-dependent. Group V atoms remain on these substitutional sites (even at concentrations above their solid solubility) throughout the remainder of the anneal sequence, or for a wide range of implantation temperatures. At anneal temperatures of 900–1000° C, diffusion to precipitation sites or to surface sinks tends to reduce the substitutional concentration toward solid-solubility values. In hot-substrate implantations, the replacement mechanism tends to drive Group III elements toward interstitial sites and diffusion processes toward nonregular (precipitation) sites. Both of these mechanisms reduce the substitutional component, but even so, concentrations in excess of the solid solubility can be attained. At implantation temperatures of 350–450° C, the Group III elements indium, gallium, and thallium tend to have roughly equal substitutional and interstitial components. Possibly, the interstitial component is governed by charge-state effects, since it is suppressed when a high concentration of substitutional Group V atoms is present. The behavior of the Group III elements is sufficiently complex and species-dependent that further generalization at this point is not warranted. Boron, for example, does not exhibit the *tetrahedral* interstitial component, even though the replacement reaction has been clearly established.

4.4 Lattice-Location Results in More Detail

4.4.1 Group V Implants in Silicon

It is clear from Table 4.4 that the Group V elements arsenic, antimony, and bismuth show high substitutional components in silicon following implantation at 350–450° C. Some of the parameters influencing the lattice location of the implanted Group V ions have already been summarized in Part 4.3, viz. implantation temperature (Fig. 4.16) and subsequent anneal treatment (Fig. 4.17). The dose dependence of the substitutional fraction and of the lattice disorder are compared in Fig. 4.19 for implant temperatures of 350–450° C. For all three elements, the substitutional fraction is essentially

TABLE 4.4

Lattice Location of Group III and Group V Elements in Silicon[a]

Implanted atom[b]	Implantation temperature (°C)	% along row[c]		Solubility at 1100° C [(atoms/cm³) × 10^{-19}]	Diffusion constant[d] at 1100° C [(cm²/sec) × 10^{14}]
		⟨111⟩	⟨110⟩		
As	350	60	58	200[e]	2
	450	56	55		
Sb	350	89	87	4.5[e]	2
	450	90	88		
Bi	350	—	—	0.02[e]	2
	450	87	86		
Ga	350	55	27	3.5[e]	50
	450	<10	<10		
In	350	52	25	0.2[f]	8
	450	61	34		
Tl	350	84	46	0.02[g]	8
	450	53	27		

[a] From Eriksson *et al.*[174]

[b] Implantation conditions: 350–450° C, ~1.5×10^{14} ions/cm², 40 keV. Assuming a mean depth of ~0.05 microns, 1.5×10^{14} ions/cm² corresponds to an average concentration of ~3×10^{19}/cm³.

[c] %⟨110⟩ ≈ substitutional component; % ⟨111⟩ − % ⟨110⟩ ≈ interstitial component. These data represent an average over measurements on several samples. The experimental errors are 3–5%.

[d] From Boltaks.[231]

[e] Equilibrium solubilities given by Trumbore.[228]

[f] From the maximum solubility calculated by Fischler.[229]

[g] From diffusion studies by Fuller and Ditzenberger.[230]

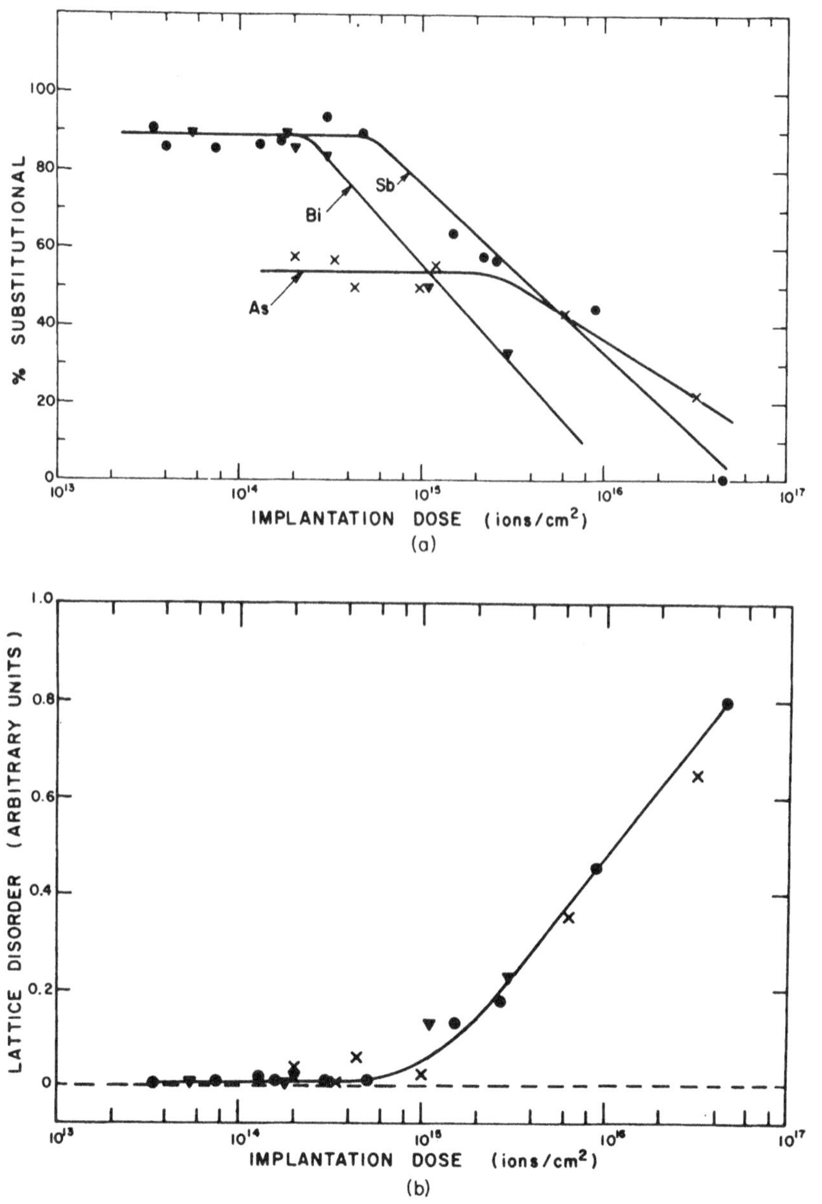

Fig. 4.19. (a) Dose dependence of the substitutional component for "hot" implantations of Group V dopants in silicon: ×, As; ●, Sb; ▼, Bi. Implantation conditions: 40 keV and 350–450° C. (From Eriksson *et al.*[174]) (b) Dose dependence of the lattice disorder accompanying the Group V implantations in (a): ×, As; ●, Sb; ▼, Bi. (From Eriksson *et al.*[174])

independent of dose up to a certain point. At higher doses, a drop-off in substitutional level occurs. As can be seen from Fig. 4.19(b), this is associated with a rapid increase in the accumulated lattice disorder. From the data in Fig. 4.19(a), assuming an implant depth of ~500 Å, the following maximum substitutional concentrations are obtained: for arsenic, 1.5×10^{21} ions/cm^3; for antimony, 0.8×10^{21} ions/cm^3; and for bismuth, 0.2×10^{21} ions/cm^3. In the case of bismuth, this is roughly three orders of magnitude above its published solid solubility at 1100° C (Table 4.4). For arsenic, on the other hand, 1.5×10^{21} ions/cm^3 is approximately equal to the maximum equilibrium solubility.

The accumulation of disorder in Fig. 4.19(b) appears to be independent of ion species. The dose at which the substitutional level starts to decrease, however, is definitely species-dependent.

The fact that at lower doses the substitutional level is independent of dose indicates that the probability of an implanted ion being trapped on a nonlattice site is constant and is independent of dopant concentration at these temperatures. Figure 4.19(a) also shows the surprising fact that arsenic (the element with the highest solubility) has a lower substitutional level (50–60%) than either antimony or bismuth. Since no significant dose dependence is observed below 10^{15} ions/cm^2, one may assume that the low substitutional level for arsenic persists at lower doses. This peculiar behavior compared to the other Group V implants is not understood at present.

The dose dependence of bismuth (and thallium) implants at room temperature is illustrated in Fig. 4.20(a). A strong correlation is seen between the decrease in the substitutional dopant level and the increase in lattice disorder with increasing dose (as was also the case in the hot implants, Fig. 4.19). However, the dose required to form an amorphous layer at room temperature (~10^{14} ions/cm^2) is at least two orders of magnitude lower than at 350–450° C. The anneal behavior of a low-dose bismuth implant (Fig. 4.20(b)] shows that the high substitutional level is maintained up to 800° C. The anneal of high-dose room-temperature Group V implants is shown in Fig. 4.17. A high substitutional level is observed as soon as the amorphous layer recrystallizes, and no dramatic change occurs up to ~900° C. Again, as in the hot implants, arsenic appears to exhibit a significantly lower substitutional level than antimony or bismuth.

Measurements on phosphorus implantations using the ^{32}P isotope and the electron-emission technique [Fig. 4.5(b)] have been reported by Gibson *et al.*[168] These measurements were too preliminary to give quantitative information on the lattice location; however, they indicated that a major fraction of the implanted phosphorus atoms were on substitutional sites. Recently, Mitchell,[225] using the ^{31}P(p, α) ^{28}Si reaction, has shown that ~85% of the implanted phosphorus atoms are substitutional following an 800° C anneal

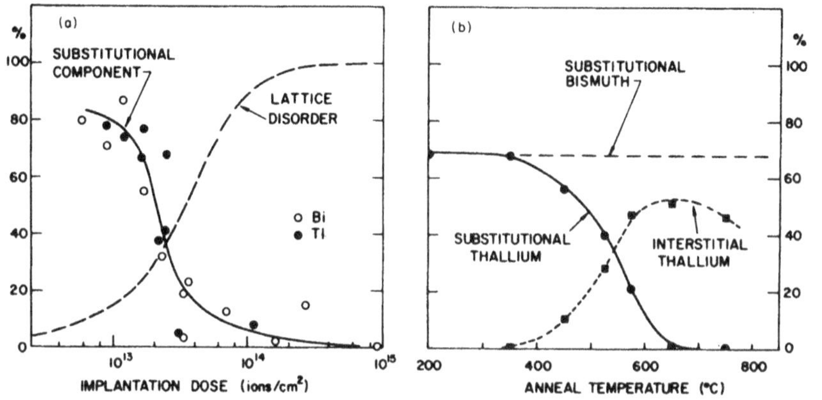

Fig. 4.20. (a) Dose dependence of the substitutional level of thallium (●) and bismuth (O) implants in silicon *before* annealing. Implant conditions: 40 keV at 25° C. The broken line indicates the total amount of lattice disorder in the implanted region, expressed as the percentage of silicon atoms displaced more than 0.2 Å from their equilibrium lattice sites. (From Eriksson *et al.*[171]) (b) Anneal behavior of a low-dose (1.1×10^{13} ions/cm²) thallium implant from (a). The broken line indicates, for comparison, the behavior of a corresponding low-dose bismuth implant. (Derived from Eriksson *et al.*[174])

of 40-keV room-temperature implants (at doses of 10^{15}–10^{16} ions/cm²).

Thus, there is no major difference in behavior for the Group V ions between 350–450° C and 25° C implants, provided the gross lattice disorder in the latter case has been subsequently annealed. All elements show large substitutional fractions and no evidence of any tetrahedral interstitial fraction.

4.4.2 Group III Implants in Silicon

In this section, the heavy Group III elements are treated in detail first. Some recent measurements for boron are then discussed in subsection d.

In contrast to the relatively simple behavior of the Group V elements, the lattice location of the Group III elements (gallium, indium, and thallium) depends strongly on implantation temperature and on subsequent anneal treatment. The substitutional fractions are generally lower for the Group III elements than for the Group V. Nevertheless, the substitutional concentrations of gallium, indium, and thallium that can be obtained following implantation at 350° C are much greater than their estimated equilibrium solubilities at the implantation temperature (cf. Table 4.4). The most remarkable observation (Table 4.4) is that these elements exhibit an interstitial component, which in many cases is comparable in magnitude to the substitutional one. The ratio of the interstitial to substitutional component depends on implantation temperature and also upon subsequent anneal. There is also a dependence on ion species.

a. *Implantation Temperature.* Detailed measurements by Fladda *et al.*[216] on the temperature dependence of thallium implantations (Fig. 4.21) indicate that the interstitial to substitutional ratio is strongly temperature-dependent. A 1 : 1 relation between substitutional and interstitial components is maintained for thallium only in a narrow temperature region near 400° C. Note that a 600° C implant leads to a pure tetrahedral interstitial component, whereas at 25° C and low doses the thallium atoms occupy substitutional sites, in agreement with Fig. 4.20. The behavior of thallium implantations at 350° C has been investigated also as a function of dose.[174] It is found that the substitutional to interstitial ratio is essentially independent of dose and close to unity over the whole dose range studied, viz. 10^{13}–10^{15} ions/cm^2.

For indium, both substitutional and interstitial components decrease towards zero for implantations at ~600° C, and a similar result is seen for gallium at 450–500° C.[174] Whether or not this decrease occurs via an intermediate interstitial component (as for thallium) has not been entirely resolved, but such an effect does not appear to be supported by the anneal measurements described below.

b. *Anneal of Elevated-Temperature* (>300° C) *Implants.* During subsequent annealing of 350–450° C Group III implants, the relation between

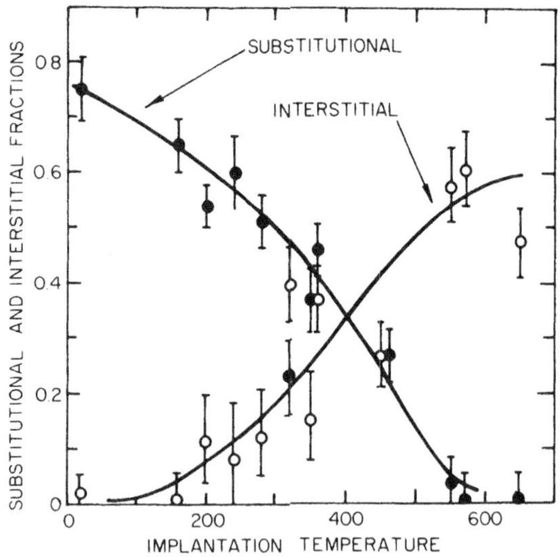

Fig. 4.21. Lattice location of implanted thallium atoms in silicon as a function of implantation temperature: ●, substitutional fraction; ○, tetrahedral interstitial fraction. The data at 20° C are for a dose of 2×10^{12} ions/cm^2, the data between 150 and 350° C are for 1–3 $\times 10^{13}$ ions/cm^2, and for temperatures higher than 350° C the dose is ~10^{14} ions/cm^2. (Data from Fladda *et al.*[216] and Eriksson *et al.*[174])

substitutional and interstitial fractions is not always maintained. A typical anneal sequence for a "hot" thallium implant was shown in Fig. 4.18(b). Below ~525° C, the substitutional and interstitial fractions are nearly equal. By 575° C, however, the substitutional thallium component has disappeared, and there is a corresponding increase in the interstitial component. This behavior is consistent with the ~600° C implantation results of Fig. 4.21. At even higher anneal temperatures (>600° C), the measurements indicate that the interstitial thallium atoms are moving to random locations.

The anneal behavior of "hot" gallium and indium implantations has also been studied [Fig. 4.18(a)]. As for the thallium case, a gradual motion to random positions is noted in the temperature range 550–800° C, but a measurable substitutional component still exists at temperatures where the thallium dopants had all moved to interstitial sites, (i.e., at 575–650° C).

It is clear from the gallium and indium data that a subsequent annealing treatment does not always produce the same result as an increase in implantation temperature: For example, a gallium implant at 450° C or above exhibits essentially no detectable substitutional or interstitial components, whereas a 350° C implant on annealing appears to retain both these components to $\gtrsim$700° C.

c. *Anneal of Room-Temperature Implants.* For implantation doses such that an amorphous layer is formed (i.e., typically 2–5 × 10^{14} ions/cm^2), the anneal behavior of Group III elements exhibits some of the same characteristics as those noted for Group V dopants.[174] At the temperature (~550–600° C) where the lattice reorders, typically 40–60% of the implanted atoms are on regular lattice sites—in this case, however, both interstitial and substitutional sites. For thallium, *only* an interstitial component (~50%) was noted [Fig. 4.18(b)], while for gallium [Fig. 4.18(a)], nearly equal numbers (~25%) of substitutional and interstitial components were present. For indium, the evidence for an interstitial component is not as well established. The anneal behavior also indicates that the Group III dopants move from substitutional sites at significantly lower temperatures than the Group V dopants. The anneal behavior of low-dose (~10^{13} ions/cm^2) thallium implantations is illustrated in Fig. 4.20(b).

With the exception of thallium, the lattice location of Group III elements implanted at room temperature has not been as well characterized as those of the Group V species. In particular, the dependence on ion dose needs further attention.

d. *Boron Implantations.* Preliminary measurements using the ^{11}B(p, α) ^{8}Be reaction (Section 4.2.4) had indicated that a major fraction of implanted boron atoms occupied substitutional sites following a 40-keV implantation at 25° C.[231a] Gibson *et al.*,[168] on the other hand, reported measurements on boron implantations at 400 keV, 25° C, and 10^{15} ions/cm^2 in which they found that the boron atoms were *non*substitutional. Later measurements by Fladda *et al.*[231b] and by North and Gibson,[231c] using 56-keV and 150-keV boron implants, respectively, have shown that a large substitutional component is indeed obtained following a room-temperature implantation. No explanation has yet been given for the anomalous behavior of the earlier 400-keV study.

The dependence on implantation temperature (25–500° C) has been investigated by Fladda *et al.*[231b] for implantation doses of 10^{15} ions/cm^2. At room temperature the substitutional component is 30–40% but it falls off toward zero in the 100–300° C region. For boron, no tetrahedral interstitial component is detected, in agreement with the proposition by Watkins[227] based on EPR studies. North and Gibson's measurements[231c] also confirm that the nonsubstitutional boron atoms are *not* on the tetrahedral interstitial sites along the ⟨111⟩, and they suggest that the atoms may be occupying some specific interstitial site along the ⟨110⟩.

Figure 4.22 shows the anneal behavior of 25° C boron implantations in silicon for doses of 10^{14}–10^{15} ions/cm^2. Similar results have been observed by North and Gibson[231c] for a dose of 4×10^{15} ions/cm^2 at 150 keV. It is noted from Fig. 4.22 that the overall behavior is essentially independent of implantation dose. The initial substitutional fraction, however, does depend on the dose. Upon anneal, the substitutional level first decreases above ~200° C, reaches a minimum level around 500° C, and then at still higher temperatures rapidly *increases* again to reach 90% at ~900° C. This latter increase in substitutional level may perhaps be interpreted as a recombination of interstitial boron with thermally activated vacancies (cf. Section 5.4.2c). The same increase in substitutional level is noted[231b] during the anneal of 300–500° C implantations, in which case the initial substitutional component is zero.

The decrease in substitutional content between 200° and 500° C has been attributed to the replacement reaction,[231b, 231c] resulting from the silicon interstitials released during the anneal of the lattice disorder. As expected, Fladda *et al.* did not find a similar decrease when annealing a boron-*diffused* sample. We will return to a discussion of this in Section 4.4.6, after some post-bombardment experiments of boron-implanted silicon have been described.

The major difference between boron and the heavier Group III elements is the large increase in the substitutional level of boron at higher anneal temperatures (cf. Figs. 4.22 and 4.18). This is probably associated with the fact that in these lattice-location studies boron is the only Group III element whose solubility is significantly greater than the concentration achieved during implantation. Hence, a similar effect might be expected for a low-dose gallium implantation. It is also possible that the lower diffusion constant of boron plays a role.

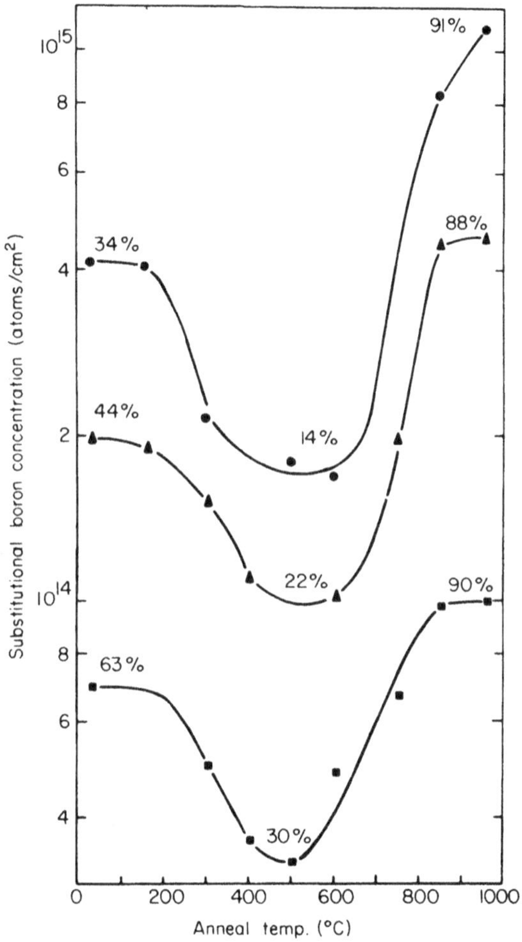

Fig. 4.22. Anneal behavior of 56-keV room-temperature boron implants in silicon. Data are shown for three different doses: ● 1.2×10^{15} ions/cm²; ▲ 4.6×10^{14} ions/cm²; ■ 1.1×10^{14} ions/cm². The percentage of substitutional boron atoms are indicated at several appropriate places on the figure. (From Fladda *et al.*[231b])

4.4.3 Implantation of Other Ion Species

At present, investigations of species other than Group III and Group V dopants have been carried out only for a rather restricted range of implantation conditions, so that it is difficult to present a general picture of the lattice-location behavior. The following summarizes† some of the lattice-location studies in silicon for ions other than Group III and Group V elements

† A more complete list of the ion species investigated is given in Table 4.3.

(Unless otherwise stated, doses were between 1 and 5×10^{14} ions/cm^2 at 30–50 keV—i.e., an average concentration of 2–8×10^{19}/cm^3.)

a. *Group IV.* Tin in silicon[174] and also lead in germanium[226] were found to be 90–95% substitutional for 350–450° C substrate temperatures during implantation. We suggest that Group IV species will in general exhibit the same behavior as the Group V.

b. *Group I.* Gold implantations in silicon illustrate the importance of keeping the implantation temperature low enough, so that appreciable diffusion does not occur. For implantations at 450° C,[174] less than 5% of the gold atoms are found on substitutional sites (Fig. 4.3), while at 200° C (and sufficiently low dose) approximately 40% are substitutional.[224]

c. *Group II.* Zinc,[231d] cadmium,[55,174] and mercury[224,231d] implanted in silicon† at 350° C exhibit interstitial components. In the zinc and cadmium cases ~20–30% were on the tetradheral interstitial sites, and for mercury ~40–50%. In all cases the substitutional component was less than ~5%. One may conclude from this that the Group IIB elements exhibit common features.

Anneal of a 350° C cadmium implantation[55] shows that the decrease in the interstitial component occurs in the same temperature range as that for the heavy Group III elements, i.e., at 600–700° C. This decrease occurs at slightly lower temperatures for zinc and at higher temperatures (800° C) for mercury.[231e]

When cadmium is implanted into silicon to a sufficiently high dose that an amorphous layer is created, outdiffusion effects are observed[55] on annealing to 600° C (see also Section 3.4.4). This is also observed for mercury implantations.

For low-dose (~10^{13} ions/cm^2) mercury implantations in silicon at 25° C which are annealed to 400° C, a substitutional component of 30% is observed[231d] with no interstitial component. Similar results were obtained for thallium and bismuth (Fig. 4.20). These results support the proposition that will be made later: viz., that the interstitial component is due to effects occurring after the particle has come to rest.

d. *Group VI.* For tellurium implantations at 450° C, approximately half of the implanted atoms were on substitutional sites, and in high-dose room-temperature implantations approximately 60% were substitutional after the lattice reordered.[55] Similar results have recently been obtained for selen-

† Mercury in germanium at an implantation temperature of 300° C had approximately equal numbers (~30%) on substitutional and interstitial sites.[224]

ium.[231d] In none of these cases could any significant tetrahedral interstitial component be observed. Thus, one again finds that elements within the same group of the periodic table exhibit a regular behavior.

4.4.4 Implantations into Germanium

In order to obtain a broader base for a discussion of implantation phenomena in semiconductors, studies of lattice-location behavior have been extended also to germanium to see whether the features observed in silicon—in particular, the interstitial component—would appear. These studies are not yet as extensive as those in silicon.

Table 4.5 summarizes the data obtained on germanium samples implanted with 1–5 × 10^{14} ions/cm^2 at 300–350° C and 30 keV. Since there was no drastic dependence on either temperature or dose within these limits, only average values are presented. Values of diffusion constants and equilibrium solubilities are included where possible. Comparison of these germanium data with those for silicon in Table 4.4 indicates that the implantation behavior of Group IV and Group V elements is similar in both crystals in that the implanted atoms are predominantly on substitutional sites. Similarly, the two sets of thallium data are similar in that the interstitial and substitutional components are roughly equal in magnitude, following implantation at these

TABLE 4.5

"Hot" Implantation Behavior in Germanium[a]

Implanted atom[b]	% along row[c]		Solubility[d] at 800° C	Diffusion constant[e] at 800° C
	⟨111⟩	⟨110⟩	[(atoms/cm^3) × 10^{-19}]	[(cm^2/sec) × 10^{14}]
Sb	90	85	1.5	2000
Bi	76	80	<0.02	—
In	75	75	0.4	20
Tl	65	35	<0.02	15
Hg[f]	60	30	—	—
Pb	87	85	0.04	2

[a] From Björkqvist *et al.*[226]

[b] Implantation conditions: 300–350° C, 30 keV, 1–5 × 10^{14} ions/cm^2.

[c] %⟨110⟩ ≈ substitutional component; % ⟨111⟩–% ⟨110⟩ ≈ tetrahedral interstitial component. Typical errors ±5%.

[d] Data from Trumbore[228] and Fischler.[229]

[e] Data from Boltaks.[231]

[f] Data from Fladda *et al.*[224]

temperatures. A further similarity is that thallium, when implanted into germanium at low dose ($\sim 10^{13}$ ions/cm^2) and low temperature (150° C), shows only a substitutional component (70–80%)[224] [cf. Fig. 4.20(a)]. The indium data in Table 4.5, on the other hand, differ markedly from those for silicon in that there is *no* interstitial indium component present in germanium. For antimony implants of 4×10^{14} ions/cm^2, the dependence on implantation temperature has been studied over a wider temperature interval (250–550° C), with little variation in the substitutional component.[224]

If an upper limit of 0.05 micron is used for the depth distribution at 30 keV, a dose of 2×10^{14} ions/cm^2 is equivalent to an average concentration of approximately 4×10^{19} ions/cm^3. On the basis of this estimate, the substitutional components for all species exceed the solid solubility at 800° C, and in the case of lead, bismuth, and thallium they exceed it by two orders of magnitude. Again this parallels the behavior observed in silicon.

Comparison of the data of Tables 4.4 and 4.5 shows that no correlation exists between the reported diffusion constants and the lattice-location behavior†—the Group V elements are largely substitutional both in silicon and germanium, although they are the slower diffusants in silicon and the faster diffusants in germanium. Further, indium and thallium have very nearly the same diffusion constants in germanium, but still show a markedly different behavior. On the other hand, during subsequent anneal of implantations in germanium, both antimony and bismuth precipitate at lower temperatures ($\sim$600° C) than does indium.[224] It therefore appears that in both silicon and germanium, a correlation to the diffusion constant can be made for the anneal behavior. When making these points, two things should be remembered. First, it can be questioned to what extent literature values of diffusion constants are applicable to implantation conditions, with the high dopant concentrations involved. Second, little is known about the diffusion *mechanisms* of the various elements.

The indium data are particularly interesting, in that indium is the only heavy Group III element for which no significant interstitial component has yet been observed, either for "hot" implantations (Table 4.5), or for annealed room-temperature implantations.[157] Both these measurements show a high ($\sim$75%) substitutional component. This high substitutional component persists even at implant temperatures of 550° C.[224] It should also be remarked here that for a high-dose thallium implantation (such that an amorphous layer is formed) the thallium atoms occupy only *substitutional* sites ($\sim$80%)

† Preliminary measurements for indium and antimony implantations in germanium[173] had indicated that the lattice location behavior in silicon and germanium might be correlated with the diffusion constant. This has not been borne out in later measurements, as is clear from the present discussion.

after the lattice has reordered.[224] This contrasts sharply with the behavior in silicon [Fig. 4.18(b)], where there is a large interstitial fraction and negligible substitutional thallium after annealing to ~600° C.

When comparing the implantation behavior in silicon and germanium, it is tempting (despite the differences noted above) to conclude that all *p*-type dopants and all *n*-type dopants, respectively, behave in similar fashion. More detailed measurements in germanium would be desirable to verify this conclusion.

4.4.5 Mixed Group III and Group V Implantations in Silicon and Germanium

The most puzzling result presented so far is the existence of an interstitial component for *p*-type dopants and the observation that this component is often equal in magnitude to the substitutional one.

Measurements on mixed implantations in silicon or germanium show[219] that if a Group III dopant is implanted into a crystal that had previously been implanted with a Group V dopant, the Group III atoms become substitutional to a high degree and show *no* measurable interstitial component, provided the concentration of the Group V dopant is sufficiently large. This can be seen in the scattering spectra of Fig. 4.14(a), where (in contrast to the behavior in Fig. 4.10) the thallium peaks along the ⟨111⟩ and ⟨110⟩ directions are identical. A summary of these mixed-implantation studies is given in Table 4.6, along with the comparable data for a Group III implantation alone. No influence of the Group III dopants on the lattice location of the Group V dopants is observed in these measurements. Measurements were also made with the implantation order reversed—i.e., with the Group III element implanted first. These measurements showed that the implantation order is of no significant importance.

Mixed implants of thallium plus indium in germanium were also studied (Table 4.6), since indium in germanium is the only case of a Group III implant having a high substitutional component and no interstitial component, (cf. Table 4.5). However, it has no effect on the thallium interstitial component in a mixed implant, suggesting that only substitutional *donors* are effective in suppressing the interstitial component.

Preliminary dose-effect measurements indicate that the suppression of the interstitial component in a mixed implant occurs only when the concentration of *n*-type dopants exceeds the concentration of *p*-type dopants. This dose dependence is illustrated in Table 4.6 by the thallium-plus-arsenic and indium-plus-bismuth implantations into silicon. For example, when the arsenic and thallium doses are comparable, an interstitial thallium component of ~15% remains, whereas for a sixfold higher arsenic dose, no interstitial

TABLE 4.6

COMPARISON OF THE LATTICE-LOCATION BEHAVIOR OF GROUP III ELEMENTS IN SILICON AND GERMANIUM AS IMPLANTED ALONE AND AS MIXED WITH GROUP V DOPANTS

Ions[a]	Group III implant alone[b,c]		Mixed implant[c,d]		Dose ratio (V/III)
	% subst.	% int.	% subst.	% int.	
Silicon					
Ga + Sb	27	28	70	0	2
Tl + Sb	46	38	66	0	1.5
Tl + As	46	38	55	15	1
Tl + As	46	38	70	15	4
Tl + As	46	38	75	0	6
In + Bi	25	27	68	19	1
In + Bi	25	27	70	14	2
In + Bi	25	27	65	0	4
Germanium					
Tl + Sb	35	30	71	0	2
Tl + In	35 75	30 (Tl) 0 (In)	38 76	30 (Tl) 0 (In)	1[e]

[a] Typical implantation conditions: 350° C, 30 keV, 1–4 × 10^{14} ions/cm^2. Because the implantation energy is kept constant, the range distributions for the two components do not completely overlap.

[b] Data from Eriksson *et al.*[174] and Björkqvist *et al.*[226]

[c] Typical error ±5%.

[d] Data from Eriksson *et al.*[219] and Fladda *et al.*[224]

[e] Indium/thallium ratio.

component is observed. (It should be recalled that for arsenic only ~60% of the implanted atoms are on substitutional sites [Fig. 4.19(a)].) When a mixed thallium-plus-arsenic implantation is annealed to 650° C, the high substitutional thallium component is retained,[224] in strong contrast to the anneal behavior of a pure thallium implantation [Fig. 4.18(b)].

It is not possible at present to specify in detail why there is no measurable interstitial component under mixed-implant conditions. The results suggest, however, that charge-state effects are important in determining the lattice location of implanted atoms. We shall return to this question in Part 4.5.

4.4.6 POSTBOMBARDMENT OF IMPLANTED SAMPLES

In Section 4.3.3, the observed lattice location behavior was attributed to the interaction of the implanted atoms with vacancies and with interstitial host atoms (and other defects) created during the implantation. Therefore, the ultimate lattice location depends on processes that occur after an implanted atom has lost its initial kinetic energy. Further, it should be possible to change the lattice location of implanted atoms by creating defects in various postbombardment processes. The mixed-implant results of the previous section support this point of view.

Direct evidence of a motion of dopant atoms from substitutional sites during proton irradiation has been observed[156] in channeling-effect measurements on boron-diffused layers in silicon at 25° C. Such measurements are illustrated in Fig. 4.23, which shows how the percentage of boron and lattice atoms located within ~0.2 Å of the axis varies as a function of proton dose. A 680-keV proton beam was used to induce the $^{11}B(p, \alpha)\ ^{8}Be$ reaction, and the yield of α-particles was used in the channeling-effect measurements to locate the boron atoms (cf. Section 4.4.2.d). Backscattering was used to monitor the lattice disorder. A rapid decrease of boron atoms along the ⟨111⟩ is observed and a similar decrease occurs along the ⟨110⟩. For the same dose interval, there is a negligible increase in lattice disorder. It was concluded therefore that an enhanced effect drives the boron atoms off the lattice sites. These measurements also indicate that the interstitial site occupied by the displaced boron atoms is not the tetrahedral interstitial site. This matter was discussed in more detail in Section 4.4.2d. Similar measurements on antimony-implanted silicon showed no significant change in the substitutional component for a 1.0-MeV helium-bombardment dose of 10^{18} ions/cm^2.

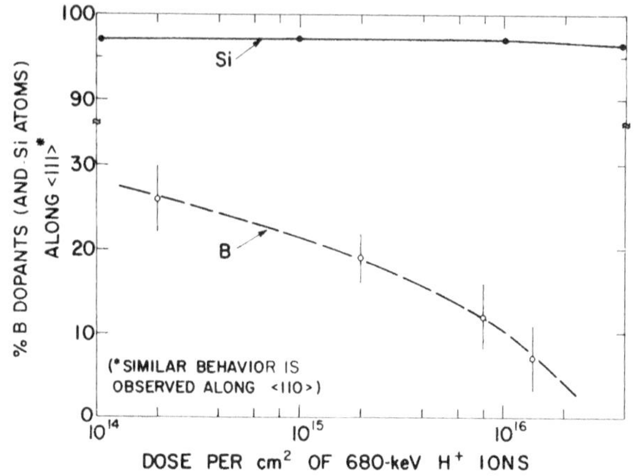

Fig. 4.23. Radiation sensitivity of boron-doped silicon to 680-keV proton bombardment at 25° C. The broken line indicates the percentage of boron atoms that are within ~0.2 Å of ⟨111⟩ atomic rows as a function of bombardment dose. The solid line shows that little permanent lattice disorder is introduced during the analysis. The sample had a surface concentration of ~10^{19} ions/cm^3 and a junction depth of ~1 micron. (From Björkqvist *et al.*[217])

The influence of post-irradiation on the lattice location of implanted thallium in silicon has been investigated by Fladda *et al.*[216] at various temperatures, using a beam of 1.8-MeV carbon ions. In order to study the induced motion of thallium atoms from substitutional to interstitial sites, implant conditions of ~200° C, 40 keV, and ~10^{13} ions/cm^2 were chosen. As seen in Fig. 4.21, this produces a high (~60%) substitutional component, a low (~10%) interstitial component, and relatively little lattice disorder. Furthermore, this disorder was largely annealed out (at 300° C) before postbombardment started. During this anneal treatment, the decrease in substitutional thallium level was less than 10%.

In Fig. 4.24 the substitutional and interstitial thallium fractions are plotted as a function of the random carbon bombardment dose at 340° C. Two different samples (●, ▲ = 1.5×10^{13} thallium ions/cm^2 and ○, △ = 2.6×10^{13} thallium ions/cm^2) are included. The data are plotted as a function of carbon dose for random orientation only, since the radiation damage in the implanted region during the measurement of the aligned spectra is almost insignificant. It is apparent from Fig. 4.24 that during the bombardment, there is a strong decrease in the substitutional fraction and a corresponding increase in the interstitial fraction, thus indicating a direct motion of thallium atoms from substitutional to interstitial sites. Further, it can be seen that the carbon dose $Q_{1/2}$ required to reduce the substitutional component to 50% of its initial value depends on the thallium concentration. Measurements on several samples at 340° C (Fig. 4.25) indicate a linear relationship between $Q_{1/2}$ and the initial content of substitutional thallium. This would imply that the defect–thallium interaction rate does not depend on the thallium content. In Fig. 4.26, carbon-bombardment measurements are shown at four different temperatures (200°, 270°, 340°, and 410° C). Only the substitutional fraction is shown, and the dotted lines indicate the best linear fits to the data points. The bombardment doses have been normalized to the same initial substitutional thallium fraction, using Fig. 4.25. The slope becomes steeper with increasing temperature, suggesting that the displacement of thallium atoms from substitutional to interstitial positions is a thermally activated process.

Carbon bombardments were also carried out at 25° C, with no detectable change in the substitutional fraction, even after bombardment to doses significantly larger than those of Fig. 4.26. Experiments were also carried out to test the possible influence of mixed-implant doping of the substrate on the displacement mechanism. For this purpose a silicon sample, implanted with both arsenic (3×10^{14} ions/cm^2) and thallium (6×10^{13} ions/cm^2) at 350° C was used. Under this mixed-implant condition, there is *no* detectable interstitial thallium fraction present initially (Table 4.6). Even when the sample

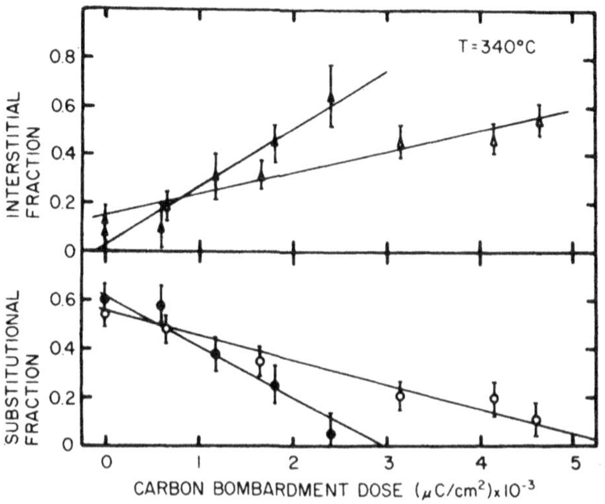

Fig. 4.24. Lattice location of implanted thallium atoms in silicon as a function of 1.8-MeV carbon bombardment at 340° C. Open symbols refer to an implanted thallium dose of 2.6×10^{13} ions/cm² and solid symbols to a dose of 1.5×10^{13} ions/cm². The solid lines are best linear fits to the data. The carbon doses are those measured without secondary-electron suppression; the actual doses may be much smaller. (From Fladda *et al.*[216])

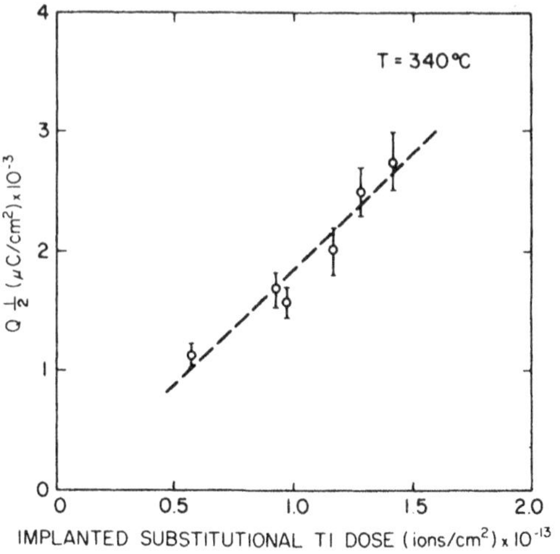

Fig. 4.25. The bombardment dose $Q_{1/2}$ of 1.8-MeV carbon ions required to decrease, in silicon at 340° C, the initial substitutional thallium component by a factor of 2 versus the implanted substitutional thallium dose. As in Fig. 4.24, the carbon doses have been measured without secondary-electron suppression. (From Fladda *et al.*[216])

was subsequently bombarded at 340° C with carbon ions to a random dose several times larger than those characteristic of Fig. 4.26, *no* decrease in the substitutional level could be detected.

Fladda *et al.*[216] also tried to verify the opposite reaction—i.e., to drive interstitial ions into substitutional sites. For this purpose, silicon samples implanted with $\sim 10^{14}$ thallium ions/cm^2 at 600° C were used; under such conditions (Fig. 4.21), the interstitial component is ~60% and the substitutional component is insignificant. Carbon bombardments were then performed both at room temperature and at 200° C, but no detectable changes in either the interstitial or the substitutional components were observed even after prolonged irradiation (10^{17} carbon ions/cm^2).

Recently, Fladda *et al.*[231b] have made postbombardment experiments with boron-*implanted* silicon. As was shown in Fig. 4.22, a room-temperature implantation results in a 40–50% substitutional boron level. These room-temperature implantations were then bombarded with protons under the same conditions as in Fig. 4.23; *no* decrease in the substitutional boron level was noted, even for very large bombardment doses. On the other hand, if the implanted sample was first annealed to 900° C (cf. Fig. 4.22), and then postbombarded with protons, a strong decrease in the substitutional boron level occurred with increasing proton dose (Fig. 4.27), indicating that in the annealed sample the replacement reaction is again operative.

As for thallium (Fig. 4.25), the postbombardment dose required to displace a certain fraction of the boron atoms from substitutional sites depends on the boron concentration. It was found that anneal temperatures of 200° C or greater are required before any evidence of the replacement reaction could be detected.

One may speculate as to the reasons why the replacement reaction is ineffective for an unannealed room-temperature implantation. We suggest that, following a room-temperature implantation, the boron atoms are either complexed or surrounded by electrically active centers in such a way as to inhibit the replacement reaction.† It is now possible to return to Fig. 4.22 and try to interpret the decrease in the substitutional level that occurs above ~200° C. We propose that the conditions restricting the boron-replacement reaction disappear around 200° C and that, as the lattice disorder anneals in the temperature interval up to ~500° C (as observed by Mitchell[225]), interstitial silicon atoms are released. These interstitials are now able to displace boron atoms and a decrease in substitutional level occurs.

† It is tempting to relate this behavior to the "mixed" implant results of Section 4.4.5 where thallium atoms (in the presence of an excess of arsenic or antimony) are *not* displaced from substitutional sites by postbombardment with carbon ions.

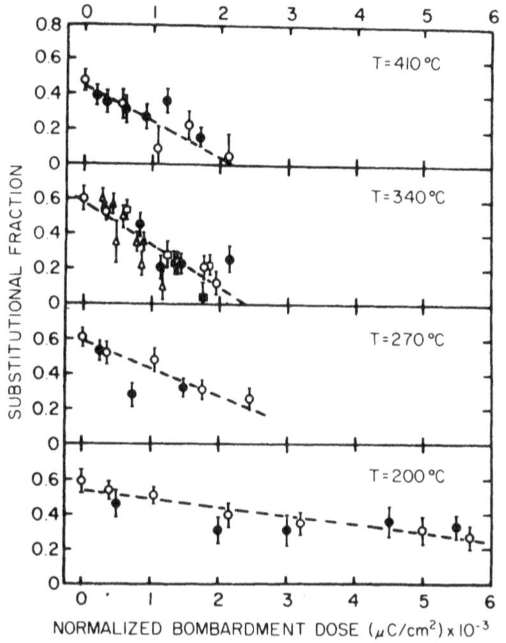

Fig. 4.26. The displacement of substitutional thallium atoms to tetrahedral interstitial sites in silicon during 1.8-MeV carbon bombardment at various temperatures. The bombardment-dose scales have been normalized to the same initial substitutional thallium level by means of Fig. 4.25. The dashed lines are best linear fits to the data. (From Fladda *et al.*[216])

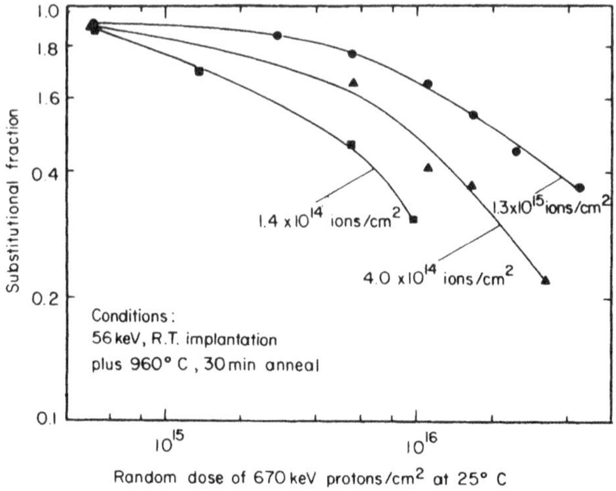

Fig. 4.27. The result of postbombardment of 56-keV room-temperature boron implantations in silicon (following anneal at 960° C for 30 min). The substitutional fraction is shown as a function of the random dose of 670-keV protons. Data are shown for three different doses. (From Fladda *et al.*[231b])

4.5 Discussion of Factors Influencing Lattice Location

4.5.1 General Model

Rather than carry out a detailed interpretation of each element that has been studied, we will try in this section to give support for the general model of the behavior of implanted atoms that was sketched out in Part 4.3. In particular, we will concentrate on the mechanisms that might lead to substitutional or interstitial components. In Table 4.7 are listed some of the "reactions" that might be involved in determining the lattice location of implanted atoms (using silicon as an example). It should be emphasized at this point that our interpretations are somewhat speculative. Until a wider range of dopant species are investigated and certain key experiments have been conducted, it will be difficult to judge the validity of our approach.

We believe that the lattice location of implanted species is not governed to any large extent by *dynamical* events during the slowing-down process, but instead is determined after the atom has lost its initial kinetic energy. The experimental results lend support to this belief. We do not wish to imply,

TABLE 4.7

Possible Reactions Determining the Lattice Location of Implanted Species X (in Silicon)[a]

Substitutional reactions	
1. Dopant interacts with vacancy	X or $X_I + V \rightarrow X_S$
2. Dopant becomes substitutional upon anneal of lattice disorder	X + lattice reorder $\rightarrow X_S$
3. Dopant becomes substitutional during slowing down (dynamic replacement)	$X + Si_S \rightarrow X_S + Si_I$
Interstitial reactions	
4. Substitutional dopant moves to interstitial site	$X_S \rightarrow X_I + V$
5. Dopant becomes interstitial upon anneal of lattice disorder	X + lattice reorder $\rightarrow X_I$
6. Replacement reaction of Watkins	$X_S + Si_I \rightarrow X_I + Si_S$
Other reactions	
7. Dopant moves to nonregular sites (for example, precipitates)	X_S or $X_I \rightarrow X$
8. Dopants interact (for example, to form pairs)	$X_S + X_I \rightarrow [X_S \cdot X_I]$
9. Dopant interacts with defect species to form complex (for example, vacancy–dopant pair)	$X_S + D \rightarrow [X_S \cdot D]$ $X_I + D \rightarrow [X_I \cdot D]$

[a] Index S means substitutional, index I means interstitial.

however, that the dynamical replacement reaction, number 3 in Table 4.7, is not an existing process, only that it is of minor importance for the final location under implantation conditions. The simplest "proof" for this is that the observed lattice location of atoms neighboring in mass is often entirely different. Further, there is often a strong temperature dependence in the implantation behavior.

The major issue is thus to evaluate which implantation reactions are important, and the factors determining these reactions. When such a knowledge has been obtained, ion implantation of semiconductors will be on a much firmer ground. To achieve this, lattice-location experiments alone are not sufficient. For example, a much more detailed knowledge of normal and enhanced-diffusion mechanisms under implantation conditions are needed. Further, the properties of the defects in implanted layers are still not known to any large extent. These two factors alone almost preclude a final description of ion-implantation phenomena.

In the present discussion, we are considering only the situation in which the implanted atom stops in a relatively crystalline region (i.e., for low-dose or elevated-temperature implantations). When the implanted atom is enveloped by a heavily disordered region (either due to the formation of an amorphous layer or the existence of severe localized disorder around the end point), the concept of lattice sites is of course meaningless. Also the behavior of well-channeled ions is not considered; as yet, there are not enough experimental data for a discussion of their behavior. Our development is primarily concerned with the *majority* of the implanted atoms, since channeling-effect measurements cannot be unambiguously interpreted when only 10–15% effects are observed.

4.5.2 Implantation Reactions

a. *Substitutional Reactions.* The basic premise is that an implanted atom occupies a substitutional site at some time immediately after the atom loses its initial kinetic energy (reaction 1 in Table 4.7). The strongest support for this is the large substitutional component observed in low-dose bismuth and thallium implantations at room temperature [Fig. 4.20(a)].† We suggest that toward the end of its track, each implanted atom displaces a sufficiently large number of lattice atoms so that there are always vacant sites nearby. It should be remarked that there is no *direct* evidence of this motion of an implanted atom to a substitutional site due to interaction with a vacancy.

† It is also seen in the recent measurements for room-temperature boron implantations described in Section 4.4.2d.

During the reordering of an amorphous layer, the implanted atoms can also compete successfully for substitutional sites (reaction 2). Concentrations may still exceed the solubility; however, because of the relatively high temperatures involved, other reactions may play a large role in determining the ultimate lattice-location behavior.

In addition, as was illustrated in Fig. 4.22, dopant atoms can acquire substitutional sites during a high-temperature anneal—presumably by recombination with thermally activated vacancies. Since at these high temperatures ($\gtrsim$ 800° C for silicon) most of the implantation-produced defects have already annealed, thermal-equilibrium considerations might be expected to hold: i.e., the values of the solid solubility and the diffusion constant might well determine the observed behavior. Indeed, channeling-effect measurements could be a useful way of determining these quantities. With boron implantations, for example, Fig. 4.22 indicates that the behavior is *not* solubility limited, since even at 950° C no effect of precipitation can be noted at the highest boron concentration ($\sim 5 \times 10^{19}$ atoms/cm^3). This value is well below the solubility of 2×10^{20} atoms/cm^3 at this temperature, as reported by Vick and Whittle.[231f] On the other hand, in cases where the solid solubility has been exceeded (cf. Mayer *et al.*,[157]) precipitation effects become clearly visible at these higher anneal temperatures.

b. *Interstitial Reactions.* For thallium in silicon, two reactions (4 and 6) have been observed in which substitutional dopants are displaced into tetrahedral interstitial positions. In both cases the decrease in the substitutional component could be correlated with an increase in the interstitial component. In the anneal studies of low-dose thallium implantations [Fig. 4.20(b)], the substitutional level decreased at temperatures around 550° C—well above that at which the isolated disordered regions anneal ($\sim$250° C). On this basis, the displacement to interstitial sites can be ascribed to thermal effects (reaction 4) without invoking defect species.

In postbombardment studies, it has been shown (Fig. 4.24) that thallium ions can be displaced from substitutional sites even at temperatures of 300° C by 1.8-MeV carbon irradiation and that the decrease in the substitutional component was correlated with the carbon-ion dose. This reaction (as well as that observed for boron: Figs. 4.23 and 4.27) is clearly associated with defects created during irradiation. We suggest that this reaction is the Watkins replacement mechanism (reaction 6), in which the substitutional dopant is displaced into one of the neighboring interstitial sites by interaction with a mobile silicon interstitial.

Several other implanted Group II and III elements also show the presence of a tetrahedral interstitial component, but a well-defined motion from substitutional to interstitial sites has not been observed. In none of these cases has there been annealing or postbombardment studies of low-dose implantations.

c. *Other Reactions.* For all implanted species, if annealing is carried out at sufficiently high temperatures, the number of dopant atoms on lattice sites decreases (reaction 7). We suggest that this occurs when the implanted atoms can diffuse sufficiently far to be able to interact with some "sink"—e.g., precipitate, or reach the surface or a dislocation. For Group V implantations in silicon, this decrease in substitutional content occurs at temperatures of 800–900° C (depending on concentration). At these temperatures, normal diffusion is significant enough to provide sufficient motion. For Group III (and also cadmium) implantations in silicon, the decrease in substitutional and interstitial content occurs well below 800° C. This lower-temperature motion could be associated with the presence of the interstitial component. For these species, the decrease is observed at even lower temperatures in the case of implantations made as a function of substrate temperature. This indicates that the defects created during implantation can accelerate the displacement to off-lattice sites, perhaps by means of an enhanced-diffusion effect (cf. Part 2.5).

The early observation that, for implantations at 350–450° C in silicon, the Group III tetrahedral interstitial component was often nearly equal to the substitutional one raised the question whether or not this could be explained as an interstitial–substitutional pairing effect (reaction 8), possibly assisted by Coulombic forces. Such effects have been seen in conventionally doped layers. Although this possibility cannot presently be entirely ruled out, it does not seem to be a major effect. The strongest support for it is the 1 : 1 relation between the interstitial and substitutional components and the often-observed [cf. Fig. 4.18(a)] simultaneous decrease of these two components during subsequent anneal. The strongest argument against this pairing effect is that in thallium implantations in silicon, the substitutional–interstitial ratio depends strongly on implantation temperature (Fig. 4.21). A note of caution should be added at this point, however: Gallium and indium may be sufficiently different from thallium that this generalization may not be valid.

No evidence for or against the importance of reaction 9 for implantation conditions has yet been reported. It is generally believed, however, that vacancy–dopant complexes are not important above approximately 300° C.

4.5.3 Charge-State Effects

It has been clearly demonstrated that the tetrahedral interstitial component of Group III species can be suppressed by implantations into material containing a higher concentration of Group V species (Section 4.4.5). These experiments showed that the lattice location of an implanted acceptor is influenced by the presence of donor species—suggesting that charge-state effects are important. Unfortunately, such experiments do not establish the nature of the charge states involved, but only give guidance for further experiments. The most difficult hurdle to surmount is the fact that for elevated-temperature implantations of both Group III and Group V elements, anneal temperatures of 800–900° C are required before the maximum electrical activity is observed (see Chapter 5). This implies that compensating defects are present. Until the nature of these defects is clarified, it is difficult to draw any firm conclusions.

5

Hall-Effect and Sheet-Resistivity Measurements in Silicon

Written in Collaboration with O. J. MARSH†

5.1 Introduction

The electrical characteristics of implanted layers do not fall into the same neat categories found in lattice-location studies. Although many dopant elements show similar behavior, it is premature to make generalizations. The reason for this is not hard to visualize. The electrical behavior of an implanted layer has been found to be dominated by radiation-damage defects up to anneal temperatures of 500° C or higher. The anneal characteristics depend on the dose, ion species, and substrate temperature during implantation. The anneal temperatures lie above those studied in fast-neutron irradiation, so that it is difficult to draw on experience gained in irradiation studies. For most implantation conditions, doses $> 10^{14}$ ions/cm^2 can result in degenerate doping conditions (carrier concentrations $\gtrsim 10^{19}$/cm^3) and also concentrations above solid solubility values. The properties of heavily doped layers differ significantly from those observed in samples with lower dopant concentrations ($\lesssim 10^{17}$/cm^3).[232]

Because of these factors, most studies have attempted to obtain only an empirical feeling for the electrical characteristics of implanted layers. The approach has been to determine (1) the anneal temperature at which the number of carriers per cm^2 reaches a maximum value, (2) the relation between the maximum number of carriers per cm^2 and the implanted dose, (3) the

† Address: Hughes Research Laboratories, Malibu, California.

mobilities and dopant activation energies in implanted layers, (4) the influence of the substrate temperature during implantation upon the subsequent anneal behavior, and (5) the depth distribution of the free carriers. In each investigation only one or two aspects of the foregoing have been studied for a given dopant (generally Group III or Group V elements) and then only under specific implantation conditions. The anneal behavior is also energy-dependent since the maximum carrier concentration is inversely proportional to the mean range straggling (ΔR_p), and hence to the ion energy (Section 2.3.5).

Comparison with studies on diffused layers in silicon illustrates some of the difficulties encountered in analysis of implanted layers. In diffusion processes, temperature is the primary factor determining the surface concentration and diffusion constant. Because of the simple form of the diffusion profile, the diffusion constant is usually deduced from the surface concentration and the junction location (the point at which the concentration of diffusants equals that of the substrate dopant species). The depth distribution is determined in more detail from sheet-resistance (four-point probe) measurements combined with successive layer removal. It is generally assumed that the carrier mobilities follow those in bulk silicon. In implanted samples, sheet-resistivity measurements are generally not sufficient because the mobility values may deviate significantly from values found in uniformly doped, uncompensated silicon. For this reason, Hall-effect measurement techniques were developed to determine carrier mobilities and concentrations. Even here, the interpretation is not straightforward because the carrier concentrations and mobilities are strongly depth dependent, and hence the number of carriers deduced from Hall-effect measurements may differ somewhat from the total number of carriers in the implanted layer. This difficulty can be resolved by combining Hall-effect measurements with layer-removal techniques.

This chapter is concerned primarily with the dopant behavior in the layers where the majority of the implanted atoms are located. Consequently, we have placed most emphasis on the use of Hall-effect and sheet-resistivity measurements to determine the carrier concentration and mobility. Junction characteristics are not discussed in detail because they reveal information only about the region where the concentration of implanted atoms is near that of the substrate doping density (i.e., near the junction location). Of course, "good" junctions are necessary in Hall-effect analysis, since the presence of a well-defined depletion layer and low junction-leakage currents are required to electrically isolate the implanted layer from the substrate. The ability to form high-quality *p-n* junctions by use of implantation techniques has been amply demonstrated by their use in devices. For a discussion of junction characteristics and device formation, see Chapter 6 and review articles by Large and Bicknell,[126] Mayer and Marsh,[58] and Large.[232a]

Some of the major points of this chapter are illustrated by the data shown in Fig. 5.1 which gives the number of carriers N_s per cm^2 determined from Hall-effect measurements as a function of anneal temperature. For silicon samples implanted at room temperature with high doses ($\sim 10^{15}$ ions/cm^2) such that an amorphous layer is formed, a marked increase in the number of carriers occurs between 500° and 600° C [Fig. 5.1(a)]. This increase is associated with the reordering of the amorphous layer. The maximum number of carriers at 600° C is a function of ion species, viz. $N_s(\mathrm{Sb}) > N_s(\mathrm{Ga}) > N_s(\mathrm{Al})$. The anneal behavior of N_s in both the antimony- and gallium-implanted samples is analogous to that of the number of substitutional atoms observed in lattitice-location studies (Chapter 4).

This same correlation is *not* found in either hot-substrate or low-dose implantations—i.e., where an amorphous layer is not formed [Fig. 5.1(b)]. For an implantation of 2×10^{14} antimony ions/cm^2 at a substrate temperature

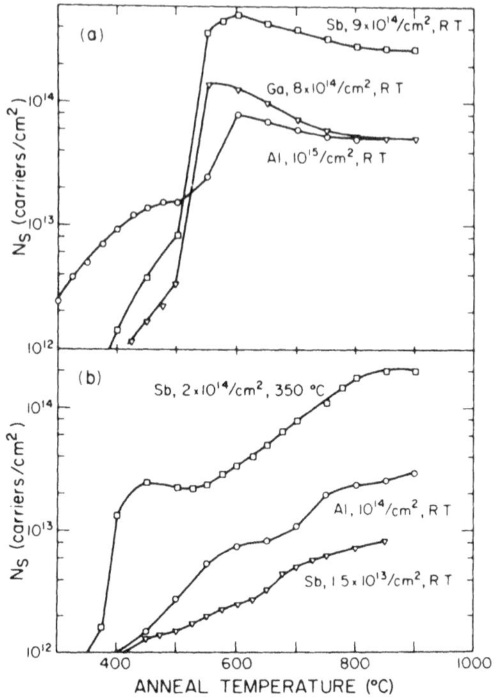

Fig. 5.1. Number of carriers/cm^2 measured in implanted layers in silicon versus anneal temperature: (a) Anneal behavior for samples implanted at room temperature to a dose of $\sim 10^{15}$ ions/cm^2 such that an amorphous layer was formed. (b) Anneal behavior of a hot-substrate (350° C) antimony implantation and of lower-dose room-temperature antimony and aluminum implantations under conditions such that an amorphous layer was *not* formed. Implantation energies were between 40 and 50 keV, and anneal times between 10 and 15 min.

of 350° C, over 90% of the implanted atoms are on substitutional sites, whereas the number of carriers after anneal at 500° C is only 10% of the implant dose N_D. In such cases, anneal temperatures of 800° C are required before $N_s \approx N_D$. In room-temperature antimony-implanted samples of low dose [Fig. 5.1(b)], anneal temperatures of 700° C or greater are required before $N_s \approx N_D$, even though the gross lattice disorder has annealed and most of the antimony atoms are substitutional after anneal at 300° C. For cases in which an amorphous layer is not formed, the anneal behavior is species- and dose-dependent (cf. Section 5.4.2). Generally, as the implantation dose is increased, the anneal temperature required for full electrical activity increases. At high anneal temperatures, the maximum values of N_s are set by equilibrium conditions.

Thus the observed electrical anneal behavior depends very strongly on ion species, substrate temperature, and total dose. Other factors, such as ion energy and beam-to-substrate alignment during implantation, may also play a role. The approach we have adopted is to consider the types of behavior (anneal stages, mobilities, diffusion, etc.) that have been found in implanted layers, rather than to describe in detail the behavior of each dopant species. This approach was chosen because we believe that there is insufficient evidence to describe the electrical behavior of a given species under all implantation conditions. Of course, it is possible to describe the electrical behavior for rather specific conditions. For boron and phosphorus, a much broader range of experience has been obtained (there have been approximately ten publications for each of these species). But even for these two species, there are many questions—e.g., the influence of channeling or enhanced diffusion—that have not been completely resolved.

Tables 5.1 and 5.2 list the various elements whose electrical characteristics have been evaluated in implanted layers in silicon. The table heading *Observation and measurement* indicates experimental results and the measurement technique used. Since the electrical characteristics cannot, at present, be described over a general range of implantation conditions, we have listed the specific implantation conditions that apply to each observation. The references cited do not cover all the investigations that have been made of a particular dopant, but refer only to representative studies. A more complete citation is given in the individual sections. In terms of the observations listed in Tables 5.1 and 5.2, anneal behavior as a function of implantation dose and implanted element is discussed in Section 5.4.2, substrate temperature during implantation in Section 5.4.4, and channeling and the presence of an interstitial component in Sections 5.4.5 and 5.4.3. The relationship between N_D (the number of implanted ions/cm^2) and N_s (the number of carriers/cm^2) is presented in Part 5.6. We have limited the discussion to the case of silicon, for which electrical behavior has been studied in detail. Although implantation

TABLE 5.1

ELECTRICAL BEHAVIOR OF DOPANT SPECIES THAT EXHIBIT *n*-TYPE BEHAVIOR IN IMPLANTED LAYERS IN SILICON

Ion	Observation and measurement	Implantation conditions	References
P	$N_s \approx N_D$[a] (Hall, ρ_s)	R.T.,[b] 13–400 keV, to $10^{15}/cm^2$	168, 238
	600° C anneal stage (Hall, ρ_s)	R.T., 30–300 keV, to 10^{15}	253, 254
	Anneal, T_I dependence[c] (ρ_s)	−195 to 420° C, 300 keV, 10^{15}	254
	High-temperature anneal (Hall, ρ_s)	R.T., 420° C	168, 254
	Anneal, dose dependence (Hall)	R.T., 50 keV, 10^{13}–10^{17}	252, 253
	Channeling, N_s (Hall)	R.T., 400 keV, 10^{13} and 10^{15}	168
	Channeling, $n(x)$ (strip, Hall)	R.T., 8–200 keV	76
	Enhanced diffusion (strip, tracer)	$T_I > 650°$ C	88
As	$N_s \approx \frac{1}{2}N_D$ (ρ_s)	78° K, 1–1.7 MeV, 10^{15}	83
	High-temperature anneal (Hall)	500° C, 20 keV, 10^{14}–10^{15}	236
Sb	$N_s \lesssim N_D$ (Hall)	500° C, 20–50 keV, 10^{13}–10^{15}	237
	Exceeds solubility (Hall)	500° C, 20–50 keV, 10^{13}–10^{15}	237
	600° C anneal stage (Hall)	R.T., 40 keV, 10^{14}	246
	High-temperature anneal (Hall)	R.T., 40 keV, 10^{14}	246
Bi	$N_s \approx \frac{1}{8}N_D$ (Hall)	R.T., 40 keV, 3×10^{13}–2×10^{14}	240
	Exceeds solubility (Hall)	R.T., 40 keV, 3×10^{13}–2×10^{14}	240
	600° C anneal stage (Hall)	R.T., 40 keV, 3×10^{13}–2×10^{14}	240
N	$N_s \ll N_D$ (Hall)	R.T., 50 keV, 6×10^{14}	244
	High-temperature anneal (Hall, ρ_s)	R.T., 50 keV, 10^{14}: 2 MeV, 10^{16}	141
Li	$N_s \ll N_D$ (Hall)	R.T., 400 keV, 5×10^{15}	168
Cs	$N_s \ll N_D$ (ρ_s)	~500° C, 10 keV, $>10^{16}$	264
	Exceeds solubility (ρ_s)	~500° C, 10 keV, $>10^{16}$	264
	Anneal, T_I dependence (ρ_s)	200–600° C, 10 keV, $>10^{16}$	264
Na	$N_s \ll N_D$ (ρ_s)	~500° C, 10 keV, $>10^{16}$	264, 265
	Exceeds solubility (ρ_s)	~500° C, 10 keV, $>10^{16}$	264, 265
Te	$N_s \ll N_D$ (Hall)	R.T., 20–50 keV, 10^{14}–10^{15}	55
	600° C anneal stage (Hall)	R.T., 20–50 keV, 10^{14}–10^{15}	55
Cd	Interstitial, $N_s \ll N_D$ (Hall)	350° C, 40 keV, 2×10^{14}	55

[a] N_s is the number of carriers/cm² and N_D is the number of implanted ions per cm².
[b] R.T. = room temperature.
[c] T_I is substrate temperature during implantation.

TABLE 5.2

ELECTRICAL BEHAVIOR OF DOPANT SPECIES THAT EXHIBIT p-TYPE BEHAVIOR IN IMPLANTED LAYERS IN SILICON[a]

Ion	Observation and measurement	Implantation conditions	References
B	$N_s \approx N_D$ (Hall)	R.T., 500° C, 50 keV 10^{13}–10^{15}	238
	600° C anneal stage (ρ_s)	−195° C, 200 keV, 10^{15}	254
	High-temperature anneal (Hall, ρ_s)	R.T., 500° C, 17–150 keV, 10^{15}	243, 255
	Anneal, T_i dependence (ρ_s)	−195 to 420° C, 200 keV, 10^{15}	254
	Anneal, dose dependence (ρ_s)	90° C, 25–150 keV, 10^{13}–10^{16}	247
Al	$N_s \ll N_D$ (Hall)	R.T. and 500° C, 30–60 keV, 10^{14}–10^{15}	237, 252
	600° C anneal stage (Hall)	R.T., 60 keV, 10^{15}	237
	High-temperature anneal (Hall)	R.T. and 500° C, 30–60 keV, 10^{14}	237, 238
	Enhanced diffusion (strip)	R.T., 10–30 keV, 10^{14}–10^{15}	97
	Compensation (strip)	R.T., 30 keV, 10^{13}	237
Ga	$N_s < N_D$ (Hall)	500° C, 20–75 keV, 10^{14}–10^{16}	236
	Solubility limited (Hall)	500° C, 20–75 keV, 10^{14}–10^{16}	262
	600° C anneal stage (Hall)	R.T., 40 keV, 4–8 × 10^{14}	246
	High-temperature anneal (Hall)	500° C, 30 keV, 10^{13}–10^{16}	246
In	$N_s \ll N_D$ (Hall)	R.T., 40 keV, 10^{14}	235
Tl	$N_s \ll N_D$ (Hall)	R.T. and 450° C, 40 keV, 10^{14}	237
	Interstitial, donor (Hall)	450° C, 40 keV, 10^{14}	237

[a] See Table 5.1 for explanation of terms.

processes have been investigated in germanium, gallium arsenide, zinc telluride, silicon carbide, and other materials, it is not yet possible to make any general comparisons (for a review of this work, see Mayer and Marsh[58]).

5.2 Interpretation of Hall-Effect Measurements

Hall-effect measurements are used to determine carrier concentrations and mobilities in semiconductors.[233] There are two major problems to consider when evaluating Hall-effect measurements on implanted samples: the quality of the junction, and the carrier density and mobility distributions. The junction requirement is illustrated in Fig. 5.2, which indicates an implanted region (shaded) etched to form a van der Pauw-type[234] Hall pattern.† In the Hall-effect measurement, the current passing through the sample must be confined

† We have chosen this geometry because of its simplicity and the fact that most of the data discussed in this chapter were obtained from van der Pauw configurations. Other Hall geometries have been used.[168]

to the implanted layer. To accomplish this, one must isolate the implanted layer from the substrate by the formation of a low-leakage junction [or use extremely high resistivity ($\rho \gtrsim 10{,}000$ ohm-cm) silicon so that N_s is much greater than the number of dopant atoms in the substrate]. To obtain meaningful data, all junction leakage currents (both bulk, I_2, and surface, I_3, components) should be less than 5% of the measurement current. This is a difficult criterion to meet in the region of low anneal temperatures ($\lesssim 400°$ C), where the junction characteristics are often poor (high leakage currents) and the resistance of the implanted layer is high. The measurement technique is described in detail by Johansson *et al.*[235]

The sheet Hall coefficient R_s is determined for this configuration (Fig. 5.2) by measuring the voltage change ΔV_{13} normal to the current path I_{24} when a magnetic field B is applied perpendicular to the sample. R_s is given by

$$R_s = 10^8(\Delta V_{13}/BI_{24}) \tag{5.1}$$

where measured values are in units of volts, amperes, and gauss.[233]

The sheet resistivity ρ_s is obtained for a symmetrical pattern from the potential difference V_{34} between two adjacent contacts occurring when a current I_{12} is passed between the other two contacts:

$$\rho_s = (\pi/\ln 2)(V_{34}/I_{12}) \tag{5.2}$$

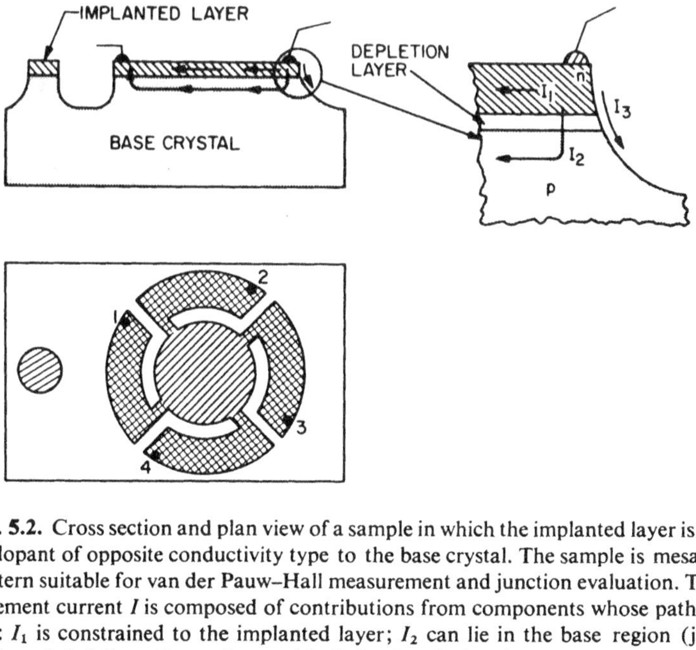

Fig. 5.2. Cross section and plan view of a sample in which the implanted layer is formed with a dopant of opposite conductivity type to the base crystal. The sample is mesa-etched in a pattern suitable for van der Pauw–Hall measurement and junction evaluation. The Hall measurement current I is composed of contributions from components whose paths are as follows: I_1 is constrained to the implanted layer; I_2 can lie in the base region (junction leakage); and I_3 follows the surface and bulk (surface leakage).

If the pattern is not symmetrical, correction for geometry is made by determining the sheet resistivity from

$$\rho_s = (\pi/2 \ln 2)(R_1 + R_2)\, F(R_1/R_2) \tag{5.3}$$

where $R_1 = V_{34}/I_{12}$, $R_2 = V_{41}/I_{23}$, and $F(R_1/R_2)$ is a geometrical correction factor.[234] Sheet resistivity can also be obtained by the conventional four-point probe measurement as long as the probe spacings are appreciably greater than the layer thickness.[235a]

From the measurements outlined above, the number of carriers N_s per cm^2 and the Hall mobility μ_H can be obtained. For a sample with a uniform concentration of carriers n to a depth t the value of N_s is given by

$$N_s = nt = r/R_s e \tag{5.4}$$

where $r = \mu_H/\mu$ is the ratio of Hall mobility ($\mu_H = R_s/\rho_s$) to conductivity mobility ($\mu = 1/eN_s \rho_s$).

The value of r depends on scattering mechanisms and impurity concentration. For lightly doped silicon (impurity concentration $\lesssim 10^{16}/cm^3$), experimental data indicate that $r \approx 1.2$–1.3 for electrons and $r \approx 0.7$–0.8 for holes.[232,235b] For electrons the theory gives a value of 1.93 for the case where scattering by ionized centers dominates, and predicts that the mobility ratio approaches unity when degenerate doping levels are reached (i.e., when the Fermi level lies several kT within the conduction band). For heavily doped silicon, the experimental situation is not well defined.[232] The problem is that drift-mobility measurements are extremely difficult on heavily doped materials, and conductivity-mobility measurements require a knowledge of carrier concentration. To meet the latter requirement, one can determine the impurity concentration (e.g., by neutron activation[235c]); however, it is not certain that all the impurity atoms are electrically active. For electrons, a mobility ratio of unity for impurity concentrations $> 10^{18}/cm^3$ seems valid. For holes, on the other hand, we have found only one recent measurement of conductivity mobility,[235c] and it is difficult to predict a lower concentration at which r is unity. It is generally assumed that the conductivity and Hall mobilities are equal in implanted layers. Use of this assumption might lead to a 20–30% error in N_s (the values of N_s would be smaller than the number of carriers/cm^2 for electrons and greater for holes).

Figure 5.3 shows values of N_s, ρ_s, and μ_H as a function of anneal temperature for two n-type silicon samples implanted at 40 keV with 4×10^{14} gallium ions/cm^2 (dashed curves) and 3×10^{14} indium ions/cm^2 (solid curves). The values of N_s vary between 10^{11} and $10^{14}/cm^2$, and with a magnetic field of 4000 G and a current of 4 μA, the corresponding voltage changes, ΔV,

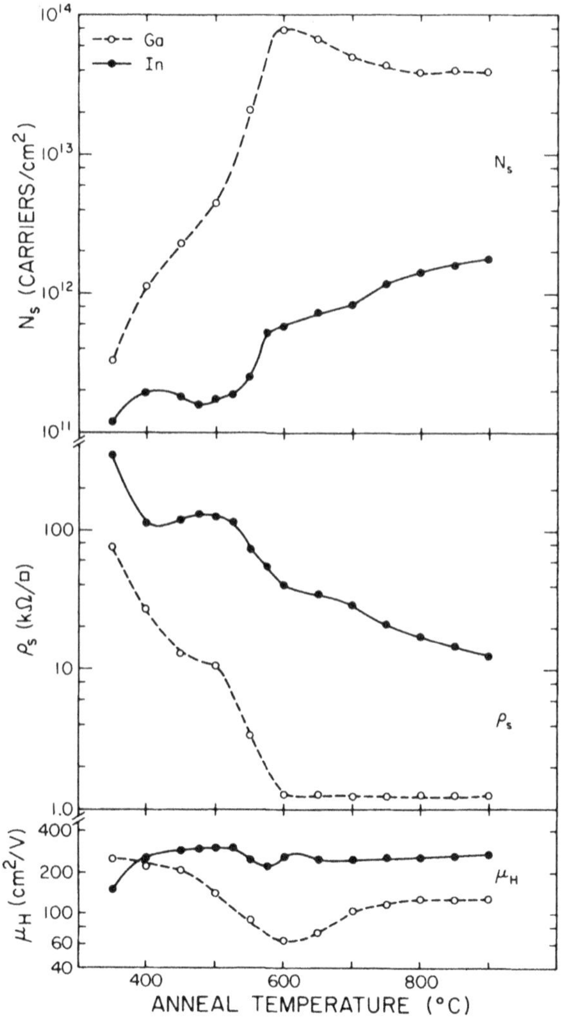

Fig. 5.3. Values of N_s, ρ_s, and μ_H versus anneal temperature for 40-keV room-temperature implantations of 4×10^{14} gallium (dashed curves) and 3×10^{14} indium (solid curves) ions/cm^2. Anneal time, 15 min. (Data from Johansson *et al.*[235])

range from 10 to 10^{-2} mV.† In practice, to compensate for the decrease in ΔV as N_s increases, higher currents are used so that the ΔV values typically are in the range 0.1–10 mV. The sheet resistivity can vary from several hundreds of k$\Omega/\square$ down to small values. For a current of 1 μA, the corresponding measured voltages have values between ~0.5 volts and some millivolts. As can be seen from Fig. 5.3, sheet-resistivity measurements alone do not give sufficient information about the number of carriers in the implanted layer. For example, the decrease in N_s for the gallium-implanted sample at high anneal temperatures is not apparent in the sheet-resistivity values because of the change in mobility.

Although the two directly determined parameters are R_s and ρ_s, values of N_s (assuming $r = 1$) and μ_H are generally given. In interpreting Hall-effect and resistivity measurements on implanted samples, the fact that both the concentration n and mobility μ can vary with depth must be considered.[236–239] It has been found[236,237] that the mobility (averaged over the entire implanted layer) may be a factor of 2 or 3 higher than the carrier mobility in the region containing the majority of implanted atoms. If both n and μ are functions of the distance x perpendicular to the surface of the layer, Hall-effect measurements give weighted averages[237,239] of the carriers and mobilities. Assuming now that $\mu_H = \mu$ and that there are no circulating currents in the layer,[239a]

$$R_s = \frac{\int_0^t n(x)\,\mu^2(x)\,dx}{e[\int_0^t n(x)\,\mu(x)\,dx]^2} \tag{5.5}$$

and

$$\mu_H = R_s/\rho_s = \frac{\int_0^t n(x)\,\mu^2(x)\,dx}{\int_0^t n(x)\,\mu(x)\,dx} \tag{5.6}$$

The influence of $n(x)$ and $\mu(x)$ on the measured quantities can be seen more easily by considering (in analogy with the two-carrier model for the Hall coefficient)[233] an implanted sample consisting of two layers, one containing N_1 carriers/cm^2 with mobility μ_1 and the other N_2 carriers/cm^2 with mobility

† The Hall effect is in general not the only source for the voltage change ΔV observed between the probes when a magnetic field is applied. Other galvanomagnetic and thermomagnetic effects (Nernst, Rhigi–Leduc, and Ettinghausen effects) exist,[233] and all can contribute to ΔV. Moreover, thermal gradients may be present, resulting in thermoelectric voltages; also an *IR* drop (offset-voltage) can exist between the probes. The influence of these effects and other sources of error are discussed in detail by Johansson *et al.*[235]

μ_2. Then, for the case in which the conductivity and Hall mobilities are equal,

$$N_s = (N_1\mu_1 + N_2\mu_2)^2/(N_1\mu_1^2 + N_2\mu_2^2) \tag{5.7}$$

and

$$\mu_{eff} = (N_1\mu_1^2 + N_2\mu_2^2)/(N_1\mu_1 + N_2\mu_2) \tag{5.8}$$

For $N_1 > N_2$ and $N_1\mu_1^2 < N_2\mu_2^2$, the value of N_s will be significantly less than $N_1 + N_2$ and the value of μ_{eff} will lie between μ_1 and μ_2.

In general the value of N_s will be *less* than the total number of carriers in the implanted layer. This is caused by the influence of the less heavily doped regions where mobility values are higher than those near the maximum in the dopant concentration. Figure 5.4 shows $n(x)$ and $\mu(x)$ for a 40-keV bismuth implantation annealed at 700° C.[240] The carrier and mobility distributions were obtained by combining Hall-effect and layer-removal techniques. The value of N_s (arrow in Fig. 5.4) measured before layer removal

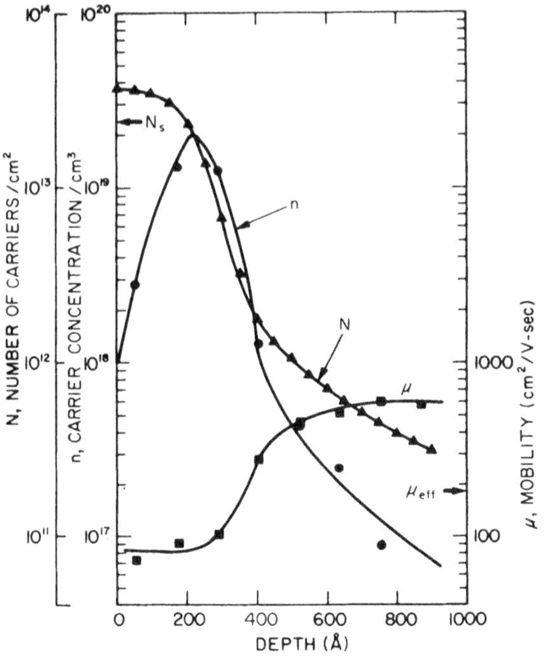

Fig. 5.4. Carrier concentration and mobility distribution for a sample implanted at room temperature with 2×10^{14} bismuth ions/cm^2 at 40 keV and annealed to 700° C. The arrows show the effective values of mobility and N_s (2.4×10^{13}/cm^2) measured before stripping. The total number of carriers/cm^2, determined from integration of the carrier concentration, is 3.6×10^{13}/cm^2. (Data from Marsh *et al.*[240])

was $2.4 \times 10^{13}/cm^2$, and the value of N obtained by integrating the smoothed curve of $n(x)$ was $3.6 \times 10^{13}/cm^2$. The effective mobility μ_{eff} lies between the mobility values in the heavily and lightly doped regions.

The interpretation of Hall-effect measurements, then, must be based on some knowledge of the carrier distribution. The anneal treatment must also be considered in evaluating Hall data, as indicated by the following:

1. In well-annealed samples, the relationship between mobility and carrier concentration follows that established in bulk silicon (Part 5.3). In this case, the difference between N_s and the number of carriers/cm^2 depends on the distribution of dopant centers. The analysis of Baron *et al.*[237] indicates that the number of carriers can be 30–50% greater than N_s for cases in which most of the implanted atoms stop in a narrow region around the amorphous range R_p and only a small percentage are in a channeled tail in the distribution. On the other hand, where there is a broad distribution of implanted atoms such that $n(x)$ and $\mu(x)$ do not vary strongly with depth, the value of N_s will closely approximate the total number of carriers.
2. In partially annealed samples, the mobility values are influenced by the presence of defect centers. Under these conditions, the *distribution* of disorder can have a strong influence on the anneal characteristics of N_s and μ_{eff}. That is, various portions of the implanted-atom distribution may have different anneal characteristics due to the amount of disorder and defect centers in each layer. For the case in which part of the distribution of implanted atoms lies in an amorphous layer, dopant centers in that layer will make only a small contribution to the measured value of N_s up to anneal temperatures where the amorphous layer reorders. In general, the relative contribution that each region in the implanted layer makes to N_s will change during annealing, because of changes both in the fraction of electrically active centers and in the amount of disorder. For example, at low anneal temperatures those regions with a low concentration of disorder may dominate.

The factors just discussed make interpretation of Hall-effect and resistivity measurements as a function of temperature very difficult. Because of the strong manner in which the temperature dependence of carrier mobility varies with impurity concentration,[241] the relation of N_s to the total number of carriers may be strongly temperature dependent. This could lead to errors in the determination of, for example, the dopant ionization energy.

Most of the problems in the interpretation of electrical measurements are a direct result of the shape of the implanted-atom distribution. They are minimized when $n(x)$ and $\mu(x)$ are not strongly depth-dependent. The carrier and mobility distributions can be determined[236,237] by measuring R_s and ρ_s, removing a thin layer of the implanted samples, and remeasuring. This cycle is repeated until most of the implanted layer is removed. The number of carriers in the ith layer and their mobility can be obtained from

$$(R_s)_i/(\rho_s)_i^2 - (R_s)_{i+1}/(\rho_s)^2_{i+1} = en_i \mu_i^2 d_i \tag{5.9}$$

and

$$1/(\rho_s)_i - 1/(\rho_s)_{i+1} = en_i \mu_i d_i \tag{5.10}$$

giving

$$\mu_i = \Delta(R_s/\rho_s^2)_i/\Delta(1/\rho_s)_i \tag{5.11}$$

and

$$n_i = \Delta(1/\rho_s)_i/ed_i \mu_i \tag{5.12}$$

$(R_s)_i$ and $(\rho_s)_i$ are the sheet Hall coefficient and the sheet resistivity that are measured after removal of the ith layer of thickness d_i. The number of carriers per cm^2 is obtained from $\sum_i n_i d_i$.

Layer-removal techniques generally involve anodic oxidation and oxide-layer removal by hydrofluoric acid (as in Section 2.2.1). This introduces two possible errors—uncertainty in the thickness of the removed layer, and degradation of the junction characteristics. The latter effect arises because of both the increased surface-leakage currents that often occur after many process steps and the increasing resistance of the implanted region as successive layers are removed. One should determine directly the thickness removed (by use of weight loss, optical interferometry, or other techniques) because high concentrations of dopants or disorder may influence the removal rate.

A major point in this section is that the profile of implanted atoms must be considered in evaluating Hall-effect data. In general, the value of N_s is somewhat less than the total number of carriers/cm^2, and μ_{eff} is greater than the mobility value at the maximum in the carrier concentration. These differences become less significant in cases where the depth distribution of carriers is broad or where $n(x)$ and $\mu(x)$ are relatively constant over the major portion of the implanted region. A strong effect is caused by tails in the distribution where the mobility values are high.

5.3 Carrier Mobilities and Dopant Ionization Energies

In an implanted layer annealed to high temperatures, carrier mobilities over much of the layer tend to follow the dependence on dopant concentration found in bulk silicon and, at least for boron and phosphorus implants, the dopant ionization energies agree with previously published values. This agreement with bulk values does not hold in all cases; in particular, anomalously low mobility values are sometimes found in the region adjacent to the surface.

First, we will consider the existing mobility data. Baron *et al.*[237] have evaluated the carrier and mobility distributions in implanted layers by combining Hall-effect measurements with layer-removal (stripping) techniques. They found that the variation in mobility for many cases followed the same dependence on carrier concentration as is found for uniformly doped samples. In Fig. 5.5, their data on mobility versus carrier concentration is plotted for a number of *n*- and *p*-type dopants. For purposes of comparison, curves of Hall mobility versus impurity concentration are also given for bulk, uncompensated *n*- and *p*-type silicon.† This agreement with mobility values in bulk silicon has also been observed in samples implanted with antimony at 100 keV[241a] and phosphorus at 280 keV[76b, 241b] after annealing to 850° C or higher.

In Fig. 5.5(a) (*n*-type doping), the three curves for implanted samples have several features in common. The carrier mobility in the region closest to the surface (circled) is much lower than that of the bulk curve. As the depth from the original surface increases, the measured points approach and then approximate the bulk curve. Despite the wide range of doses and maximum carrier concentrations, the point at which the measurements break away from the bulk curve seems related to the depth at which the measurements were taken, rather than to some given carrier concentration. Similar results were found in phosphorus-doped silicon by Galaktionova *et al.*[76] The low values of mobility observed near the surface indicate that there is an additional source of carrier-scattering located in this region. This additional scattering of carriers might be caused by defect clusters, dislocations, or charged defect centers.

The same trends are apparent in Fig. 5.5(b) for *p*-type dopants. Note that the aluminum-implanted sample (open triangles) was annealed only to 400° C (the mobility values in this sample are discussed in Section 5.4.3). The points identified as boron diffusion in Fig. 5.4(b) were measured on a deep (>1 micron) diffusion by the same stripping techniques that were used to evaluate the implanted samples.

The ionization energies determined from analysis of Hall-effect measurements as a function of temperature are consistent with values obtained in bulk silicon‡: phosphorus, 0.045 eV[242] and 0.027 eV[168] (lower value attributed to high doping concentration); boron, 0.045 eV[243]; and nitrogen 0.045 eV.[244] The Hall-effect measurements of Seidel and MacRae[243] on a sample

† As Baron *et al.* point out,[237] it may not be justified to equate carrier concentration with impurity concentration. However, the agreement between the bulk curve and implant data is sufficiently close to warrant this comparison.

‡The ionization energies (eV) for the Group III and Group V dopants in silicon are: boron, 0.045; aluminum, 0.057; gallium, 0.065; indium, 0.16; thallium, 0.26; antimony, 0.039; phosphorus, 0.044; arsenic, 0.049; bismuth, 0.069.[241c]

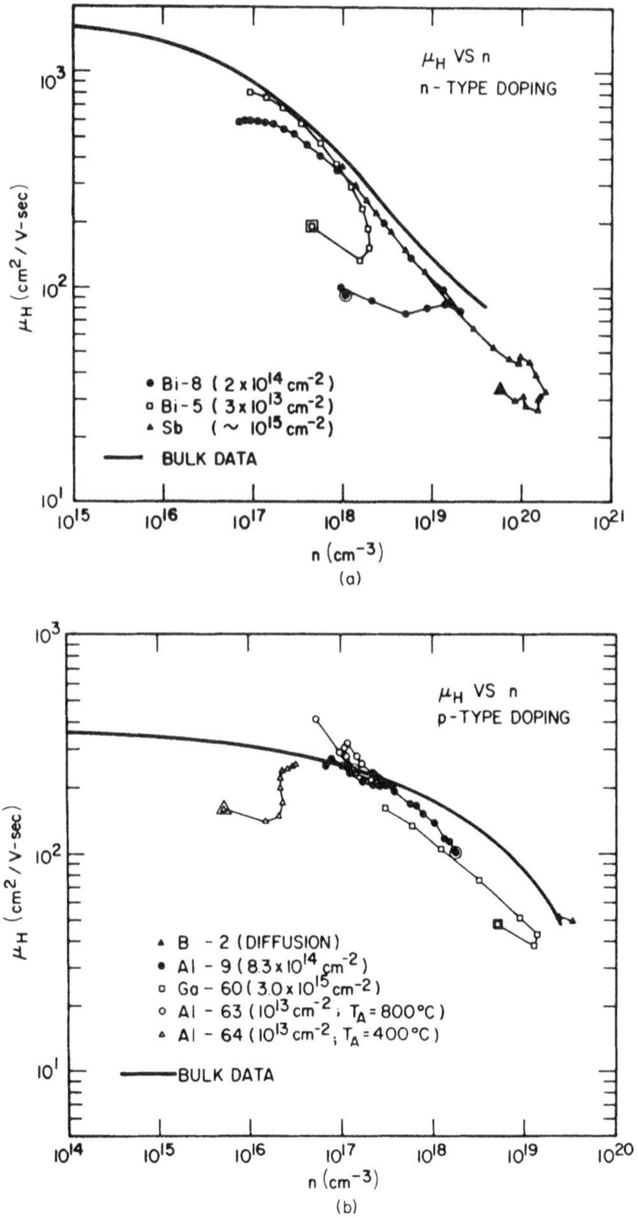

Fig. 5.5. Values of the Hall mobility μ_H versus the carrier concentration n determined from Hall-effect and layer-removal measurements. The points nearest the surface are encircled; points adjacent in depth are connected. The heavy curves labeled *bulk data* were calculated for uniformly doped, uncompensated silicon. Data for (a) n-type dopants, and (b) p-type dopants. (Curves taken from Baron *et al.*[237])

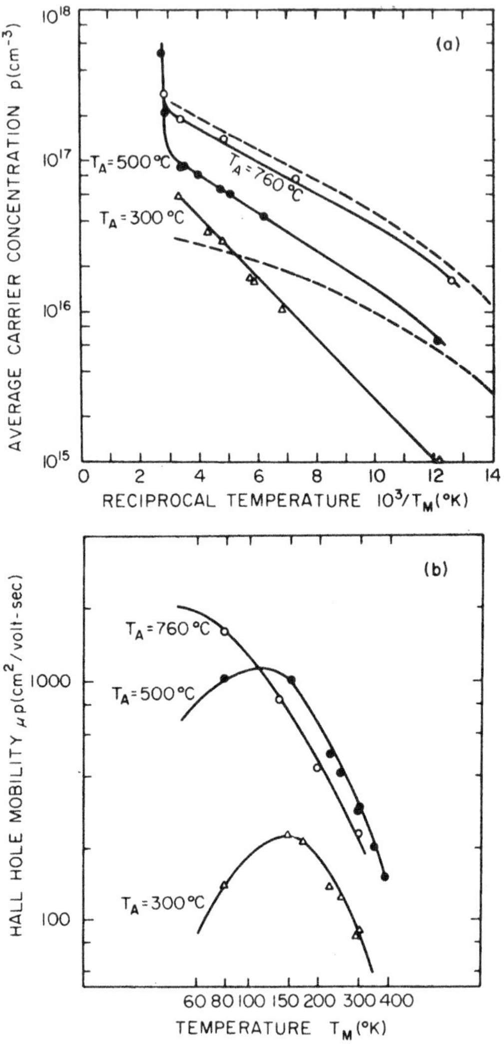

Fig. 5.6. Hall data as a function of measurement temperature for a 150-keV room-temperature boron implantation to a dose of 2×10^{13} ions/cm^2 for anneal temperatures of 300°, 500°, and 760° C: (a) carrier concentrations assuming a layer thickness of 5000 Å for all anneal temperatures (dashed curves are from Morin and Maita[241] for uncompensated boron-doped silicon); (b) hole mobility. (Curves from Seidel and MacRae.[243])

implanted at room temperature with 150-keV boron ions to a dose of 2×10^{13} ions/cm^2 are shown in Fig. 5.6. A low implantation dose was used to avoid degeneracy conditions (cf. Fig. 5.7); the maximum carrier concentration was of the order of a few times 10^{17}/cm^3. They also included [Fig. 5.6(a)] the data and theory of Morin and Maita[241] (dashed line) for uncompensated boron-doped silicon. The temperature dependence of the dashed curve represents the gradual freeze-out of carriers from the 0.045-eV boron level.† Although there is evidence of a deeper-lying level in the sample annealed at 300° C, the data follow the curve of Morin and Maita after anneal at 760° C.

The data on Hall mobility as a function of temperature are shown in Fig 5.6(b). For anneals at 500° and 760° C, the mobility values measured at room temperature are close to those in uncompensated silicon (for the same carrier concentration). However, for the sample annealed at 300° C, the mobility values are a factor of 3 smaller than those in uncompensated material. From analyses of the temperature dependence of the mobility, Seidel and MacRae[243] deduce that there were about 5×10^{18} scattering centers per cm^3 after this anneal at 300° C. In phosphorus-implanted silicon,

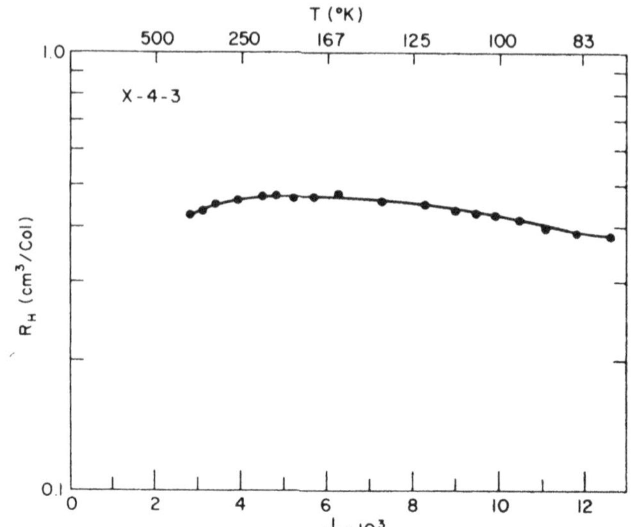

Fig. 5.7. Variation in the Hall coefficient versus measurement temperature for a sample implanted at room temperature with 10^{15} phosphorus ions/cm^2 at 80 keV and annealed at 600° C for 10 min. (Data from Clark and Manchester.[238])

† In this temperature range and assuming the absence of compensating centers, the carrier concentration is proportional to $\exp(E_A/2kT)$ where E_A is the dopant energy level (see Putley,[233] or Blakemore[245]).

Tetel'baum[242] found, in a similar analysis, that acceptors were present after anneal at 500° C.

In high-dose (10^{15} ions/cm^2) phosphorus implantations in silicon annealed to 600° C, the carrier concentrations and mobilities are not strongly temperature-dependent [168,238] This is shown in Fig. 5.7 for Clark and Manchester's[238] data on a room-temperature 80-keV phosphorus implantation. They point out that this behavior agrees with that observed in degenerate bulk silicon. Similar results were found by Gibson *et al.*[168]

The results presented in this section indicate that under proper anneal conditions and for common dopants,† the characteristics $n(T)$ and $\mu(n, T)$ of the implanted layer are comparable to those found in bulk silicon. This does *not* imply that the number of carriers/cm^2 is equal to the implanted dose. In fact, it is only for a relatively few cases that this is true.

5.4 Anneal Characteristics

5.4.1 General Considerations

The electrical characteristics of implanted Group III and Group V dopant atoms exhibit many common features in their anneal behavior.[240,246] Room-temperature implantations to dose levels such that an amorphous layer is formed exhibit large increases in the value of N_s (one to two orders of magnitude in some cases[246]) in the temperature region where reordering of the amorphous layer occurs. A three- to four-fold drop in mobility is often found (cf. Figs. 5.8 and 5.10), illustrating the transition from a situation where the electrical behavior is dominated by the lightly doped region with high carrier mobility to that where the heavily doped region with low carrier mobility dominates. The anneal behavior of implantations at elevated substrate temperatures (350–500° C) exhibits a smooth increase in values of N_s, and typically anneal temperatures of 750–900° C are required to achieve the maximum electrical activity for doses of 10^{14}–10^{15} ions/cm^2.

There are also marked differences in behavior. For example, the ratio of N_s to N_D (the implanted dose) varies from species to species and also depends on N_D and anneal temperature. Boron implantations require anneal temperatures of 900–1000° C for maximum electrical activity. Thallium implantations exhibit n-type behavior under certain implantation and anneal conditions.

In interpreting the anneal characteristics of ion-implanted samples, one should distinguish between isochronal and isothermal anneal sequences. In the isochronal case, an anneal time is selected (generally on the basis of an

† In some cases—for example, sodium, thallium—the dopant behavior in implanted samples differs from that in diffused samples.

estimate of the sample heating and cooling rate) and then the characteristics of the layer are measured after successive anneals at progressively higher temperatures. The detailed shape of such isochronal anneal curves is influenced by the duration of the anneal time, as is shown in Fig. 5.8. In this figure, values of N_s and μ_H for a series of 15-min isochronal (I.C.) and isothermal (I.T., time values in hours) anneal cycles are shown for a similar gallium implantation to that in Fig. 5.1. For the isothermal anneal at 450° C, there is a gradual rise in the number of carriers/cm^2. For the higher-temperature anneal at 550° C, there is a tenfold increase in the number of carriers in a 30-min period and then a gradual decrease.

From these data it can be seen that the shape of an isochronal anneal curve will depend somewhat on the anneal time selected. So far, most emphasis has been placed on isochronal anneals, with the exception of boron[75,238,247] and phosphorus[238,248] implantations, for which both isochronal and isothermal anneal behavior have been investigated. Except for the high-temperature anneal stage in boron, which will be discussed later, the anneal characteristics do not seem to follow simple first-order kinetics. The isothermal anneal characteristics are also dose-dependent.[248] In spite of these difficulties, one may anticipate that a judicious combination of isothermal and isochronal anneal sequences will eventually provide more insight into the annealing

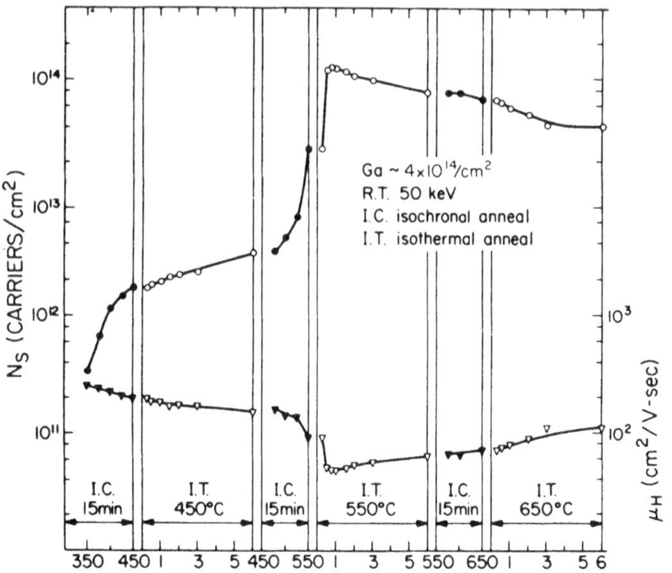

Fig. 5.8. A series of 15-min isochronal (I.C.) and isothermal (I.T.) anneal sequences on a room-temperature 50-keV gallium implantation of 4 × 10^{14} ions/cm^2. For the isothermal anneal, the scale units are in hours. (Data from Johansson *et al.*[235])

characteristics. The theory of annealing and methods of analysis of annealing curves are discussed in detail by Damask and Dienes.[158]

One aspect of the anneal behavior of implanted layers that has not been adequately investigated is the possible influence on the electrical behavior of the ambient during anneal. In general, each laboratory chooses a particular ambient (vacuum, argon, nitrogen) and adopts it as a standard condition. At present insufficient data exist to indicate whether or not the ambient plays a significant role.† Another effect is the possible removal of some of the implanted layer during the routine chemical treatments used to reduce junction leakage currents. For example, HF has an etching effect on a room-temperature implanted surface. A third possible effect is the influence that oxygen, carbon, dislocation density, etc. in the original material may have on the anneal behavior. It is generally assumed that the concentration of implanted atoms and defect centers completely overwhelms any contribution from these centers. This may not be correct, especially in the tail of the distribution where the concentration of implanted atoms is low. In the following discussion, we assume that none of these effects plays a role. The agreement in the data taken at different laboratories lends some credence to this.

5.4.2 Temperature Zones in Anneal Characteristics

a. $T < 600°$ C. Channeling-effect measurements of disorder in low-dose implantations (Part 3.4) indicate that a well-defined anneal stage is present in silicon at anneal temperatures near 300° C. This anneal was attributed to the reordering of the isolated-disorder regions around the track of the particles. In the electrical behavior of implanted samples, it is generally found that anneal temperatures of 300–350° C are required before type conversion occurs. For example, low-dose bismuth implants in 10-Ω-cm p-type silicon, where the bismuth is known to lie on substitutional sites,[171] do not show n-type behavior until an anneal of 300° C.[240] Type conversion and junction behavior are found after 300–350° C anneal‡ in samples implanted with phosphorus, lithium, gallium, and antimony.[168,237,246] Implanted layers

† In the case of compound semiconductors, such as GaAs, careful control of the ambient must be maintained to prevent the more volatile elements from leaving the sample. Layers of SiO_2, deposited by sputtering or decomposition of tetraethyl orthosilicate, have been used to prevent decomposition of implanted samples of CdTe[249] and GaAs[250,251] during anneal.

‡ The requirement for 300° C anneal to obtain type conversion is not always found. In boron implantations,[75] p-type behavior is noted after implantation at room temperature. Also in the tail of the distribution of phosphorus implantations, anneal stages at 70° and 170° C are found.[88] The anneal temperature at which junction behavior is noted may depend on the substrate resistivity.

typically show increases in the number of carriers/cm^2 for anneals above 350° C (see, for example, Figs. 5.1 and 5.8). We attribute this type of behavior to the presence of defects in the immediate vicinity of the implanted atoms—defects that must dissociate and diffuse away before dopant behavior of the implanted atoms can be measured. This is in contrast to the case with fast-neutron irradiation, where the clusters of disorder do not envelope the dopant (or impurity) species and the defects must move from the clusters to the impurities (Part 3.5).

For anneal temperatures $\lesssim 400°$ C, the presence of defect clusters is evident from low carrier-mobility values and the strong temperature dependence of the donor and acceptor concentration in these implanted layers. In phosphorus,[252] aluminum,[237] and boron[243] implantations, for example, the values of the mobility are less than for uncompensated material with the same carrier concentration. At anneal temperatures of 500° C, the mobility values do not deviate too strongly from the values that would be calculated from the carrier concentration, suggesting that the major portion of the defect cluster has disappeared.

Compensating defects and deep levels are still present, however, even after anneal at 500° C. These defects do not have a strong influence on carrier mobility at room temperature but do influence the *number* of carriers. The presence of a deeper-lying level is clearly evident in Hall-effect measurements made as a function of temperature in low-dose phosphorus implantations ($\sim 10^{13}$ ions/cm^2). Analysis[168] of a sample after 480° C anneal indicated strong compensation. The presence of a deeper-lying level in boron implantations is also evident from comparison of the experimental and theoretical (dashed line) curves in Fig. 5.6. Seidel and MacRae[243] point out that the slope of the Hall-effect curve should not be interpreted as the ionization energy of any one level, but may be associated with the presence of one or more deeper-lying levels whose ionization energies are greater than that of boron. The observed activation energy depends upon the compensation ratio between the concentration of shallow- and deep-level impurities.[168]

b. $T \approx 600°$ C. In general, it is found that as the dose of implanted ions increases, a higher anneal temperature is required for N_s to reach its maximum value.[247,252,253] However, in room-temperature implantations, as the dose is increased to a point at which an amorphous layer is formed (also indicated by the strong milky appearance), the major increase in N_s occurs at temperatures near 600° C.† This increase is shown in Fig. 5.9 for room-temperature aluminum implantations at 30 keV (see also Figs. 5.1 and 5.8). The sample

† More exactly, this stage occurs between 500° and 650° C, depending on the isochronal time interval and the implantation conditions. For convenience, we will use "600° C" to indicate this range.

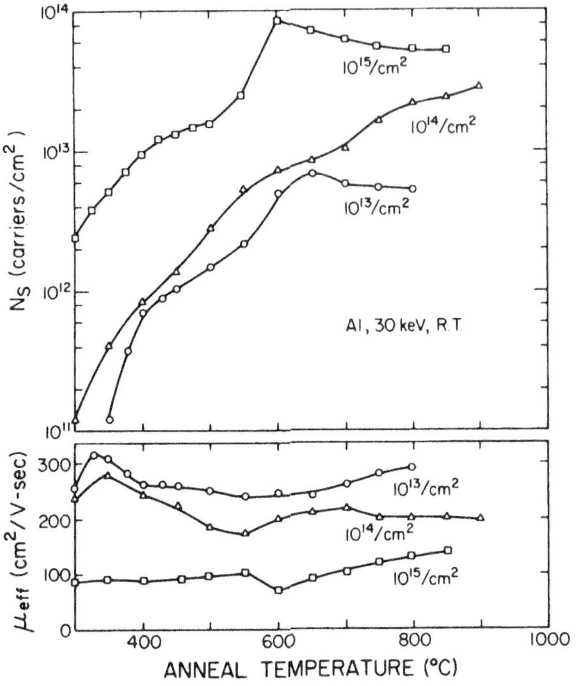

Fig. 5.9. Values of N_s and μ_{eff} as a function of anneal temperature for room-temperature aluminum implantations at 30 keV and at dose levels of 10^{13}, 10^{14}, and 10^{15} ions/cm^2. (From Baron *et al.*[237])

implanted with 10^{15} ions/cm^2 exhibits a nearly fivefold increase in the number of carriers/cm^2 as the anneal temperature is increased from 550 to 600° C. For a lower-dose implant (10^{14} ions/cm^2), the anneal characteristics indicate that N_s is increasing even at temperatures >800° C. Similar results to those shown in Fig. 5.9 were found by Clark and Manchester[238] for implantation of 10^{14} aluminum ions/cm^2.

The anneal stage near 600° C has also been observed in gallium[246] (Fig. 5.8), phosphorus,[238,252–255] antimony,[236,237] and bismuth[240] implantations at room temperature.† Because of the strong correlations with the anneal of the amorphous layer noted in channeling-effect measurements (Chapters 3 and 4), we identify this anneal stage in the electrical characteristics with the reordering of the amorphous layer. For every room-temperature implantation

† Measurements on room-temperature phosphorus, arsenic, and antimony implants at 260–300 keV[76b, 241b, 255a] and antimony implants at 100 keV[241a] have also shown this 600° C anneal stage whenever the dose was sufficiently large to form an amorphous layer.

reported to date in which a large increase in N_s was found near 600° C anneal temperatures, the total ion dose was sufficiently high to have formed an amorphous layer. In boron implants (40 keV, 10^{15} ions/cm^2), Blamires *et al.*[256] found that the sheet conductivity after a 630 °C anneal was a factor of 5 higher if the sample had also been bombarded with 10^{16} neon ions/cm^2 (to form an amorphous layer overlapping the boron distribution) as compared to the value obtained with the boron implant alone at that anneal temperature. In the latter case, the disorder created by 10^{15} boron ions/cm^2 is insufficient to form the amorphous layer. For low-temperature boron implantations to dose levels ($\sim 10^{15}$ ions/cm^2) such that an amorphous layer is formed, an anneal stage at 600° C is again observed.[176a] For room-temperature boron implantations of the same dose, an amorphous layer is not formed (Section 3.4.1) and this anneal stage is absent.

The 600° C stage is often the major feature in the anneal behavior, and generally the maximum number of carriers/cm^2, N_s(max), is observed immediately after this stage.[246] The value of N_s(max) is often significantly less than the ion dose, as has been found for aluminum, gallium, bismuth, and tellurium. For the case of gallium[246] and alumium shown in Figs. 5.8 and 5.9, the maximum number of carriers/cm^2 is approximately 10–20% of the implanted dose. This limitation in the maximum number of carriers/cm^2 may be associated with the number on substitutional sites and the presence of compensating centers (cf. Section 5.4.3). The decrease in the number of carriers for the gallium implantation in Fig. 5.8 is similar to the decrease in the substitutional component found in channeling-effect measurements [Fig. 4.18(a)].

For phosphorus implantations[76,168,238,255] to doses of 10^{15} ions/cm^2, the number of carriers/cm^2 was approximately equal to the implanted dose after $\sim$600° C anneal. Similar results were noted in antimony implants, as indicated in Fig. 5.10. Channeling-effect measurements[156] on comparable antimony implantations indicated that after anneal at 600° C, approximately 90% of the antimony was on substitutional sites. For bismuth implantations, the electrical characteristics exhibited a markedly different effect than that for antimony or phosphorus.[237] As shown by comparing Fig. 5.10 to Fig. 4.17, the effective number of carriers/cm^2 and the number of bismuth atoms on substitutional sites exhibit a similar dependence upon anneal, but the number of carriers/cm^2 is only 10–15% of the number on substitutional sites. The decrease in the number of carriers/cm^2 observed at high anneal temperatures follows roughly the same pattern as the decrease in the number of bismuth atoms on substitutional sites (Fig. 4.17) noted in channeling-effect measurements. As for antimony and gallium,[246] the decrease in N_s for bismuth implantations may be attributed to a relaxation toward equilibrium solubility values. Consequently, the magnitude of the decrease will depend on the maximum dopant concentration and hence on the implantation energy. For a

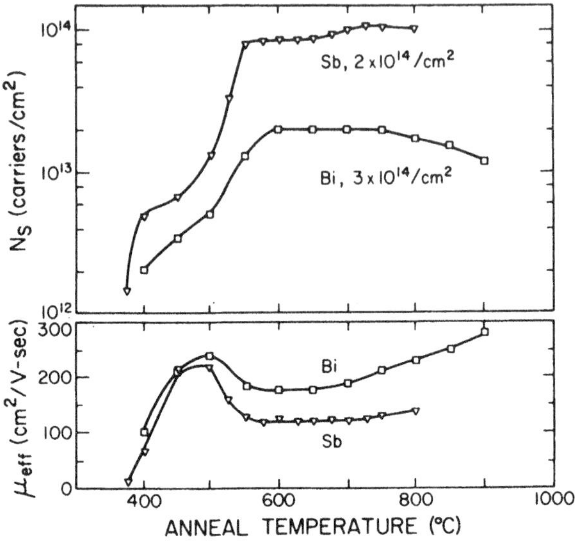

Fig. 5.10. Values of N_s and μ_{eff} as a function of anneal temperature for 40-keV room-temperature implantations of antimony and bismuth at doses of 2 and 3 × 10^{14} ions/cm², respectively. (Antimony data from Johansson,[246] bismuth data from Marsh *et al.*[240])

given species, the value of N_s at high anneal temperatures can therefore be increased by increasing the implantation energy.

Stripping data[240] on these bismuth implants indicate that the true value of N_s, determined by integrating the carrier concentration, is ~1.5 times the effective number of carriers/cm²; therefore, N_s is about 20% of the number of substitutional bismuth atoms. In spite of the fact that N_s is much less than the number of implanted ions/cm², the maximum carrier concentration of $2 \times 10^{19}/cm^3$ found in stripped samples is still 10 times greater than the maximum solid solubility of bismuth in silicon (2×10^{18} atoms/cm³ at 1300° C). The dependence of the mobility on the carrier concentration for these samples (Fig. 5.4) indicates that the mobility values are significantly low. Baron *et al.*[237] suggest that compensating acceptor centers are present, although the nature of these centers is not yet known. At higher implantation energies (240 keV), Crowder[256a] found that the values of N_s were considerably larger, viz. about 50% of the number of implanted bismuth ions. This indicates that in the lower-energy implants, compensating centers were located in the region near the surface.

It was pointed out in Chapter 3 that the reordering of the amorphous layer does not always occur in the same fashion. In room-temperature implantations at 40 keV, the anneal temperature increases if the implantation dose exceeds 10^{15}–10^{16} ions/cm^2 (this critical dose level depends upon the ion species). A similar influence of dose on anneal temperature is also noted in the electrical characteristics of phosphorus implantations.[252,253] As the dose of implanted ions is increased above $\sim 10^{15}$ ions/cm^2, increasingly higher anneal temperatures are required and a sharp stage in the anneal characteristics is no longer found.

c. $T > 600°$ C. In early investigations of implantation processes, it was hoped that the use of elevated substrate temperatures during implantation to reduce the residual lattice disorder would eliminate the requirement for anneal temperatures greater than 500° C. This was suggested by high-dose ($> 10^{15}$ ions/cm^2) antimony implantations at substrate temperatures of 450–500° C in which the maximum number of carriers was observed following a 500° C anneal.[236] In this work, the carrier concentration was well above the value set by the thermal-equilibrium solubility (cf. Part 5.6), and the decrease in N_s due to precipitation at still higher anneal temperatures tended to mask any increase due to the annealing of defects.

In almost every case where an amorphous layer is *not* present [except possibly for high-dose ($\gtrsim 10^{15}$ ions/cm^2) hot-substrate implantations], anneal temperatures of at least 700–750° C are required for N_s to reach its maximum value.† This indicates that defect centers are present up to these temperatures. The high-anneal-temperature requirement cannot be attributed to a lack of dopant atoms on substitutional sites. For example, in 350° C antimony implantations at dose levels up to 5×10^{14} ions/cm^2, at least 90% of the atoms are substitutional [Fig. 4.19(a)]. Also, in boron implantations, the temperature dependence of the Hall-effect measurements indicated that anneal temperatures of 760° C were required to remove the influence of deep levels—i.e., to produce the normal value for the dopant activation energy.[242,243,257] In channeled phosphorus implantations[168] at room temperature, it was found that defect centers were present after anneal to temperatures of 700–740° C. The origin or nature of the defects responsible for the high-anneal-temperature requirement are not known. Unfortunately, there are no anneal data on the electrical behavior of fast-neutron-irradiated samples that can be used for comparison. Also, Johnson[258] has demonstrated the need for anneal temperatures greater than those predicted from fast-neutron irradiations. He implanted samples at 600° C and then introduced damage by implanting at room temperature with silicon ions at doses of $\sim 5 \times 10^{12}$ ions/cm^2. In the anneal characteristics of phosphorus and arsenic implants,

† Samples implanted at room temperature with doses of 10^{13} ions/cm^2 or less tend to require lower anneal temperatures for the value of N_s to approach its maximum. In Fig. 5.9, for the sample implanted with 10^{13} aluminum ions/cm^2, N_s reached approximately one-half its maximum value after anneal at 600° C.

Johnson found a relatively steep anneal at about 200° C, which he attributed to dissociation of dopant–vacancy pairs, and then a slower change at higher anneal temperatures. Typically, the values of N_s after anneal at 600° C were 50–70% of those measured before implantation with silicon ions.

It should not be inferred that all the high-temperature anneal characteristics are due solely to the presence of defect centers. In the case of boron, for example, there is strong evidence to suggest that another mechanism is playing a role. Figure 5.11(a) shows the anneal characteristics of boron-implanted samples as a function of implantation dose and temperature. Figure 5.11(a) shows the values of N_s for 500° C implants and Fig. 5.11(b) the relative number of holes and boron atoms on substitutional sites for room-temperature implants.

These curves show that anneal temperatures of ~900° C are required before the number of carriers/cm^2 becomes equal to the implanted dose. Clark and Manchester[238] have measured the isothermal anneal characteristics of a room-temperature boron implantation (5×10^{14} ions/cm^2). They estimate an activation energy of ~4.4 eV from the anneal kinetics. Seidel and MacRae[243] also find an activation energy of the order of 5 eV associated with this high-temperature annealing. They suggest capture of an interstitial boron atom by a vacancy as a mechanism that would account for the increased electrical activity at 800° C. This suggestion has been born out by the channeling-effect measurements of Fladda *et al.*[231b] and North and Gibson[231c] (Section 4.4.2d). They observed an increase in the number of boron atoms on substitutional sites, following anneal at temperatures greater than 600° C. After anneal at temperatures of 900° C or greater, approximately 90% of the boron atoms were on substitutional sites. For anneal temperatures $\gtrsim$600° C, North and Gibson[231c] found a close correlation between the number of carriers per cm^2 and the number of boron atoms on substitutional sites [Fig. 5.11(b)]. Note that in this work the maximum boron concentration did not exceed published solubility values.[231f]

In evaluations of room-temperature boron implantations,[58, 238, 255, 258a] a negative anneal behavior was observed; i.e., for temperatures less than 600° C, the number of carriers decreased with an increase in anneal temperature. This negative anneal behavior can now be understood in terms of the decrease in substitutional content found for anneal temperatures between 200° and 500° C. (See Figs. 4.22 and 5.11b.)

The various anneal regimes are summarized in Table 5.3, along with our suggested interpretation. Note that much of this interpretation must be considered as speculative. Although we are confident of the existence of isolated disordered regions and the formation of an amorphous layer, we can only surmise their effect on electrical behavior. Much of this same caution also applies to the interpretation of fast-neutron irradiation data.

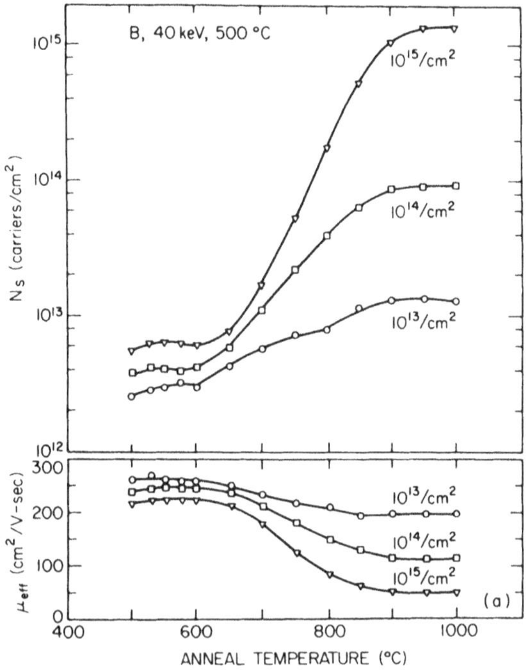

Fig. 5.11(a). Variation of N_s and μ_{eff} with anneal temperature for 50-keV boron implantations (substrate temperature, 500° C) to dose levels of 10^{13}, 10^{14}, and 10^{15} ions/cm². (From Baron *et al.*[237])

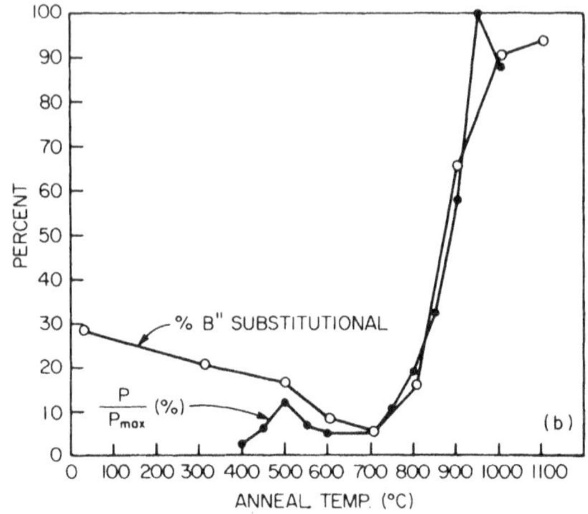

Fig. 5.11(b). Anneal behavior of silicon samples implanted at room temperature with 150-keV boron ions. Open data points are the percentage of boron atoms which occupy substitutional sites as determined by (p, α) reactions (Section 4.4.2d) for a sample implanted with 4×10^{15} ions/cm². Solid data points are the percentage of holes per cm² (relative to the maximum number of holes measured after anneal at 950° C) as determined by Hall-effect measurements for a sample implanted with 2×10^{15} ions/cm². (From North and Gibson.[231c])

TABLE 5.3

DEFECT-ANNEALING BEHAVIOR THAT MAY PLAY A ROLE IN EVALUATION OF ROOM-TEMPERATURE IMPLANTATIONS IN SILICON

Anneal temperature	Observation	Mechanism
70–200° C	Changes in N_S in tail of distribution	Isolated defects dissociate
300–350° C	Type conversion	Defects anneal in the vicinity of implanted atoms
~400–450° C	Mobility increases, $\mu(n)$ agrees with bulk value	Dissociation of defect clusters, which influence mobility
$<600°$ C	$n(T)$ does not agree with bulk values	Deep levels present
550–650° C	Large increase in N_S for R.T. implants (amorphous layer)	Amorphous layer reorders
$\gtrsim 750°$ C	$n(T)$ and $\mu(T)$ agree with bulk values	Deep levels absent
800–900° C	N_S increases (boron)	Interstitial boron and vacancy interaction

5.4.3 INTERSTITIAL COMPONENTS

One of the striking features of the channeling-effect measurements discussed in Chapter 4 is the presence of an interstitial component of magnitude equal to or greater than the substitutional component of the implanted species. At present there are a number of dopants whose electrical behavior has been attributed to the presence of an interstitial component.

Thermal-probe, Hall-effect, and channeling-effect measurements have been made on n- and p-type silicon samples implanted with thallium at substrate temperatures of 350–450° C. The channeling-effect measurements have been discussed in Chapter 4, and an example of the anneal data is indicated in the upper portion of Fig. 5.12. From this figure it can be seen that there are equal numbers of thallium atoms on substitutional and tetrahedral interstitial sites in the as-implanted condition (450° C). As the anneal temperature is increased to about 600° C, the fraction of thallium atoms on interstitial sites increases from about 30% to about 60% and the fraction of substitutional thallium atoms decreases to less than 5% of the total number of implanted atoms. The lower portion of Fig. 5.12 (using the same temperature scale) summarizes the results obtained from thermal-probe measurement made as a function of anneal temperature on both n- and p-type crystals For both types of crystals, the thermal-probe measurements show an n-type region in the implanted layer over the same annealing range in which a large interstitial component is present. When the implanted layers were of opposite

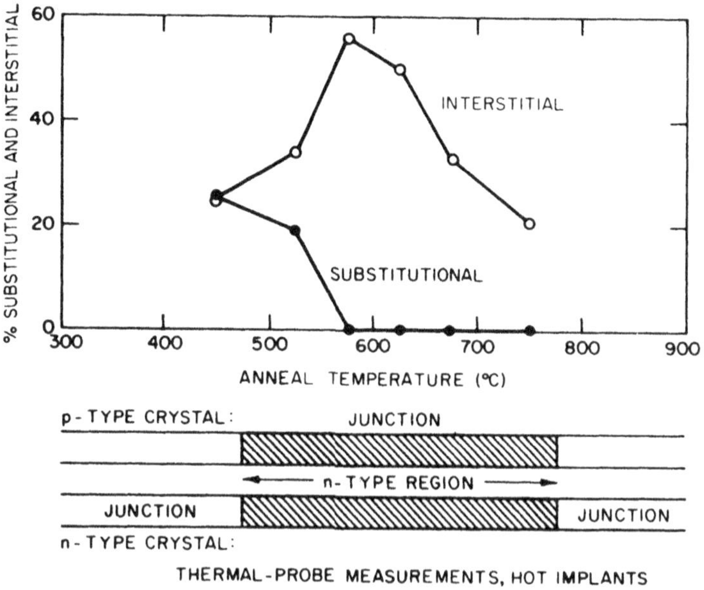

Fig. 5.12. Upper: Anneal-temperature dependence of the substitutional and interstitial components in a 450°-C thallium implantation as determined from channeling-effect measurements. Lower: Results of thermal-probe measurements on comparable thallium implants in both *n*- and *p*-type substrates. The shaded portions indicate the anneal-temperature region over which *n*-type behavior was noted. The legend *junction* indicates the existence of *p-n* junction behavior. (From Baron *et al.*[237])

conductivity type to that of the base crystal, it was possible to make Hall-effect measurements. The observed surface carrier concentration was about $10^{10}/cm^2$ when the implanted layer was *n*-type, and about $10^{11}/cm^2$ when the implanted layer was *p*-type; i.e., in both cases, N_s was very much less than the number of implanted atoms. The small number of active carriers in the *p*-type condition may be associated with the fact that thallium as an acceptor in silicon has a level located about 0.26 eV above the valence band.

Similar behavior was observed in *n*- and p-type silicon implanted with cadmium at 350° C substrate temperatures.[55] Junction evaluation and Hall-effect measurements showed *n*-type behavior for anneal temperatures between 500 and 750° C, as shown in Fig. 5.13. In this case, however, in contrast to the thallium case, no *p*-type dopant action was observed in similar implants in *n*-type silicon. Channeling-effect measurements[55] indicate that the substitutional component was $\lesssim 5\%$, and that at least 25% of the cadmium atoms were on tetrahedral interstitial sites along the $\langle 111 \rangle$ rows. For anneal temperatures $>700°$ C, the interstitial cadmium component decreased. This decrease corresponds to the decrease in the number of carriers/cm² shown in

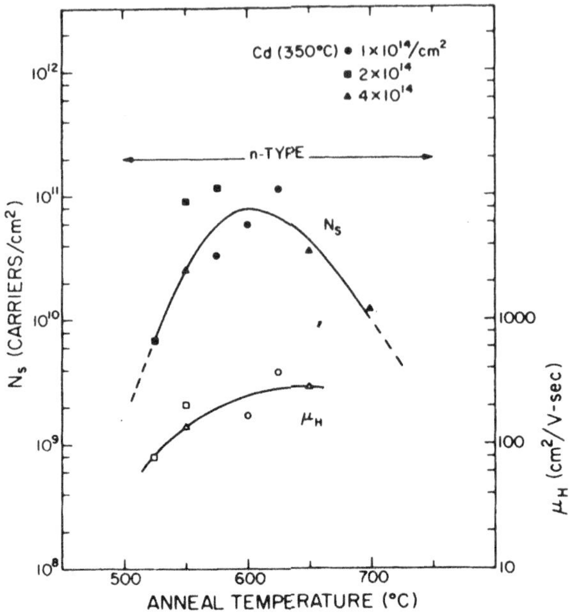

Fig. 5.13. Variation of N_S and μ_H with anneal temperature for 40-keV cadmium implantations into *p*-type silicon at 350° C substrate temperature. (From Picraux *et al.*[55])

Fig. 5.13. The maximum value of $10^{11}/\text{cm}^2$ for the number of carriers corresponds to an average concentration $\bar{n} \approx 5 \times 10^{16}/\text{cm}^3$. This is lower by 2–3 orders of magnitude than the concentration of cadmium on interstitial sites. The bulk mobility for a carrier concentration of $\sim 5 \times 10^{16}/\text{cm}^3$ in *n*-type silicon (Fig. 5.3) would be about 1000 cm²/V-sec. The effective Hall mobility in these samples (as shown in Fig. 5.13) is much lower ($\sim$250 cm²/V-sec), suggesting that the low number of carriers is due to strong compensation effects. Meyer *et al.*[231d] have found that zinc- and mercury-implanted layers in silicon also exhibit much the same electrical and lattice-location behavior as that of cadmium-implanted layers.

It should be pointed out that the *n*-type behavior in both the thallium and cadmium implantations could also be due to the presence of radiation-damage defects. Until more complete investigations of the temperature dependence of the Hall coefficient are completed, one can only tentatively identify the *n*-type behavior with the presence of the interstitial component.

Another possible interstitial case can be inferred from the anneal characteristics of aluminum-implanted layers in silicon. Here a strong *n*-type region is not formed; however, Hall-effect measurements on stripped samples strongly suggest the presence of an interstitial component. Baron *et al.*[237] have analyzed in detail the mobility and carrier-concentration distributions in aluminum implantations at a dose level of 10^{13} ions/cm^2 (the anneal behavior is shown in Fig. 5.9). Two identical samples were annealed, one to 800° C (nearly complete electrical activity, as shown in Fig. 5.9) and the other to 400° C. The depth distribution of the free-carrier concentration and mobility were measured by the combination of Hall-effect and layer-removal techniques. The data are shown in Fig. 5.14, in which the circles represent the carrier

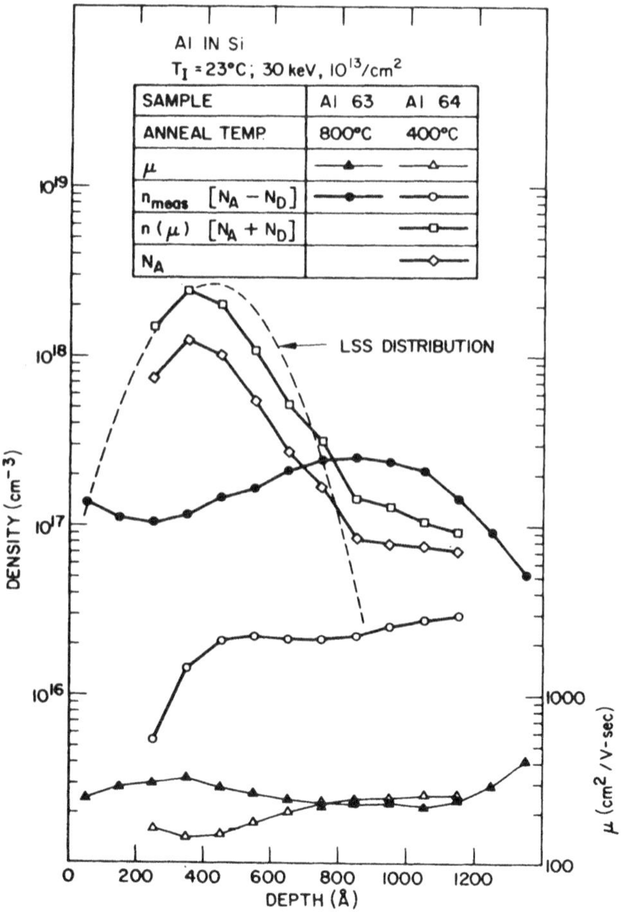

Fig. 5.14. The depth distributions of carrier density and mobility as measured by Hall-effect and layer-removal techniques for two 30-keV room-temperature aluminum implants at a dose of 10^{13} ions/cm^2. Solid points refer to the sample annealed at 400° C. For the latter sample, the depth distribution of the total density of ionized impurities, $n(\mu) = N_A + N_D$, was determined from the mobility data in Fig. 5.5(b). The dashed curve is the distribution of aluminum atoms predicted from the theory of Lindhard *et al.*[2] (From Baron *et al.*[237])

density and the triangles the mobility. The relationship between mobility and carrier concentration is shown for these samples in Fig. 5.5(b), where it can be seen that the mobility values of the well-annealed sample (Al-63) fit the bulk curve within experimental error. On the other hand, the sample that was annealed to only 400° C (Al-64) has a value of N_s that is only 5% of the dose; and [as shown in Fig. 5.5(b)] the values of the mobility are significantly below those values for the bulk curve.

Baron *et al.*[237] have used these mobility data to investigate the properties of the implanted layer. They suggested a model in which the measured value of the mobility might be used to determine the number of ionized donors (N_D) and ionized acceptors (N_A). If the implanted layer is compensated, the observed value of the carrier concentration is equal to the difference in the doping concentrations, i.e., $N_A - N_D$. The mobility value, on the other hand, is dominated by the total number of charged impurity centers, i.e., $N_A + N_D$, since at these concentration levels impurity scattering probably dominates over lattice scattering.† The values of $N_A + N_D$ are found from the relationship of the measured mobility curve [Fig. 5.5(b)] to the expected bulk value—i.e., from the concentration of ionized centers (assumed equal to $N_A + N_D$) required to give the observed mobility value. The curve of $N_A + N_D$ for the sample annealed at 400° C is shown in Fig. 5.14. It can be seen that both the shape and the maximum concentration of carriers is close to the distribution predicted from the LSS range theory (Chapter 2), indicating that nearly all of the implanted aluminum ions are electrically active. The low value of the free-carrier concentration would then imply that the aluminum acceptors are almost exactly compensated. Baron *et al.*[237] suggest that the compensating donors are interstitial aluminum ions.

If the model is correct, room-temperature aluminum implantations (at a dose of $\sim 10^{13}$ ions/cm^2) annealed to 400° C have ~95% of the aluminum atoms located in an LSS distribution; half of these will be on substitutional sites and the other half on the tetrahedral interstitial sites with compensation to a few percent. The remaining 5% of the aluminum atoms are found in a deeply penetrating tail.

5.4.4 Influence of Implantation Temperature

One of the striking features of channeling-effect measurements is the absence of an amorphous layer and the presence of a large fraction of atoms on well-defined sites, when the implantation is made into substrates held at elevated temperatures.

† The assumption that the observed mobility values are due to impurity scattering rather than to defect scattering centers is reasonable in view of the fact that in boron and phosphorus implantations the mobility values at room temperature are close to those deduced from uncompensated material.

The most significant electrical results in hot-substrate implantations are the absence of the 600° C annealing stage and the requirement of relatively high anneal temperatures to obtain a number of carriers/cm^2 comparable to the implanted dose. These results are illustrated for antimony implantations in Fig. 5.1. In channeling-effect measurements of similar hot-substrate implantations, it was shown that approximately 90% of the implanted antimony atoms were on substitutional sites.[174] Consequently, the gradual increase in the number of carriers/cm^2 [shown in Fig. 5.1(b)] cannot be attributed to a movement from interstitial to substitutional sites, and would therefore suggest the presence of defect centers. It is not known whether these defects are in the immediate vicinity of the antimony ion, or are spatially well separated and act as compensating centers. The requirement for high anneal temperatures has also been noted in hot-substrate implantations of arsenic,[236] phosphorus,[254] aluminum,[237] and gallium.[236, 246] This difference in anneal characteristics between silicon samples implanted to the same dose at room temperature (in which case an amorphous layer is formed) and at elevated temperatures has also been pointed out by Crowder *et al.*[241b, 255a]

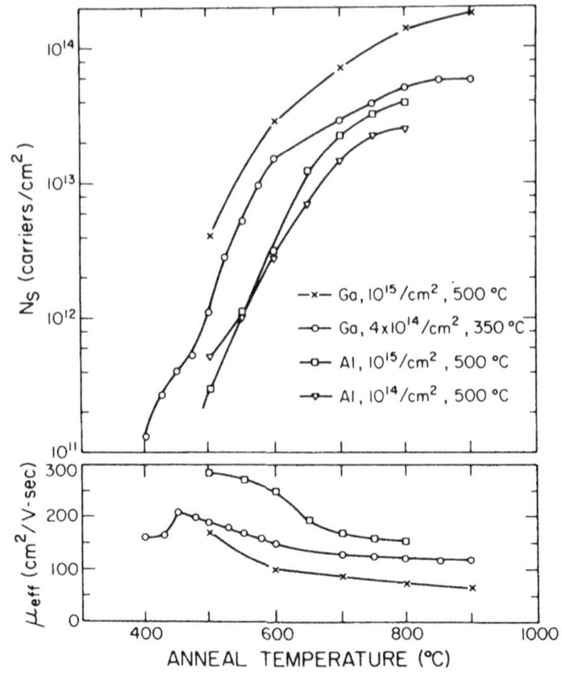

Fig. 5.15. N_s and μ_{eff} as a function of anneal temperature for *n*-type silicon samples implanted with gallium and aluminum at 30 keV and at elevated substrate temperatures. (From Baron *et al.*[237])

The anneal characteristics of gallium and aluminum implantations into silicon at substrate temperatures of 500° C are shown in Fig. 5.15 (see Figs. 5.8 and 5.9 for comparable room-temperature implantations). It is clear from the hot-implantation data in Fig. 5.15 that there is no well-defined anneal stage at 600°C and that only after annealing to ~900° C does the electrical activity approach that observed in the room-temperature implantations annealed to 600° C (10^{15} ions/cm^2). At present there are no definite indications as to the mechanisms responsible for the high anneal temperature requirement of hot implants, although for aluminum and gallium it could be the vacancy–interstitial-dopant reaction similar to the boron case in Section 5.4.2.

Boron implantations have also shown that the implantation temperature can have a large effect on the anneal characteristics.[254] Figure 5.16 shows the anneal characteristics of boron implantations at 200 keV for several substrate temperatures. Curves I and II indicate that implantations performed at or

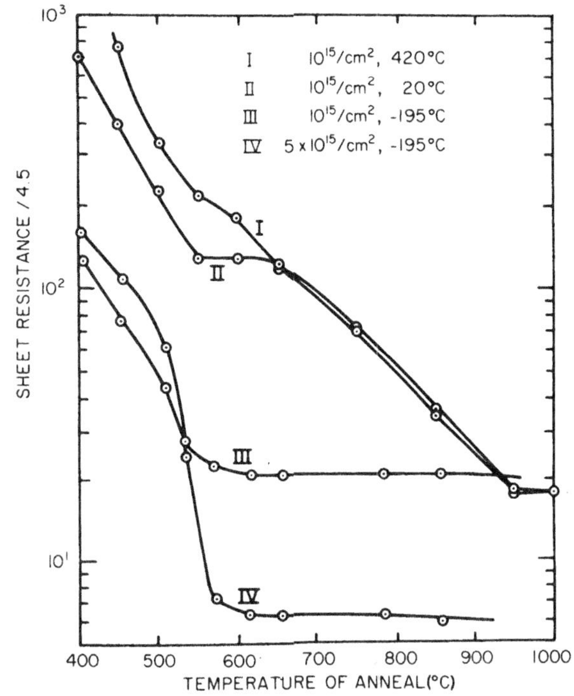

Fig. 5.16. Isochronal anneal of the sheet resistivity of 200-keV boron implantations at dose levels of 10^{15} ions/cm^2 and substrate temperatures of −195°, 20°, and 420° C. A fourth implantation was made at −195° C at a dose level of 5×10^{15} ions/cm^2. (From Davies.[254])

above room temperature require anneal temperatures of nearly 900° C. On the other hand, the implantations at substrate temperatures of −195° C show nearly complete electrical activity after an anneal of 600° C. Channeling-effect measurements[176] indicate that in low-temperature implantations, an amorphous layer is formed at boron doses $\gtrsim 10^{15}$ ions/cm^2. Thus the nearly complete anneal at 600° C reflects the reordering of the amorphous layer.[176a] It is not due solely to the use of low implantation temperatures. Hall-effect measurements on low-temperature, low-dose boron[176a] and antimony[235] implantations do not show this anneal stage.

5.4.5 Influence of Channeling

It has been shown in the discussion in Chapter 2 that channeling effects can play a dominant role in determining the distribution of implanted atoms. The electrical evaluation of channeled phosphorus[168] and gallium[259] implantations indicates significant differences in anneal characteristics between channeled and random conditions. This difference perhaps may be due to the reduction in the amount of lattice disorder during a channeled implantation, and consequently to the absence of an amorphous layer.

In spite of the paucity of experimental data, it is evident that anneal characteristics are influenced by channeling effects. At the present time it is not possible to describe the mechanisms responsible for these differences in the anneal characteristics.

5.5 Diffusion Effects

Using range-straggling calculations and values of N_s to estimate the average carrier concentration ($\bar{n} = N_s/2.5\,\Delta R$), one must always consider whether or not diffusion effects may have influenced the dopant distribution. In Table 5.4 we compare the mean range straggling ΔR for 100-keV implantations to the diffusion length $(Dt)^{1/2}$ for 10-min anneals at 900° C. The diffusion constants are not well defined at this temperature, and a considerable extrapolation is involved. In general, the highest published values have been used in Table 5.4. It can be seen from the ratio of $(Dt)^{1/2}/\Delta R$ that gallium and aluminum are the only species where appreciable broadening of the profile might occur during an anneal at 900° C, assuming that conventional diffusion effects are dominant.† However, enhanced diffusion effects (Part 2.5) are present for some ion species; therefore, one cannot *a priori* exclude the influence of diffusion processes.

† Fairfield and Crowder[19a] have shown that no appreciable broadening of the dopant distribution occurs in room-temperature arsenic- and phosphorus-implanted layers when annealed to 850° C for 30 min.

TABLE 5.4

INFLUENCE OF DIFFUSION PROCESSES ON THE DISTRIBUTION OF IMPLANTED ATOMS FOR 100-keV ION ENERGIES AND 10-min ANNEAL AT 900° C

Ion	R_p (Å)[a]	ΔR (Å)[a]	D (cm^2/sec)[b]	$(Dt)^{1/2}$ (Å)	$(Dt)^{1/2}/\Delta R$
B	3280	726	2×10^{-15}	110	0.15
Al	1440	430	4×10^{-14}	490	1.1
P	1230	380	$\sim 10^{-14}$[c]	250	0.7
Ga	610	190	$\sim 10^{-14}$	250	1.3
As	575	180	$\sim 10^{-15}$	80	0.45
Sb	450	120	$\sim 10^{-15}$	80	0.65

[a] Data from LSS theory (Part 2.3).
[b] Data from B. I. Boltaks.[231]
[c] Highest value of phosphorus diffusion constant as given by W. R. Runyon.[235b]

Gas-release measurements (see Section 3.4.4) have demonstrated that significant motion of implanted inert-gas atoms to the surface can occur at temperatures well below normal diffusion temperatures. This is most strikingly shown in the cases where an amorphous layer is formed (Jech and Kelly[54]), indicating that during the reordering of an amorphous layer a significant redistribution of the implanted atoms can take place. In considering the influence of diffusion, the boundary condition imposed by the surface will play a role. There are two extreme possibilities: The surface could act as a sink, or else as a perfectly reflecting interface. In the case of antimony implantations,[156] it was found that the surface acted as a sink for antimony atoms during high-temperature (900° C) annealing. It has also been found[259a] that for mercury, cadmium, and zinc implantations, a large fraction of the implanted atoms leave the crystal during the reordering of the amorphous layer. In the case of rubidium-implanted layers, a direct correlation was noted[259b] between the decrease in the number of carriers per cm^2 and the decrease in the number of implanted atoms in the sample.

In addition to redistribution within the implanted layer, enhanced-diffusion effects have also been shown to occur. This has been shown most directly for aluminum implantations in silicon. Figure 5.17 shows the carrier and mobility distribution for a 30-keV aluminum implantation at a substrate temperature of 500° C and a subsequent anneal temperature of 900° C. Shown on the figure for comparison is the carrier distribution that would have been obtained under these implantation conditions if all the atoms resided in an LSS distribution. It has been shown by Itoh *et al.*[97] that these deep distributions cannot be attributed to channeling. Similarly, Glotin[88] finds that

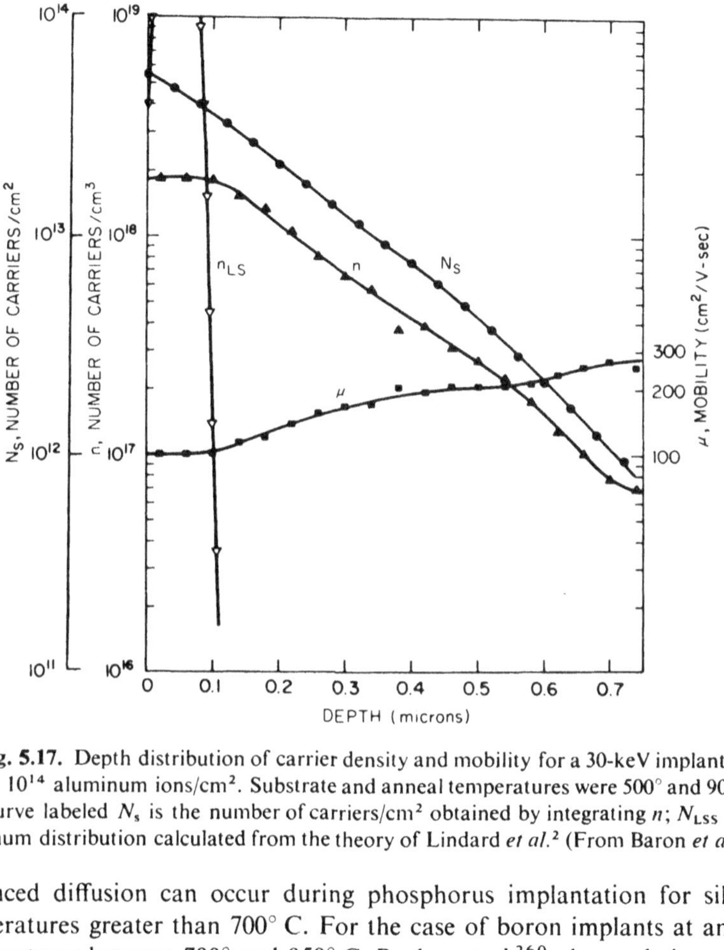

Fig. 5.17. Depth distribution of carrier density and mobility for a 30-keV implantation of 8×10^{14} aluminum ions/cm^2. Substrate and anneal temperatures were 500° and 900° C. The curve labeled N_s is the number of carriers/cm^2 obtained by integrating n; N_{LSS} is the aluminum distribution calculated from the theory of Lindard *et al.*[2] (From Baron *et al.*[237])

enhanced diffusion can occur during phosphorus implantation for silicon temperatures greater than 700° C. For the case of boron implants at anneal temperatures between 700° and 950° C, Pavlov *et al.*[260] observed changes in the junction location that did not follow standard diffusion relations. They suggested that the observed motion of the junction might be due only to an increase in the electrical activity of boron atoms already present in the vicinity of the junction location. In this case, there would not be any gross diffusion of the implanted species.

Thus, one can state with some certainty that enhanced-diffusion effects do sometimes occur during implantation (see also Part 2.5) or during the subsequent anneal treatment. However, there have not been sufficient experimental investigations to describe all the conditions under which enhanced diffusion takes place. In particular, for semiconductors, it is not certain whether the enhanced diffusion represents a motion of an interstitial species or whether it is associated with an enhanced vacancy concentration.

At anneal temperatures of 900° C or higher, diffusion effects can be expected to influence the anneal behavior. The value of N_s could decrease if the dopant species diffused to the surface or to some precipitation centers, or it could increase if the dopant atoms were driven off precipitation centers. Thermal equilibrium and solid solubility should be the dominant effects. At these anneal temperatures, the dopant distribution might be quite analogous to that found[261] in "drive-in" diffusions following a predeposition diffusion. Except for boron,[260] the influence of high anneal temperatures has not been investigated in detail.

5.6 Limitations on Electrical Behavior

The channeling-effect measurements discussed in Chapter 4 showed that the number of implanted atoms on substitutional sites could exceed the values given for conditions of thermal-equilibrium solubility. It is clear that an ion implanted at keV energies has sufficient energy to overcome any diffusion barriers at the surface (of the order of electron volts), and also to provide sufficient vacancies along its track so that substitutional sites are available. Concentrations in excess of equilibrium solubility cannot be maintained, of course, when the sample is annealed to temperatures such that appreciable diffusion occurs and implantation-produced defects disappear. Here, thermal-equilibrium conditions establish the upper limit on the concentration of dopants on lattice sites. Gallium implantations include cases where the concentrations can exceed or be limited by equilibrium solubility. For 50-keV, heated-substrate (500° C) implantations annealed to 800–900° C, the number N_s of carriers/cm^2 never exceeded $\sim 10^{14}/\text{cm}^2$, irrespective of implantation dose N_D. This is indicated in Fig. 5.18, where the solid line indicates the case for $N_s = N_D$. This maximum measured value of N_s is comparable to that set by equilibrium considerations. On the other hand, for room-temperature implantations where an amorphous layer has been formed, the value of N_s can exceed the solubility limit when annealed to temperatures around 600° C so that the amorphous layer has reordered.

Table 5.5 is a summary of the concentration of atoms on substitutional sites (from data in Chapter 4) and the maximum value of the carrier concentration. In some cases the maximum carrier concentration had to be estimated from the calculated range distribution. The concentrations of atoms on lattice sites were determined from the widths of the peaks in the spectra of backscattered particles; in many cases, this provides only a *lower* limit to the concentration of the implanted species. It should be noted in comparing the maximum carrier concentration and the maximum number on substitutional sites that these two sets of data may not always refer to identical implantation

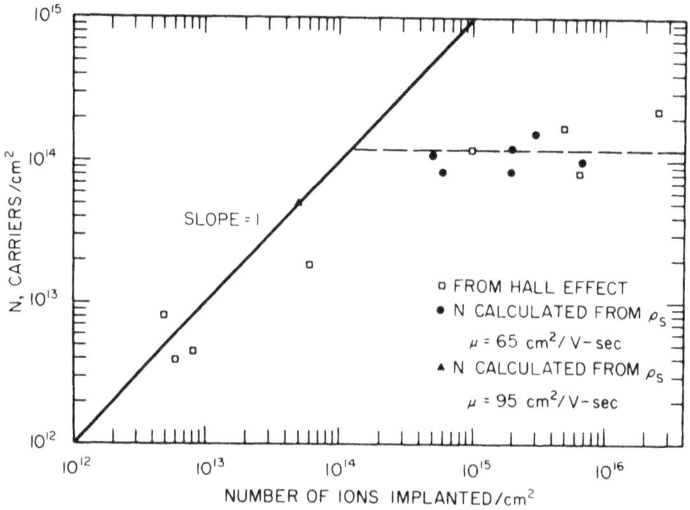

Fig. 5.18. The number of carriers/cm^2 as a function of the number of implanted gallium ions/cm^2. The substrate temperature and 10-min anneal temperature were 500° and 800–900° C, respectively. Implantation energies of 20–75 keV were used The open squares represent values obtained from Hall data; solid points, those from sheet-resistivity measurements. (From Mayer *et al.*[236])

conditions. It is clear from the data listed in Table 5.5 that the number of atoms on substitutional sites can exceed the value of the equilibrium solubility. Furthermore, the maximum carrier concentration, although generally smaller than that observed on substitutional sites, also exceeds the maximum solubility value in certain cases (tellurium, antimony, and bismuth). The implantation energies for the data given in Table 5.5 were of the order of 50 keV. By using higher-energy implantations to increase the width (ΔR_p) of the implanted layer, the values of N_s can be larger than those given. To obtain the largest values of N_s, one should use a variable-energy implant (Section 2.3.8) to obtain as uniform a dopant concentration as possible.

For 40–60 keV implantations in the dose range of 10^{14}–10^{15} ions/cm^2, one can make the following comparisons between the values of N_s (anneal temperature of 900° C) and the implanted dose N_D for the Group III and Group V dopant species:

a. $N_s \lesssim N_D$; B, P, As, *and* Sb. Except for antimony, these dopant species have solubilities in excess of 10^{20} atoms/cm^3 and the average carrier concentrations are less than the solubility values. For arsenic implantations, the value of N_s is approximately $\frac{1}{2}N_D$.[83] This is in agreement with the results found in channeling-effect measurements (Chapter 4) ,where about 60% of the implanted arsenic atoms were on substitutional sites.

TABLE 5.5

MAXIMUM VALUES[a] OF THE NUMBER OF IMPLANTED ATOMS ON SUBSTITUTIONAL SITES AND THE NUMBER OF CARRIERS/cm²

	Lattice location		Electrical		
Ion	N_{max}/cm²	n_{max}/cm³	$N_{S\ max}$/cm²	n_{max}/cm³	Solubility at 1100° C/cm³
B	2×10^{15}	2×10^{20}	2×10^{15}	2×10^{20}	5×10^{20}
Al	—	—	10^{14}	1.5×10^{19}	2×10^{19}
P	$>4 \times 10^{15}$	$>4 \times 10^{20}$	6×10^{15}	10^{20}	10^{21}
Ga S[b]	2×10^{14}	4×10^{19}	2×10^{14}	2×10^{19}	3.5×10^{19}
Ga I	2×10^{14}	4×10^{19}	—	—	
As	5×10^{15}	10^{21}	10^{15}	10^{20}	2×10^{21}
Sb	4×10^{15}	8×10^{20}	10^{15}	10^{20}	4.5×10^{19}
Tl S	10^{14}	2×10^{19}	2×10^{11} (*p*-type)	10^{17}	—
Tl I	2×10^{14}	4×10^{19}	2×10^{10} (*n*-type)	10^{16}	—
Bi	10^{15}	2×10^{20}	2×10^{13}	10^{19}	2×10^{17}
Cd S	—	—	—	—	—
Cd I	10^{14}	10^{19}	10^{11} (*n*-type)	5×10^{16}	—
Te	5×10^{14}	10^{20}	10^{13}	4×10^{18}	$\sim 10^{17}$

[a] The concentrations were estimated from range data for implantation energies of 20 to 50 keV.

[b] S and I are the substitutional and interstitial components, respectively.

b. $N_s < N_D$; Ga *and* Al. For these species the value of N_s is generally lower than the implanted dose. For high anneal temperatures, equilibrium solubility values determine the upper limit on N_s. For aluminum implantations, enhanced-diffusion effects are generally found.

c. $N_s \ll N_D$; N, In, Tl, *and* Bi. For indium and thallium† implantations, the values of N_s at room temperature should be considerably reduced because of the relatively large ionization energies (0.16 and 0.26 eV, respectively). An indication of this was found in Hall-effect measurements on indium implantations, where the value of N_s was strongly dependent on measurement temperature.[235] Note, however, that at these high dopant concentrations, the

† Here we refer to the *p*-type behavior found at high anneal temperatures.[237] The *n*-type behavior noted at intermediate temperatures is discussed in Section 5.4.3.

values of the ionization energies may be substantially smaller than the values found in lightly doped samples.[232] In analysis of bismuth implantations, Marsh *et al.*[240] found that about 20–25% of the substitutional atoms were electrically active at room temperature. This difference is too large to be explained on the basis of ionization energy. From mobility data determined by use of layer-removal techniques, Baron *et al.*[237] suggest that compensating centers were present.

5.7 Analysis of Unconventional Dopant Species

One interesting aspect of ion implantation is the possibility of introducing unconventional dopant species. By combining channeling-effect measurements with other experiments such as optical or electrical measurements, one has available a powerful method to study the behavior of foreign atoms in semiconductors. There are, of course, limitations in this type of analysis that arise from lattice-disorder and radiation-damage effects produced by the incident ion and from the high dopant concentrations encountered for typical doses and energies. On the other hand, ion implantation offers a particularly simple sample-preparation technique for studies of heavily doped semiconductors. Also, the effects of radiation damage can be greatly reduced by use of high anneal temperatures.

Essentially one would like to know lattice location (substitutional, interstitial, etc.), solubility, diffusion constant, electrical behavior (donor, acceptor, or neutral), and ionization energy. Lattice location and an upper limit to the solubility can be determined from channeling-effect measurements as a function of anneal temperature. Diffusion constants can be deduced from radiotracer measurements (or in favorable cases, from backscattering measurements) if enhanced-diffusion effects are not dominant. Electrical behavior can be determined from Hall-effect measurements and junction characteristics.

Electrical measurements can be difficult to interpret because of radiation-damage effects. However, the results presented in Section 5.4.2 indicate that anneal temperatures of 750–850° C should be sufficient to dissociate most of the damage centers. Anneal temperatures of 600° C are often sufficient for silicon samples implanted at room temperature to dose levels such that an amorphous layer is formed. Since the lattice location of implanted atoms may differ for room-temperature and hot-substrate implantation conditions (Chapter 4), it is advisable to evaluate the electrical behavior of samples implanted under both conditions.

Ionization-energy levels should be determined from samples implanted at low doses and annealed to 800° C or higher. In order to avoid degenerate conditions where the dopant levels merge with the conduction or valence bands, concentrations well below 10^{19} atoms/cm^3 are required.[232] Implantation doses of approximately 10^{13} ions/cm^2 give doping concentrations of 10^{18} atoms/cm^3 or less. This represents an implantation-dose level that is near the lower limit that can be easily analyzed by channeling-effect measurements. However, the dopant ionization energies should be determined for a series of implantation doses, since these energies may still be concentration-dependent down to concentrations of 10^{16} atoms/cm^3. These energy-level data are determined from Hall-effect measurements taken as a function of temperature, and therefore the depth distributions of the carrier concentration and mobility should be considered in the analyses. Some of the difficulties in interpretation can be overcome by orienting the sample to minimize channeling and by varying the ion energy during implantation to produce a nearly uniform distribution. A more complete investigation would combine Hall-effect measurements as a function of temperature with layer-removal techniques.

At present the electrical behavior of only a few implanted unconventional dopant species have been investigated.† It has been established that nitrogen acts as a donor[263] with an ionization energy of 0.045 eV.[244] Cesium and sodium, Group IA elements, exhibit donor action.[95,264,265] For all three of these elements, the values of N_s were significantly less than the implanted dose. In the case of cadmium,[55] the observed donor behavior was attributed to the presence of an interstitial component (Section 5.4.3).

Picraux *et al.*[55] used channeling- and Hall-effect measurements in their analysis of tellurium implantations. They found that, following the anneal of room-temperature 50-keV tellurium implantations (at doses of 4 and 8×10^{14} ions/cm^2), approximately 60 % of the tellurium atoms occupied substitutional sites—i.e., a concentration of $\sim 10^{20}$ atoms/cm^3. Hall-effect measurements (Fig. 5.19) indicated that tellurium acts as a donor and that N_s increases strongly in the temperature region where the lattice is reordering. The number of carriers/cm^2 then levels off (in a manner similar to the observed anneal behavior in the lattice-location measurements), but at a value lower by a factor of ~ 30 than the number of tellurium atoms on substitutional sites (i.e., a concentration of $\sim 4 \times 10^{18}$ carriers/cm^3). The measured effective-mobility value of 320 cm^2/V-sec is a factor of 1.5 higher than the carrier mobility calculated from the carrier concentration. As pointed out in Part 5.2, this high effective mobility is due to the mobility variation in the implanted region. The mobility value for the tellurium implant is consistent with the concentration of 4×10^{18} scattering centers/cm^3 rather than the much greater

† Two review papers by Schibli and Milnes summarize the properties of deep impurities in silicon. They discuss solubilities and ionization energies[262a] and lifetimes and capture cross sections.[262b]

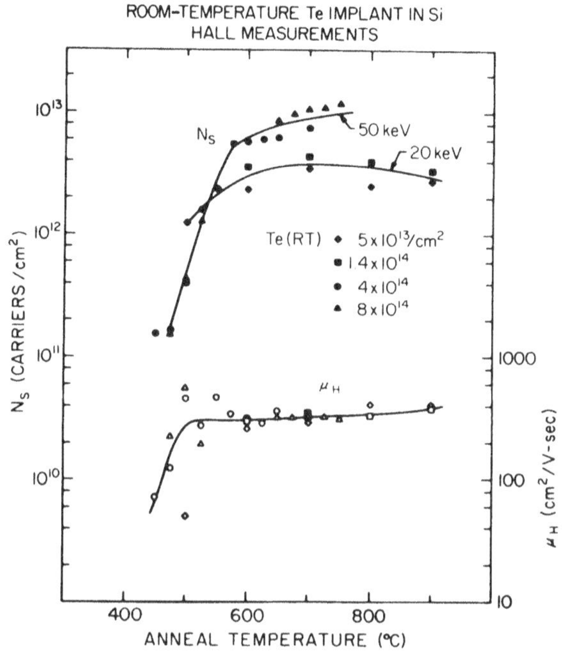

Fig. 5.19. Values of N_s and μ_H as a function of isochronal (15-min) anneal temperature for 20- and 50-keV tellurium implantations at room temperature. (From Picraux *et al.*[55])

concentration of substitutional tellurium atoms. On the basis of these data, Picraux *et al.* suggested that the implanted tellurium has a deep level. This is in agreement with the reported measurement of the energy level of tellurium in silicon of ~0.14 eV. However, temperature-dependent Hall measurements have not yet been made to determine whether the low concentration of carriers can be attributed solely to the energy of the tellurium. In more recent work[231d] the behavior of selenium was found to be comparable to that of tellurium. The implanted selenium atoms were predominantly on substitutional sites and acted as donors, but the values of N_s were higher than for comparable tellurium implants, suggesting that selenium has a smaller ionization energy.

It is clear that much further work is required if a general picture of the behavior of unconventional dopants is to be obtained. It should be stressed that extreme care must be exercised both in implantation and in the selection of substrate material. The implantation system must be capable of good mass resolution. A trace amount of a more electrically active dopant could easily dominate the electrical behavior in cases where the value of N_s is much less than the implantation dose. Oxygen or other impurities or defects in the substrate material might become electrically active in the implanted layer. This possibility should be considered when evaluating low-dose implantations.

6

Device Considerations and Applications

R. W. BOWER†

6.1 Introduction

Ion implantation is a process that can be used to introduce controlled amounts of impurities into a layer adjacent to a semiconductor surface. In this respect it is similar to conventional semiconductor-processing techniques such as diffusion and alloy regrowth. The critical point to determine in this chapter is whether ion implantation offers any significant advantages.

Earlier chapters described in detail the distribution, damage, lattice location, and electrical behavior of ions implanted into a semiconductor surface. Compared with doped layers produced by other techniques, ion-implanted regions are generally shallower, are formed at lower temperatures, and can be masked by a wider range of materials. However, the exact doping distribution may be difficult to predict, not all implanted atoms may be electrically active, and a high density of defects may remain in the implanted region.

This chapter will first consider the primary characteristics of ion-implanted layers in semiconductors. This will include a discussion of ion range and distribution as a function of ion species, energy, and dose. Several rules of thumb are presented at this point which allow first approximations to be made, and references are given to the appropriate chapters for more detailed discussion. Next, the subject of junctions formed by ion implantation will be considered, since many applications of the ion-implantation technique involve the formation of *p-n* junctions. A noteworthy conclusion of this

† *Address*: Semiconductor Division, Hughes Research Laboratories, Newport Beach, California.

section is that planar *diffused* junctions are almost entirely specified by two parameters, surface concentration and junction depth, while planar implanted diodes depend on a variety of parameters such as dose, energy, and temperatures of implantation and anneal. These added degrees of freedom allow more process flexibility, but also lead to more difficult process control.

The applications of ion implantation to device fabrication comprise the third major topic of this chapter. The section is divided in terms of planar and nonplanar device applications. This division is felt to be essential, since any technique applied to the silicon planar process must be adaptable to the rigid requirements of this process. Planar-process applications include the gate-masked ion-implanted MOSFET, bipolar transistors, integrated circuits, and a variety of circuit elements such as resistors and controlled-breakdown diodes. For nonplanar applications of ion implantation, great flexibility is allowed, since in this case no attempt is being made to improve upon a highly developed, established technology. The nonplanar devices include a variety of diode structures. These include point-contact, hyperabrupt, IMPATT, and nuclear-particle-detector diodes.

The applications of ion implantation given in this chapter are not intended to be inclusive, but are representative of the special features and problems of this doping technique. The final section analyzes these applications and points out the special features and problems of ion-implantation doping.

6.2 Properties of Doped Layers Produced by Ion Implantation

From a semiconductor-device point of view, the major factors that are important in a doped layer are its sheet resistivity, junction location, and junction characteristics. These will be determined primarily by the net dopant and defect distributions, which control carrier densities, mobilities, and lifetimes. While a thorough understanding of all the parameters affecting these distributions is not yet available, Table 6.1 serves as a guide to our present knowledge and indicates the appropriate chapter for a more complete discussion of each point.

As illustrated in Chapter 2, the amorphous range distribution can be approximated by a Gaussian distribution with mean projected range R_p and standard deviation ΔR_p about R_p. Under most implantation conditions, over 90% of the incident ions will be contained within this amorphous range distribution. The remaining fraction of the implanted ions contribute to a deeper-penetrating tail, caused by channeling or enhanced diffusion. In Chapter 2 methods are given for calculating the mean range R_p (Fig. 2.8, Tables 2.4 and 2.7) and the width of range distribution ΔR_p (Eq. 2.8, Fig. 2.11).

The functional dependences of R_p and ΔR_p on ion energy E, ion mass

TABLE 6.1

RULES OF THUMB FOR ION IMPLANTATION IN SILICON

Ranges (*Chapter* 2)
Majority of ions stop near $R_p : n \approx N_D/2.5\,\Delta R_p \approx N_D/R_p$ (for $M_1 > M_2$).
Tail in distribution due to channeling: influences junction location.
Enhanced-diffusion effects large for Al (can be small for other species).

Disorder (*Chapter* 3)
Low Dose: individual disordered regions; anneal by 300° C.
High Dose: amorphous layer; anneals by 550–600° C.
Amorphous dose, ions/cm^2 (R.T.): $\sim 10^{14}$ Sb, $\sim 10^{15}$ Al, $>10^{16}$ B.
Temperature dependent: above 50° C for heavy ions ($M_1 \gtrsim M_2$);
above −100° C for light ions (B, N).

Lattice Location (*Chapter* 4)
Group V: mostly substitutional, can exceed solubility.
Group III: substitutional and interstitial components.

Electrical (*Chapter* 5)
Diode behavior generally after anneal at 300–350° C.
Maximum in N_s at 600° C (amorphous layer).
High-temperature anneal (800–900° C) (no amorphous layer).
$N_s \lesssim N_D$ (B, P, As, Sb); $N_s < N_D$ (Ga, Al); $N_s \ll N_D$ (In, Tl, N).

M_1, and substrate mass M_2 are quite complex. However, it is useful to note that R_p depends primarily on the ratio E/M_1 and the full width of the distribution measured at half the maximum concentration, $2.5\,\Delta R_p$, is approximately equal to R_p. For energies between 20 and 100 keV, this approximation, $2.5\,\Delta R_p \approx R_p$, holds within $\pm 30\%$ in silicon for ion species from boron to antimony.

For cases where post-annealing is complete and carrier profile follows the implanted ion distribution, the dopant atoms in this "amorphous" region will determine the approximate sheet conductivity in the layer. Figure 6.1 shows plots of sheet resistance versus implanted dose at energies of 10, 30, 100, and 300 keV for boron and phosphorus. These curves were computed from the resistivity versus carrier concentration data of Sze and Irwin[241c]. The approximation is made that all ions are contained in the amorphous distribution and that an approximate carrier concentration $\bar{n} = N_D/(2.5\,\Delta R_p)$ can be used to determine the sheet resistance. These approximations are just those described in Table 6.1. Since full electrical activity of the implanted species is required for such validity, close attention must be given to the anneal characteristics of the implanted species. The discussion of anneal characteristics in Part 5.4 and Tables 5.1 and 5.2 serves as a guideline for the properties of a variety of dopant species in silicon. For applications

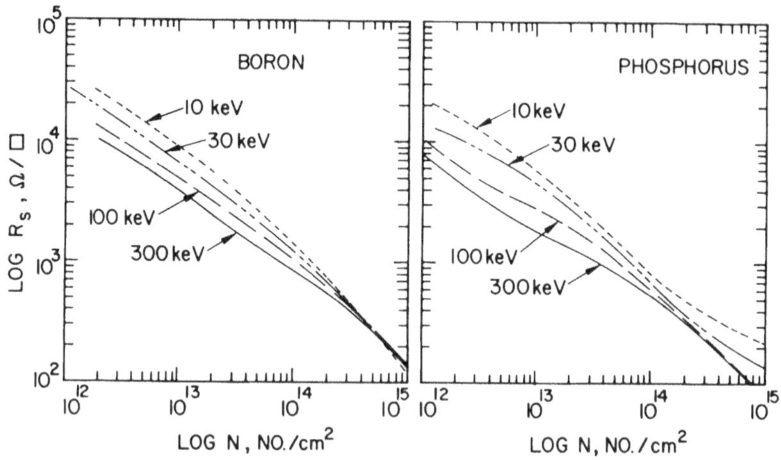

Fig. 6.1. Calculated relationship of sheet resistance R_S versus implanted dose N_D for various energies of boron and phosphorus implanted into silicon.

requiring high electrical activity, boron and phosphorus, and to a lesser extent gallium and antimony, are good choices for dopant species. However, care must be taken to choose the correct implantation conditions (dose, substrate temperature, energy) to obtain the appropriate activity at the maximum anneal temperature used during post-implantation processing.

The junction location and carrier distribution may be greatly influenced by channeling and enhanced-diffusion effects. As described in Chapter 2, it is virtually as difficult to *eliminate* channeling as it is to produce a large channeling peak. However, the junction depth may be quite consistent from one implantation to the next, as evidenced by reproducible results with implanted IMPATT diodes,[266] where junction depth is a critical parameter. For any set of implantation and anneal conditions, the best procedure is to measure the junction depth directly by angle-lap and stain techniques.

The junction characteristics of implanted diodes are not easy to describe in the general case. The junction is often located in the tail of the distribution, and the dopant distribution may be quite different at the lateral edge of the implanted layer than at the bottom.

This is quite different from the situation in diffused diodes, where the lateral distribution is the same as that at the bottom (Fig. 6.2). In this case, knowledge of the junction depth and the dopant concentration at the surface is generally sufficient to deduce the dopant distribution near the junction. Note that the radius of curvature at the "corners" of the diffused layer is an important parameter in predicting junction breakdown voltage.

In an implanted layer, the dopant distribution at the bottom interface

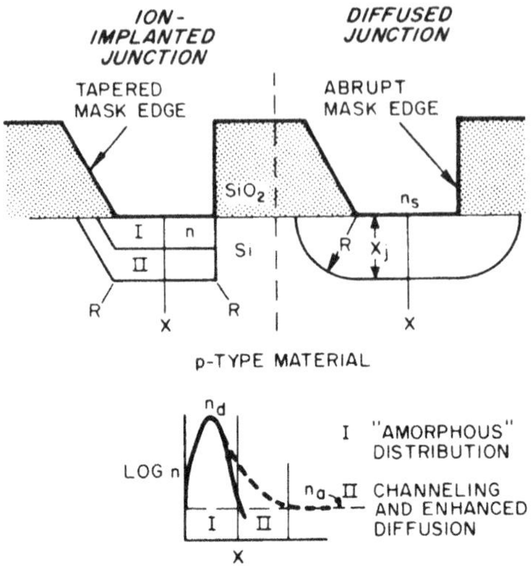

Fig. 6.2. The effect of masking on ion-implanted junctions. A perfectly abrupt mask as shown on the right will produce very abrupt lateral grading given by lateral straggling, while the junction will be graded in depth by normal straggling effects and any channeling or enhanced diffusion. Mask taper will have a major effect on the lateral junction grading, as shown on the left.

(near the junction where $N_A = N_D$) will be determined by normal range straggling and by channeling or enhanced-diffusion effects. The distribution may be quite similar to those found in graded junctions produced by diffusion. The ion distribution may be quite different on the lateral edge of the junction than on the bottom. Figure 6.2 shows the influence of the mask taper on the distribution of implanted atoms near the "sides" of the implant junction. A straight-walled (nontapered) mask will result in a very abrupt transition between the implanted and nonimplanted regions. Some lateral distribution will be introduced by the transverse range straggling (Section 2.3.5 and Sigmund and Sanders[15]) and beam divergence.

Mask taper will have a major effect[81] on the lateral distribution, as shown on the left-hand side of Fig. 6.2. The geometrical shape of the metallurgical junction surface (i.e., where $N_D = N_A$) and the dopant gradient near that surface cannot be described in as simple a fashion as for diffused structures. Consequently, the electrical characteristics (breakdown voltage, current–voltage, capacitance–voltage) of the implanted junction can be influenced by mask taper as well as range-straggling and channeling effects. The high breakdown voltage of an implanted junction with a graded dopant distribution would be lowered by a small radius of curvature and abrupt lateral distribution. However, in most implanted devices, the lateral distribution will not have a large effect on current– or capacitance–voltage characteristics because the large geometrical area of the bottom region will be dominant over that of the lateral region.

Recent work[266a] has indicated that the junction characteristics can be strongly influenced by the dose and by the temperatures of implantation and anneal. For example, in boron implantations the mode of reverse-bias breakdown can be changed by varying the dose or anneal temperature. Implanted layers with high sheet resistivity (either low dose or low anneal temperature) have high reverse breakdown, while layers with low sheet resistivity have low breakdown voltages.[266b] Much of this behavior can be attributed to the manner in which the depletion region extends into the implanted layer. Implantation and anneal temperatures have been found to have a considerable effect on forward characteristics and reverse leakage. In the forward-biased regime, diodes formed by high-temperature (500° C) implantations exhibited a marked deviation from ideal diode behavior [$I \propto \exp(eV/nkT)$, $n = 1$], while room-temperature implants followed the theory over 8 orders of magnitude. The junction leakage under reverse bias generally exhibits higher leakage currents than those found in comparable diffused junctions.

At present, the junction characteristics of implanted diodes are being investigated intensively at many laboratories. Although the diode behavior can be described under certain specific conditions, a general description is not yet available.

6.3 Applications to Planar Structures

6.3.1 Discussion of Planar Processing

The silicon planar process has several very essential features which must be considered in determining the applicability of ion-implantation doping if the economy, reliability, and reproducibility of the planar process is to be maintained. The planar process involves diffusion of donors and acceptors through oxide windows, thermal growth of silicon dioxide, vapor deposition of silicon dioxide and/or silicon nitride over thermally grown silicon dioxide, and metallization (generally aluminum) over oxide layers to provide contact to heavily doped layers. These steps are shown in Fig. 6.3. An essential feature of this process lies in the lateral spread of the diffused regions under the oxide masks to form a passivated junction region. The principle extension of the technique that is essential for bipolar devices is the epitaxial-growth process. The steps that have been evolved happened by no accident. They were forged out of severe competition by literally thousands of variations to produce reliable, economical semiconductor devices and integrated circuits. For this

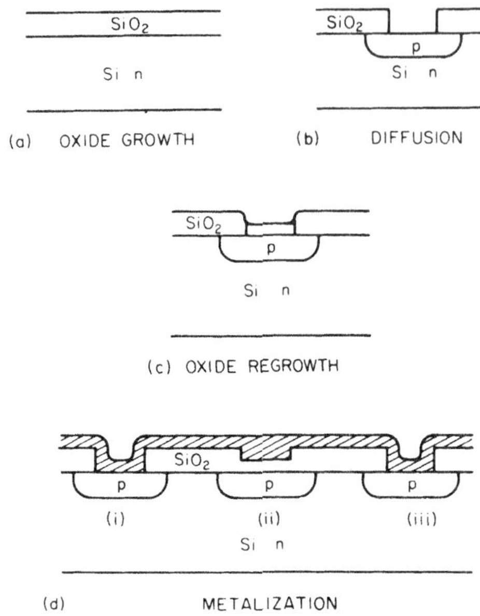

Fig. 6.3. Silicon planar process steps: (a) oxidation; (b) diffusion through oxide windows; (c) oxide regrowth, followed by reactive deposition of insulation layer; (d) metal-contact formation.

reason, it is essential to ensure that any application of ion implantation to planar technology maintains the virtues of the process.

Each of the steps described involves a temperature cycle. Diffusions generally require tens of minutes at temperatures between 1000° and 1300° C. Thermal oxide growth is typically accomplished at temperatures between 900° and 1200° C, also in tens of minutes. Similar temperature ranges are used for epitaxial growth of silicon on silicon. These three processes constitute the major high-temperature steps in the process. Vapor-deposited oxide and nitride layers are formed at temperatures of approximately 450 and 800° C respectively and thus are moderate-temperature steps. Contacts and interconnections are formed by evaporating a metallic layer at room temperature or slightly above, and then processing (sintering) at a temperature slightly below the alloy point (aluminum metallizations are generally processed at about 550° C for 10 min). Temperatures above the alloy point must be strictly avoided after the metal has been deposited on the slice.

It must be appreciated that the process sequence developed to produce a given device must pay strict attention to the time–temperature products for each step to ensure that no previous step is significantly altered. This generally requires that each successive step be accomplished at a lower temperature than the preceding one. It should be noted that for complex processes (such as are required for complementary bipolar monolithic structures where several high-temperature epitaxial-growth and diffusion steps are required) no effective means has been devised to avoid fairly serious interaction between steps. This interaction generally results in devices with seriously compromised characteristics. While certain steps such as vapor deposition of oxides or nitrides can be followed by a high-temperature step, it is generally not possible to follow metallization by a high-temperature process. These temperature-sequence constraints pose a most serious restriction on the latitude available in fabricating device structures.

6.3.2 Application of Ion-Implantation Doping to the Planar Process

Ion implantation can be applied to planar-process semiconductor doping either in terms of introducing a fixed surface density of a given dopant in a localized area or of producing a controlled concentration profile in a localized area. In the first case, ion implantation is used essentially as a predeposition step and may be followed by a high-temperature diffusion to establish the desired dopant-concentration profile. Here, direct application of normal-diffusion and thermal-oxide-growth procedures may be applied quite naturally, and no conflict with the planar process develops. In the latter case, when it is desired to establish the dopant profile by the implantation process itself, process steps involving high temperatures must be specifically avoided to prevent diffusion broadening of the profile. Therefore the standard passivation techniques of lateral diffusion under an oxide or thermal oxidation after junction formation cannot be used if the essential features of the implanted layer are to be preserved. Thus, passivation techniques for implanted layers pose an entirely different problem in some cases than those in diffusion processes.

One solution to this problem has been found by prepassivating the silicon surface with a thermally grown SiO_2 layer and subsequently implanting ions through this layer into the underlying silicon. When this prepassivation technique is used, the dopant ions must be accelerated to energies such that their mean range is well in excess of the passivation-layer thickness to ensure adequate doping of the semiconductor and to prevent severe contamination of the layer. This restricts the choice of ions to those with small mass numbers for implant energies in the range 100–300 keV. Boron, phosphorus, nitrogen, and aluminum all have a large enough range at these energies, but only boron, aluminum, and phosphorus have thus far shown suitable electrical activity for use as shallow donors and acceptors.

Adequate junction passivation and contamination control requires passivation layers of at least 1000 Å. At this thickness it is more desirable to use a

SiO_2–Si_3N_4 sandwich rather than a pure SiO_2 layer for adequate protection. Boron- and phosphorus-implanted junctions formed by this prepassivation technique have thus far exhibited excellent characteristics.[267] Leakage and breakdown of these junctions are consistent with good planar diffused diodes when consideration is taken for the radius of curvature and concentration profile.

However, there is evidence that ions implanted through a Si–SiO_2 interface can cause variations in the oxide characteristics.[268] Implantation of substantial doses of oxygen and neon, where the peak of the distribution is near the interface, leads to displacement damage, fast surface states, and doping of the semiconductor beneath the oxide. While these effects appear to anneal out at temperatures of the order of 500° C, some evidence suggests that this anneal is not complete and long-term instability may result from such radiation-damaged interfaces. Doses of less than 10^{14} boron ions/cm^2 implanted through the oxide appear to have a small effect on the interface, compared to that caused by the neon and oxygen mentioned above, although the density of fast states does increase. Although more detailed investigations are required to specify the influence of implanted ions on the characteristics of oxide-passivation layers, this prepassivation technique has thus far provided an adequate solution to this problem of forming ion-implanted junctions that are compatible with the rest of the planar process. As a rule, the thin oxide-prepassivation layers used to define the implanted region can be incorporated into the standard integrated-circuit masks, especially in the case of MOSFET circuits where the process has been applied extensively.

For most species, when the passivation step precedes metallization, anneal temperatures as high as 900° C can be used without introducing appreciable diffusion broadening. Such anneals result in virtually 100% electrical activity for species such as boron and phosphorus. However, when the implantation follows the metallization, high-temperature anneals are not generally possible. For instance, aluminum, which is commonly used for metallization, has a eutectic temperature of 575° C with silicon, and thus anneal temperatures must be kept below about 550° C. In this case, implanted dopant species such as boron or phosphorus may be far from fully electrically active† leading to high values of sheet resistivity in the implanted layers. Another consequence of low anneal temperature is that trapping and recombination levels may be present. Such levels cause the layers to have high temperature sensitivity and noise, and may create excess junction leakage. For some applications, such as the ion-implanted MOSFET, the conductance and junction characteristics of these low-temperature annealed layers ($T \approx 550°$ C) is adequate.

† This depends greatly on the implant dose and temperature. Electrically active in this context means uncompensated and fully ionized acceptors or donors.

Junction passivation is a major problem in ion implantation, although some techniques have given satisfactory results for specific device applications. Further work is required if leakage currents in implanted layers are to match those of the best diffused diodes.

6.3.3 Gate-Masked Ion-Implanted MOSFET

The application of ion implantation doping to the MOSFET technology was first described at the Electron Device Conference[268a] in October of 1966. The device, though very crude, demonstrated the concepts of low-temperature *p-n* junction formation and special masking techniques to produce a low-parasitic-capacitance structure. The gate electrode was used as the source and drain masks to form an autoregistered device. Subsequent efforts produced a stable structure completely compatible with the planar process integrated circuit technology.[267] Figure 6.4 illustrates the evolution of this device. Illustration (a) shows a standard MOSFET. The distance from source to drain is the channel length. Device performance improves as channel length is shortened; therefore high-performance devices typically have channel lengths of 5–8 microns. When conventional processing is used, the gate electrode is placed over the channel after source and drain formation. An overlap tolerance between 4 and 7 microns must be allowed to ensure that the gate always overlaps the entire channel region. This overlap adds a parasitic capacitance to the gate that can be twice as large as the active

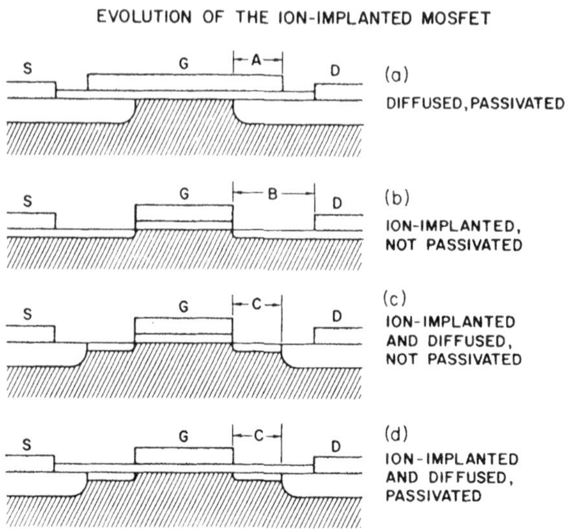

Fig. 6.4. Development steps of the gate-masked ion implanted MOSFET.

gate-to-channel capacitance. Furthermore, the Miller effect can increase effective parasitic capacitance and further reduce device performance.

Part (b) of the figure illustrates how the ion-implantation process is applied to the fabrication of this device to reduce parasitic capacitance. The gate electrode is first formed and then used as a mask to allow perfect alignment of the source and drain with respect to the gate. Although this device nicely illustrates the autoregistration technique, it is only experimental, since planar-process practice is violated by implanting into bare silicon, thus leaving unpassivated junctions. Moreover, this configuration requires separate gate and source–drain metal depositions, which is undesirable.

The next two illustrations, (c) and (d), show how the salient features of this technique are preserved while the undesirable features are eliminated. First, diffused source and drain-contact areas are formed prior to implantation to allow simultaneous deposition of the gate and source–drain contact metallization. Then, finally, the ions are implanted directly through the gate oxide layer to allow a fully passivated structure to be formed in the manner described in the previous section. This final form allows use of this technique in the whole range of planar-process applications with virtually no change in standard processing. Following this procedure, *p*-channel MOSFETs have been made with implanted boron and aluminum, and *n*-channel MOSFETs with antimony and phosphorus; however, to date, *p*-channel devices implanted with boron have been studied most extensively. Boron implantations of 10^{13} to 10^{15} ions/cm^2 at energies of 60–120 keV with implant temperatures of $-195°$ to $500°$ C have all resulted in MOSFETs with reasonably good characteristics when annealed to 545° C for 10 min.[267]

6.3.4 Application of the Gate-Masked Ion-Implantation MOS Technique to Integrated Circuits and Large-Scale Arrays

The factor-of-3 or larger reduction in total input capacitance accomplished by the ion-implantation technique relates only to the active device. In general these devices are used as integral parts of complex integrated circuits where parasitic capacitance from other sources including metal interconnections and *p-n* junction regions must also be considered. Thus, the advantage to the total circuit will be less than that for the active device. A model of MOS–NOR circuitry was studied to determine the overall circuit advantage of an ion-implanted circuit compared to one made by the conventional process. Although a full development of this analysis is not warranted in this book, Fig. 6.5 is given to illustrate the essential conclusion of the study. In this figure the pair delay τ_1, which is the time required for a digital signal to propagate through two inverter stages or logic levels, is plotted as a function of *d*, the square root of the average active-device area,

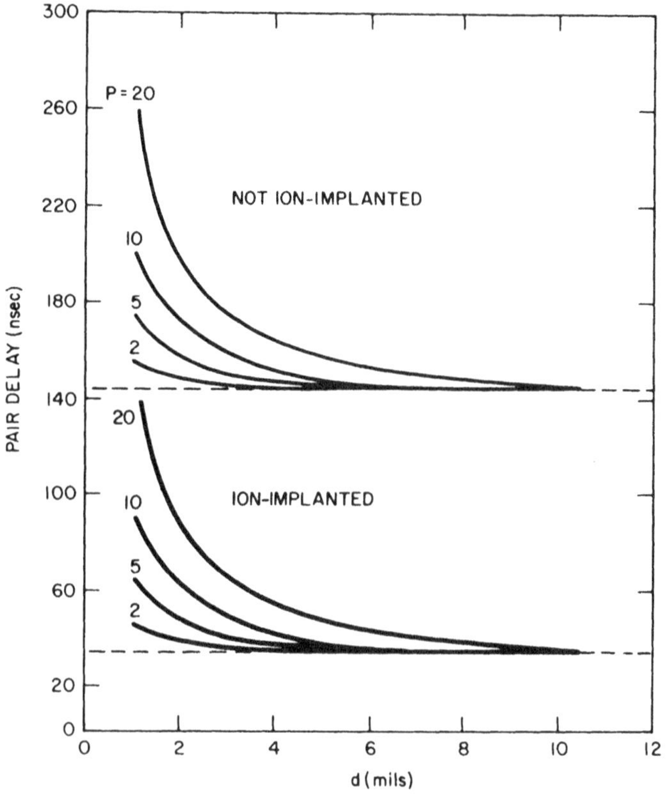

Fig. 6.5. Pair delay as a function of device size and interconnection capacitance for ion-implanted and nonimplanted devices. (From Bower and Dill.[269])

where d is measured in mils (0.001 in.). A parameter p, which is the average device-interconnection length normalized to d, is used to show the effect of interconnection capacitance. Several useful conclusions can be drawn from these curves. First, the ion-implantation technique can be used to increase conventional MOS IC circuit speed or power by approximately five times when device size is large enough for interconnection capacitance to be ignored. Also, performance can be traded for device density by making highly compact ion-implanted arrays with performance equal to that of less dense conventional arrays.

This latter consideration provides a way of producing greater functional complexity per unit area at equal performance, which directly relates to greater yield per function and thus lower cost. This analysis thus shows that the ion-implantation technique improves not only device performance but also results in improved integrated circuit performance or possibly yield.

As of this writing, shift registers, amplifiers, memories, and multiplexers have been made to advantage with ion-implanted MOS circuit techniques.[270] In Fig. 6.6† an experimental MOSFET integrated circuit fabricated with ion-implanted and conventional structures is shown. The circuit consists of a nine-stage ring oscillator and is used to evaluate the propagation delay per stage of the ion-implanted and conventional inverters. The oscillograph on

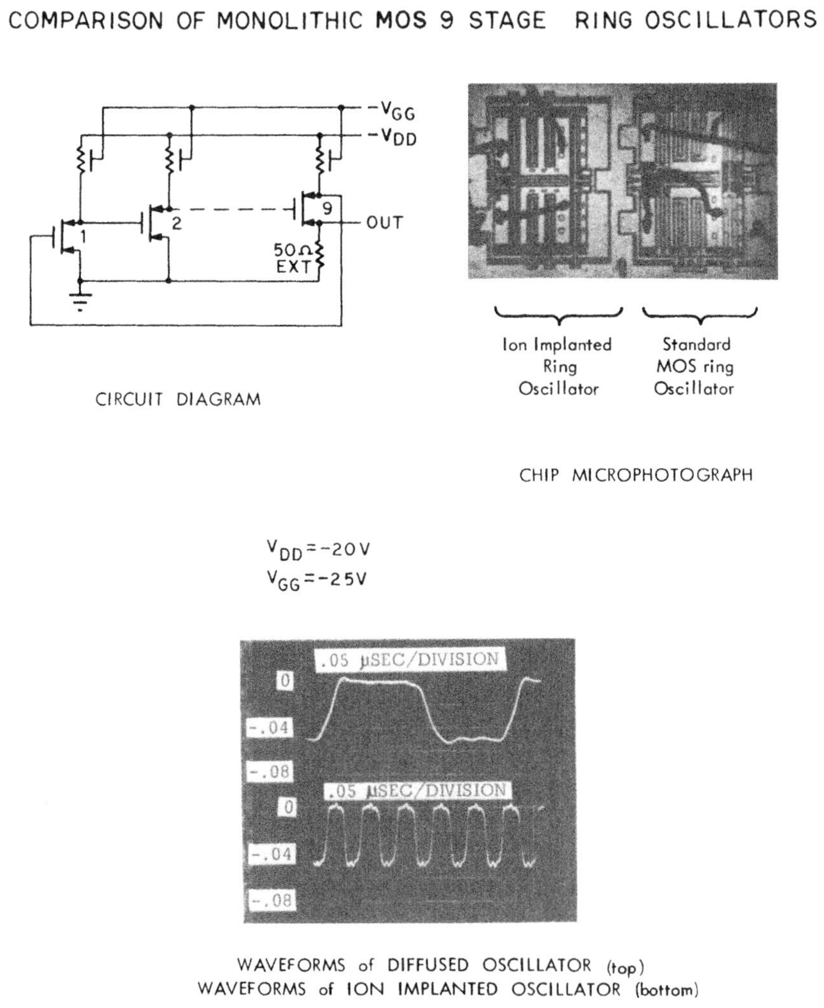

Fig. 6.6. Comparison of ion-implanted and conventional MOSFET chain oscillators.

† Circuits for Figs. 6.6–6.9 are designed and built by members of the field-effect device section of the Hughes Research Laboratories.

the right shows that the ion implanted circuit has a propagation delay of approximately $4\frac{1}{2}$ nsec per stage, whereas the diffused oscillator has a propagation delay of approximately 23 nsec per stage. This 5-to-1 speed advantage is predicted theoretically for these structures, and thus the experiment is consistent with the theory.[269] A 64-bit ion-implanted shift register, designed with the two-phase ratioless precharge technique, is shown in the photomicrograph in Fig. 6.7. This high-speed shift register has all ion-implanted

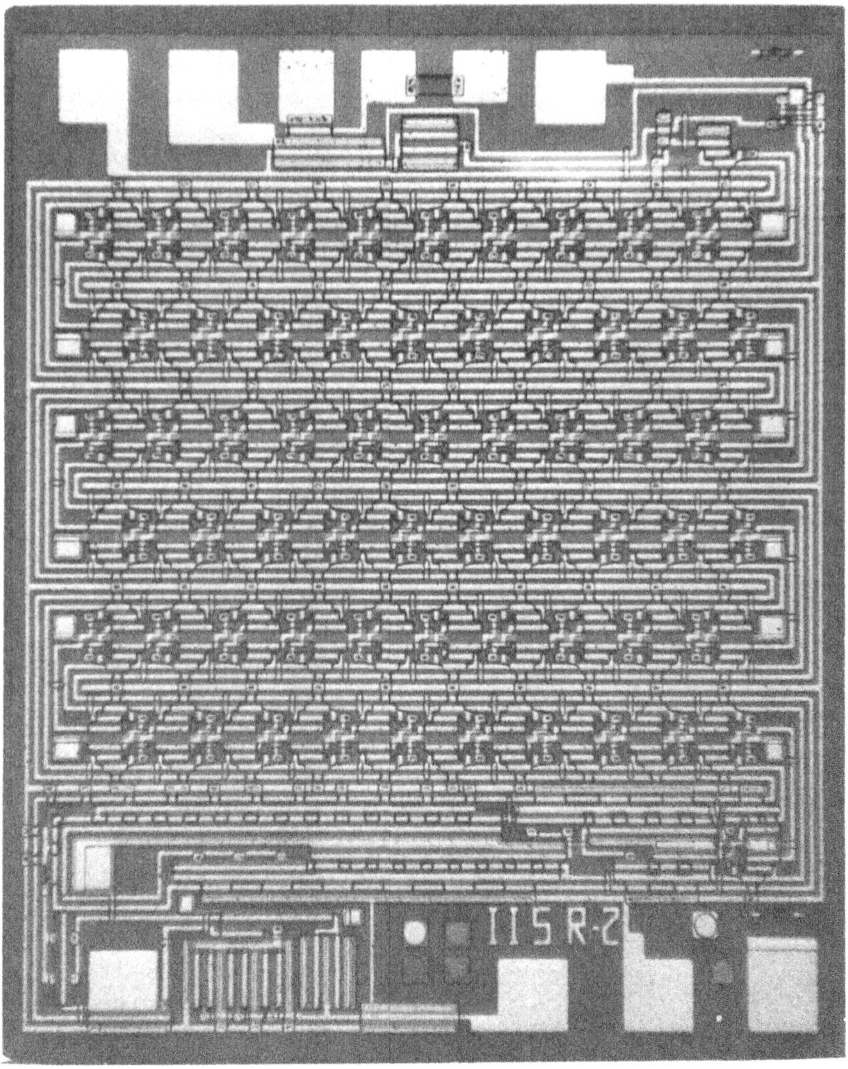

Fig. 6.7. Photomicrograph of a 64-bit ion-implanted shift register. Dimensions of the chip shown are 65×75 mils.

MOSFETs and operates at data rates up to 30 MHz with clock and power supply voltages of 20 volts and 15 volts, respectively. A dual 64-bit shift register utilizing ion-implanted resistors and MOSFETs has also been built. This circuit utilizes a two-phase ratio design and operates at essentially the same data rate and supply voltages as the shift register previously mentioned. The specifications of these ion-implanted shift registers may be contrasted with the 10-MHz data rate at 25 volts specified for the highest-performance conventional MOSFET shift register currently on the market. A ten-channel fully monolithic multiplexer utilizing ion-implanted MOSFETs operates at speeds up to 14 MHz. The shift-register cell used in this multiplexer is patterned after the two-phase ratioless cell shown in Fig. 6.7. All three of these circuits demonstrate performance significantly better than that currently available

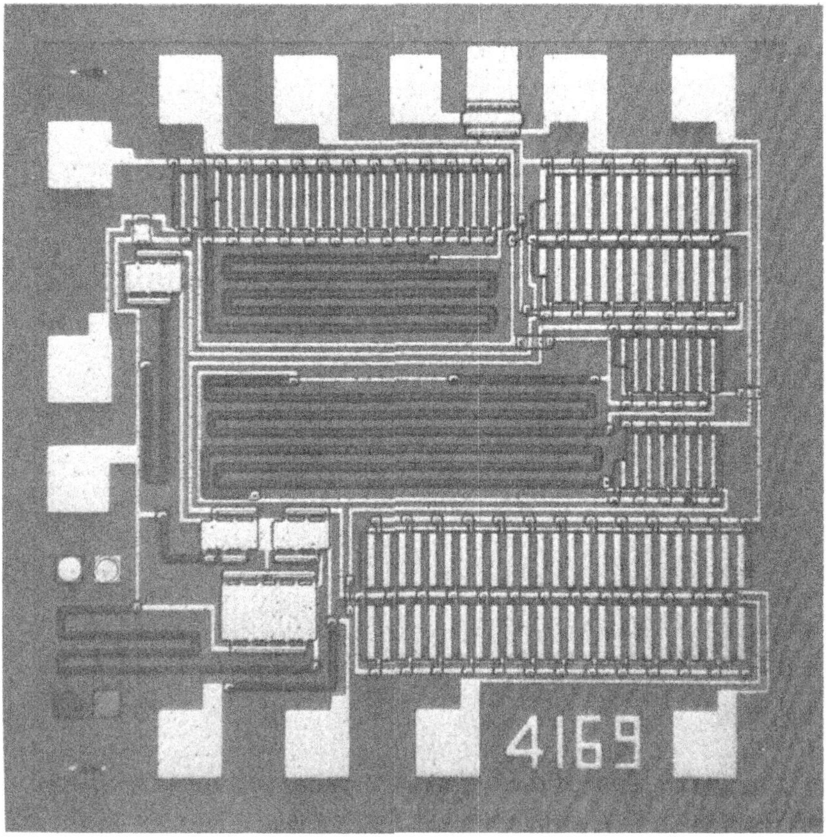

Fig. 6.8. Photomicrograph of an all-ion-implanted MOS operational amplifier. Dimensions of the chip shown are 60 × 60 mils.

with conventional MOSFETs. The 64-bit ion-implanted shift register was first available commercially in the latter part of 1969.

An all-ion-implanted MOS operational amplifier is shown in Fig. 6.8. In this circuit the resistors as well as the active MOS devices are formed by ion implantation. The resistors are fabricated without additional masking steps by implanting a thin high-resistivity layer beneath the thin gate oxide. These ion-implanted resistors provide greater linearity, lower noise, and larger dynamic range than active MOS devices used for this purpose. The performance of this amplifier is limited by basic MOS constraints, although the ion-implanted devices provide some improvement. An open-loop gain of approximately 1000 with a unity gain offset of 20 mV is typical performance for this circuit. The amplifier is useful in applications where cryogenic temperatures are encountered, or where high input impedance is required.

Figure 6.9 shows a photomicrograph of a read-only memory. The entire circuit consists of over 3000 ion-implanted MOSFETs on a chip slightly larger than 100 mils × 100 mils and includes 2048 bits of information plus encoding, decoding, and sensing. The implantation provides high-speed operation and allows a technique of writing the information pattern into the memory. This circuit is designed for 100-nsec cycle time and illustrates the high degree of complexity that is achieved with ion-implanted MOSFET integrated-circuit design as of this writing.

6.3.5 Further Application of Ion-Implantation Techniques to the MOSFET

In addition to the uses of ion implantation to produce autoregistered source and drain regions, implantation has been applied in several other ways to the MOSFET technology. Aluminum ions are implanted into the SiO_2 gate insulator to produce electron traps, resulting in a reduced sensitivity to ionizing radiation.[271] Boron ions are implanted into the surface region of the channel, thus providing a threshold-adjustment technique.[271a] Shannon *et al.*[272,272a] have implanted boron ions into the channel region of MOSFETs to allow a small channel length while the majority of the source and drain junction has the low capacitance provided by the high-resistivity *p*-type substrate. This structure is illustrated in Fig. 6.10. An offset-gate structure is made with ion implantation which has a very high drain-to-source breakdown. This device is patterned after the stacked-gate tetrode,[273] and the high-voltage operation results from the relatively low surface field provided by the weakly implanted offset-drain region. Source–drain breakdown voltages as high as 250 V on 10 Ω-cm material have been achieved with this device. Finally, Kerr[272] has reported high-mobility, depletion-mode MOSFETs produced by submerging the implanted channel beneath the

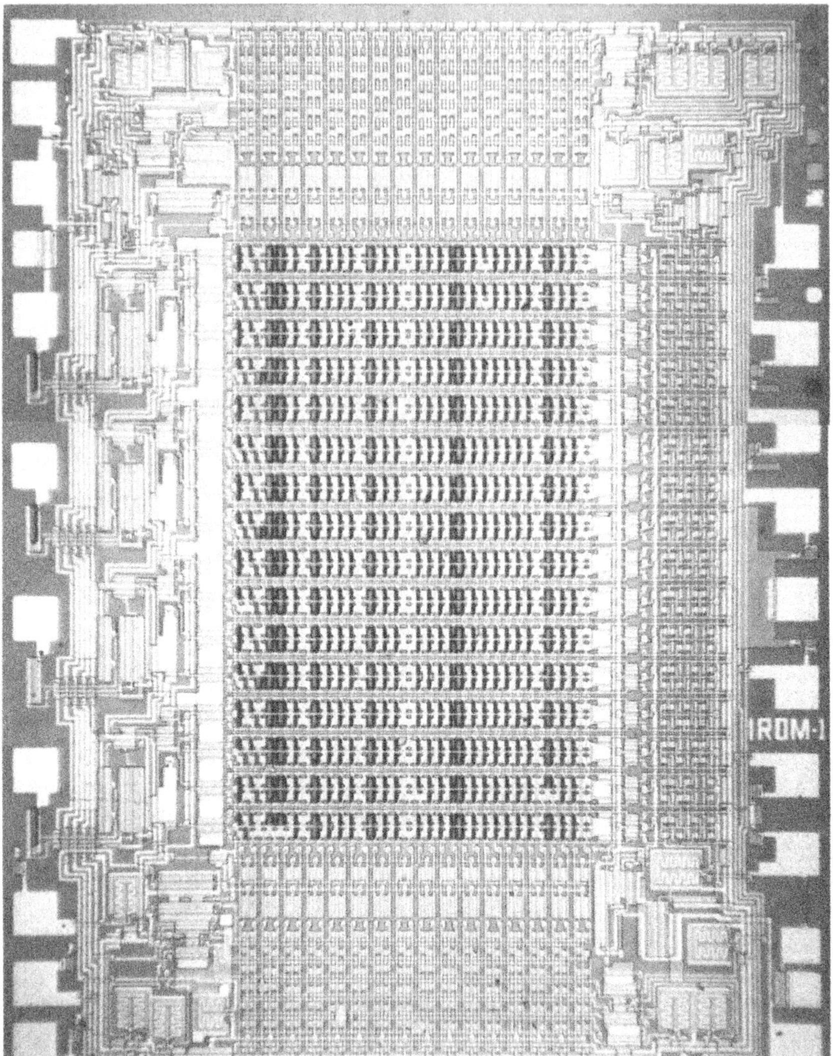

Fig. 6.9. Photomicrograph of the 2048-bit ion-implanted MOS read-only memory. Dimensions of the chip shown are 130 × 150 mils.

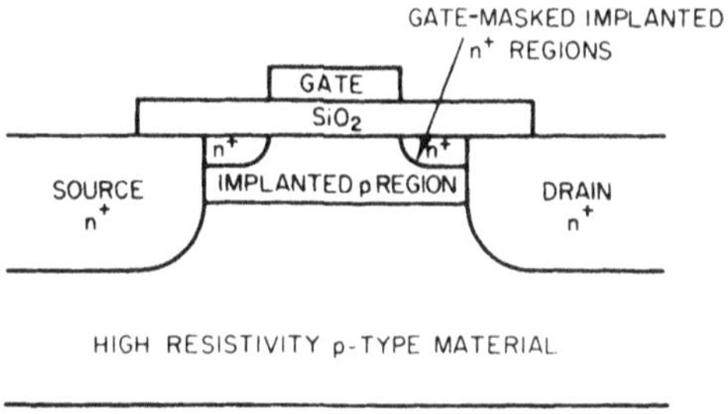

Fig. 6.10. Two ion implantations are used to construct this MOSFET. The *n*-implantation forms the self-aligned source and drain regions. The *p*-implantation is used to permit a short channel length with the high resistivity substrate material required to reduce source and drain to substrate capacitance.

Si–SiO_2 interface and avoiding surface scattering, and Lepselter *et al.*[273a] have described a self-registered MOSFET formed by ion implantation which also utilizes Schottky-barrier source and drain regions to completely eliminate high temperature diffusion.

6.3.6 Bipolar Transistors

This part has thus far treated only the application of ion implantation to MOS devices and circuits. The bipolar transistor is the major element in present-day integrated circuits. Ion-implantation doping has been applied to the fabrication of bipolar devices[274] by a number of investigators with varying degrees of success.

Numerous potential advantages appear possible in using ion-implantation doping to fabricate bipolar transistors. The low temperature of the ion-implantation doping technique should allow fabrication of *n-p-n* and *p-n-p* structures on the same slice without introducing interdependencies of multiple high-temperature diffusions. The special masking properties should allow novel geometries and process simplification to be realized. The small degree of lateral spread should provide emitters with extremely small dimensions. The reversal of process steps allows reduction or elimination of the problem of "emitter push,"[274] and profile control should allow especially high-performance devices to be made.

Although considerable progress has been made in several of these areas, the elimination of emitter push has thus far been the most convincing achievement. The emitter push arises when the heavily doped emitter junction is diffused or otherwise formed in the base region. The base is found to be "pushed" by the emitter into the collector region. This is thought to be caused[275] by the increase in vacancy concentration in the base and collector regions during emitter formation. This push is eliminated by forming the emitter region before the base is formed.[274] Ion implantation allows the base to be produced after the emitter is formed, whereas diffusion techniques do not allow such a step. This is an excellent example of process reversal that is difficult or impossible with thermal processes such as diffusion, but is easily achieved with ion implantation.

Transistors fabricated with implanted bases and/or emitters have been produced with current gains as high as 80, although 10–25 may be more typical.[274] These current gains are encouraging, although still somewhat short of the best values obtained with purely diffused devices. Implanted emitters require very high implant doses and may suffer from channeling or enhanced-diffusion tails that would reduce their injection efficiency. The implanted base, which does not require the high dose, can be used to eliminate emitter push and can be tailored in shape to produce a more ideal profile than can be produced by diffusion.[274]

This discussion indicates that ion implantation provides a promising future for bipolar-transistor fabrication; however, it will not be clear until further work is completed whether or not the technique results in a measurable advantage.

6.3.7 Applications of Ion-Implantation Doping to the Planar Integrated-Circuit Technology

The preceding two sections have dealt with the application of ion implantation to the MOSFET and bipolar devices. These two devices comprise the major three-terminal structures utilized in the planar integrated-circuit technology. A variety of other structures such as capacitors, resistors, and diodes used to complete these active devices allow a large class of linear and digital circuits to be realized. The ion-implantation technique lends itself very well to forming junction resistors and controlled-breakdown diodes.

Resistors formed by diffused *p-n* junctions have been used widely in the silicon integrated-circuit field. These resistors typically range from 100 to 1000 Ω/□ and thus are useful when the total resistor value is less than a few thousand ohms. Diffused resistors are not desirable for higher values, since large chip areas are required, and other techniques must be sought. The so called "pinch" resistors, formed by the base region of a transistor, can be made with over 20,000 Ω/□. However, the dynamic range is small and the value poorly controlled. The MOS device is also commonly used as a high-value resistor, although this approach leads to nonlinear characteristics and also tolerance problems.

High-value resistors can be formed from ion-implanted layers. These are quite reproducible, have a large dynamic range, and are linear. Such resistors can be readily made with values between 10,000 and 100,000 Ω/□. These

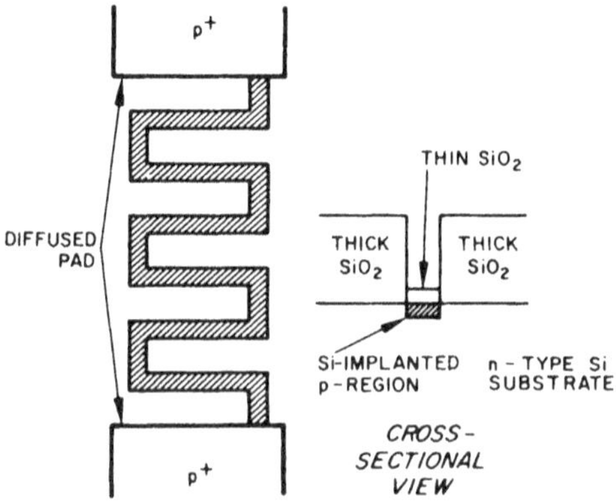

Fig. 6.11. Resistor formed by implanting ions through a thin-oxide meander line connecting two diffused pads.

resistors are formed by merely opening a controlled-dimension path through the thick oxide just before gate oxide formation, as shown in Fig. 6.11 and then regrowing the gate insulator over the exposed region. Boron ions are then implanted through the thin gate insulator. The thick insulator in the surrounding region forms a natural mask to prevent doping except in the desired thin-oxide channel.

Controlled-breakdown-voltage diodes have been formed by ion implantation. Although the junction parameters determining the breakdown voltage have not yet been specified, the most likely are sheet resistivity and gradation of the junction. Since these parameters can be greatly affected by energy, temperature of implantation, anneal, and dose, the breakdown of the resultant diode can be controlled over a wide range of values; as previously described. Diodes have been fabricated[266a] reproducibly with breakdown voltages extending from 15 to 100 V, depending on implantation conditions, by implanting boron into 1 to 3 Ω-cm silicon. Such diodes could be extremely useful as limiters, voltage references, and level shifters in MOS and bipolar integrated circuits. However, these devices have not yet been incorporated in integrated-circuit applications.

6.4 Application of Ion Implantation to Nonplanar-Process Devices

This part will deal with applications of ion-implantation doping of semiconductors to devices where the constraints of the silicon planar process are not imposed. This potentially could include the whole spectrum of solid-state devices; however, our discussion will be limited to a few of the devices that have been heavily pursued or that illustrate a particular point.

6.4.1 Point-Contact Diodes

Perhaps the first intentional use of implanted ions to improve the characteristics of a semiconductor device was reported by Ohl[276, 276a] in 1952. In this work, noble-gas ions were implanted into a low-resistivity silicon surface until the implanted layer was amorphous. This created a thin high-resistivity region on the surface. A point-contact diode was then formed by current-pulsing a wire placed on the surface until the desired *V–I* characteristics were produced. In this local region the crystal would regrow because of the heat produced by the pulsing, thus forming a low-resistance path, while the effect of the parasitic capacitance of the wire to semiconductor substrate was minimized by the large series resistance of the amorphous layer surrounding the point of contact. Although this example of ion implantation does not involve doping the crystal, it does represent a very novel and practical application of the technique. Many of the other applications may be more sophisticated than this one, but few are as simple and elegant.

6.4.2 Hyperabrupt Controlled Capacitance–Voltage Diodes

The differential capacitance per unit area versus voltage produced by an abrupt junction, such as a Schottky barrier is given approximately by $C = en(x)\, dX/dV$, where X is the depletion-layer depth and dX/dV is the rate of change of depletion-layer thickness with respect to applied voltage. If $n(x)$ is constant, then capacitance varies as the inverse square root of voltage, $V^{-1/2}$. However, if n is not constant, then the capacitance can vary more or less rapidly than $V^{-1/2}$. If n increases from the surface, the capacitance will vary more slowly than $V^{-1/2}$, and conversely if n decreases from the surface the capacitance will vary faster than $V^{-1/2}$. Diodes in which the capacitance varies faster than for $n =$ constant are known as hyperabrupt and have application as varactors. The ion-implanted doping profile is well suited for this application because the concentration can decrease very rapidly without having a surface concentration high enough to prevent Schottky-barrier formation. In the more general sense the flexibility in doping profiles that can be produced by ion implantation allows a very wide range of capacitance–voltage characteristics compared with other techniques. The rapid change in capacitance with voltage for ion-implanted layers covered with Schottky barriers was first noted in tellurium-implanted gallium arsenide in 1967.[32] More recently, several groups[277,277a] have described hyperabrupt diodes formed by ion implantation into silicon with Schottky-barrier junctions. Figure 6.12 shows the structure of these diodes. Note that the structure used by Davies[31] (Chapter 2) to determine dopant profiles is another method of making varactors. In this case Davies used a low-energy implantation instead of a metal Schottky barrier.

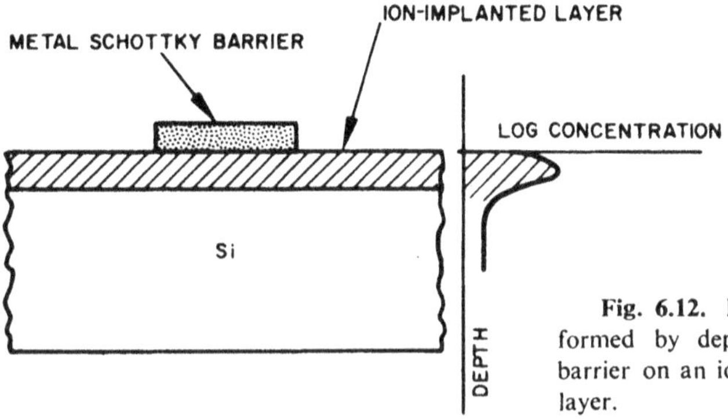

Fig. 6.12. Hyperabrupt diodes formed by depositing a Schottky barrier on an ion-implanted surface layer.

6.4.3 *p-n* JUNCTION IMPATT DIODES

As described by Ying *et al.*,[266] ion implantation has been applied to the development of microwave IMPATT diodes of the p^+-n-n^+ structure with quite positive results. Boron ions are implanted at 60 keV to a dose of 10^{15} ions/cm^2 into an n-type epitaxial layer to produce a p^+ region. After ion implantation, processing proceeds just as in the case of diffusion (e.g., mesa etch and plating). This structure is shown in Fig. 6.13. The major advantage of this technique stems from the control of the thickness of the epitaxial n-layer and the abruptness of the n^+-region resulting from the elimination of the high-temperature p^+ diffusion. This diffusion normally leads to a variation in n-layer thickness and a gradation in the n-n^+ junction. The ion implantation and anneal are accomplished in times and temperatures at which virtually no diffusion occurs. For diodes operating at frequencies of 10 GHz and below, dimensional control with diffusion is not a major problem, and ion implantation has not been proven to provide a definite advantage. However, at frequencies above 10 GHz, microwave power and efficiency are definitely improved by the ion-implantation process. High-efficiency diodes are routinely made for use at 35 and 50 GHz with implanted p^+ regions.

6.4.4 SEMICONDUCTOR NUCLEAR-PARTICLE DETECTORS

Ion-implantation techniques have been used in the fabrication of thin-window[278,278a,278b] and position-sensitive detectors.[278c,278d] In the latter application, ion implantation is uniquely suited, since a thin high-resistance (20–50 kΩ) layer about a centimeter long is required for good position definition. Laegsgaard *et al.*[278d] implanted boron ions at 60 keV into n-type silicon and adjusted the resistance between contacts by choice of ion dose (10^{13}–10^{14} ions/cm^2) and annealing temperature (400–450° C). Kalbitzer

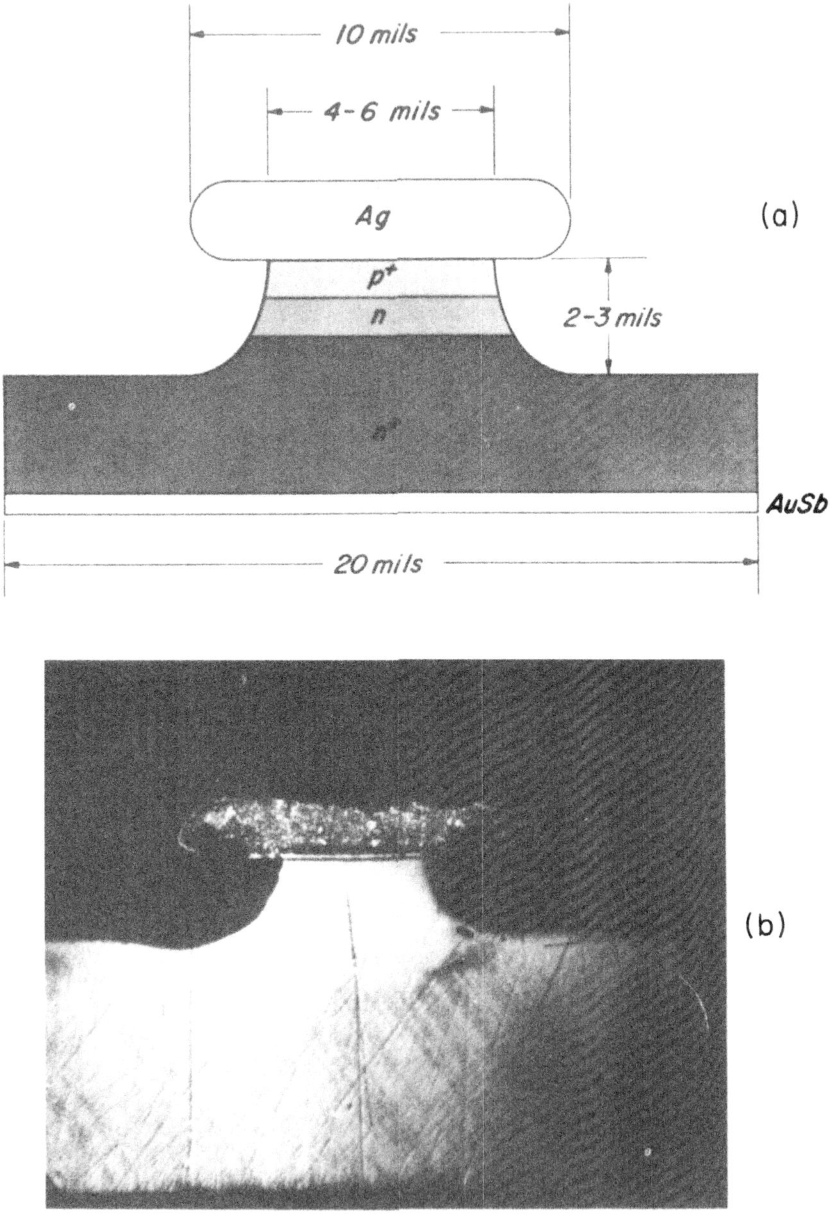

Fig. 6.13. Silicon mesa avalanche diode formed by ion implantation. (a) Illustration of the structure. (b) Angle section of an actual diode.

et al.[278c] obtained both x and y position detectors by implanting both top and bottom surfaces of the detector.

The most straightforward application of implantation is in the formation of thin, noninjecting back contacts on dE/dx detectors. In dE/dx detectors, the depletion region extends from the front surface across the device to the back contact. Blocking contacts are required. Since considerable overvoltage is generally used to ensure good charge collection, the field strength at the back contact can approach values of 10^3 V/cm. In these applications, the implanted regions must be shallow to minimize loss of carriers outside the high-field region (i.e., to minimize the window thickness). Meyer[278e,279] has found that the window thickness can extend appreciably beyond the mean range of the implanted ions (greater than a micron in some cases), and it is necessary to use low energies ($\sim$4 keV) and low doses (10^{12}–10^{13} ions/cm^2) to achieve window thicknesses less than 0.1 micron. The window thickness depends on dose, species, energy, and anneal temperature. Meyer found that boron and tellurium implantations annealed to temperatures less than 300° C give satisfactory thin-window n- and p-type blocking contacts. Implantation techniques have also been used in the fabrication of Ge (Li) detectors.[280]

Ion-implanted detectors are more stable and rugged than surface-barrier detectors, and their performance (energy resolution) is comparable to that of these devices. In applications involving exposure to large particle fluxes where detector performance degrades because of radiation damage, implanted detectors have an advantage over surface-barrier detectors in that they can be annealed to reduce the radiation damage. In Karlsruhe, it has been found that anneal treatment at 150–180° C for approximately 100 hr is sufficient to restore detector performance to its pre-irradiation value.[231e] The maximum anneal temperature is set by the epoxy used to passivate the junction.

6.5 Analysis of Ion-Implantation Device Applications

The application of ion implantation to the device structures just described illustrates many of the most important features of this doping technique.

In the case of planar devices, the gate-masked MOSFET illustrates the special masking and process-reversal features of this technique. Both the thick SiO_2 and metal areas are used as masks for the selective doping of the substrate. Furthermore, p-n junctions are formed after all other structures are completed. These special features directly result from the fact that an anneal temperature of 550° C for 10 min is adequate to produce good junctions by implanting boron through a prepassivation layer of SiO_2, and that both SiO_2 and aluminum act as an implantation mask since only material density affects ion-stopping power. The use of ion implantation to control MOS threshold vividly demonstrates the precise dopant control possible with this technique. An implanted layer with approximately 10^{11} ions/cm^2 is located within a few hundred angstroms of the Si–SiO_2 interface to properly adjust the threshold.

The use of ion implantation in bipolar-device construction illustrates the virtues of process reversal by eliminating emitter push. In this case the formation of the base region after emitter construction is the key factor, and once again results mainly from the relatively low time–temperature product necessary to form a doped layer. While precise control of the dopant profile has not yet been fully proven, the flexibility and control that are possible suggests potential improvements in bipolar devices. The unique profiles and dose control obtained by implantation have led to demonstrated advantages in the characteristics of capacitance–voltage diodes and the *p-n* junction IMPATT structures. Flexibility in the shape of doping profiles is the basis of the controlled capacitance–voltage diodes produced by implantation. This control results from the fact that doped layers can be produced by a composite of several implants at various doses, energies, and even species. This should be contrasted with diffused layers, which have a monotonically decreasing doping concentration as a function of distance into the crystal. The *p-n* junction IMPATT utilizes the low time–temperature doping sequence and shallow junctions possible with implanted layers to produce tight-tolerance junctions. This same advantage should relate to the high-frequency bipolar transistor, but has not yet been demonstrated for this use.

The high-value junction resistors result from the precise dose control that can be achieved with the ion-implantation technique. The controlled-breakdown diodes producible by implantation have their origin in the junction geometry. The combination of low lateral spread of implanted ions and small radius of junction curvature result in a wide range of possible breakdown voltage. Both energy and dose can be utilized to control the value of the breakdown voltage.

The nuclear-particle detectors produced by ion implantation demonstrate the uniformity of junction characteristics and control of dopant concentration that can be achieved by this technique. The low temperature of the implantation and anneal steps has produced detectors with the performance of surface-barrier devices and the ruggedness of junction devices.

6.6 Conclusion

This entire chapter has been devoted to a discussion of the application of ion implantation to silicon device structures. Several advantages of this doping method have been described that should result in many useful applications of the technique. Certainly, many more applications of ion implantation will occur in the future. The applications to date clearly point to the advantages of low-temperature doping, dose control, special masking, and process reversal. For the future these advantages should continue to be important, but additional features of ion implantation leading to submicron device dimensions and maskless geometry control may eventually emerge as the most significant factors. Perhaps the implantation technique can then lead us into a whole new era of semiconductor electronics.

Appendix

Channeling Behavior of Low-Z, MeV Particles in Diamond-Type Lattices

A.1 General

In recent years, much effort has been spent in studying the motion of energetic charged particles in single crystals. A channeled particle undergoes a significantly lower energy loss relative to a *random* one; it suffers an even larger reduction in the yield of any process requiring a close encounter ($\lesssim$0.2 Å) with the lattice atoms. This latter group of processes therefore provides an ideal tool for studying the channeling effect. For such studies, the diamond-type lattice is a particularly interesting structure. Within a given crystal, one may select a direction for which the steering of the channeled particles is due either to a row of uniformly spaced atoms such as the $\langle 110 \rangle$ (Fig. A.1), or to a nonuniformly spaced row such as the $\langle 111 \rangle$. Furthermore, in a compound crystal such as gallium phosphide, the gallium and phosphorus atoms lie on separate sets of atomic rows along the $\langle 110 \rangle$, whereas along the $\langle 111 \rangle$ all rows contain both gallium and phosphorus atoms. In addition, because of the asymmetric spacing along the $\langle 111 \rangle$, the [111] and $[\bar{1}\bar{1}\bar{1}]$ configurations in gallium phosphide are *not* equivalent. Similar complexities exist also for the various planar configurations (Fig. A.1).

Close-impact processes (in particular, wide-angle scattering) have been used by several groups[215,281,282] to investigate channeling behavior in the monatomic diamond lattices of silicon and germanium. Picraux *et al.*[282] extended their measurements to include diamond and several of the compound

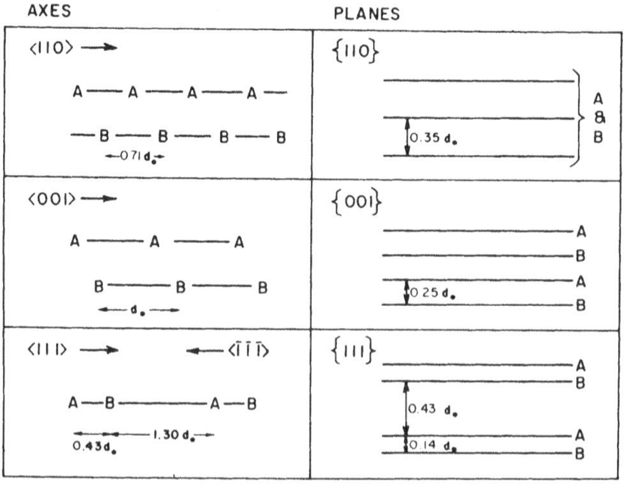

Fig. A.1. Atomic spacing in the low-index axial and planar directions of a diamond-type lattice with atoms A and B and lattice constant d_0. For monatomic lattices, $A = B$. (From Picraux *et al.*[282])

semiconductors (gallium phosphide, gallium arsenide, and gallium antimonide). There have also been quite extensive channeled energy-loss measurements in thin semiconductor crystals, particularly in silicon.[68,283–285]

The main purpose of this appendix is to compare the observed channeling behavior in various diamond-type lattices with theory. The parameters of particular importance for the analysis of implanted layers by the channeling-effect method are the *critical angle* for channeling, the *minimum yield* (i.e., the relative yield for the perfectly aligned case), the *dependence* of these two parameters on depth, and to some extent the reduced *energy loss* for channeled particles. The results show that the axial critical angles for different projectiles, energies, and semiconductor crystals have a functional dependence that is consistent with the use of average potentials—i.e., instead of treating a row of atoms as a discrete set of potential barriers, the potential may be averaged out uniformly along the row, as suggested by Lindhard.[57]

A.2 Experimental

Energy spectra of backscattered helium ions from a silicon crystal are shown in Fig. A.2(a). The energy scale of such spectra may be converted into an approximate depth scale as discussed in Section 4.2.3. Detailed orientation scans from two different depths in the silicon crystal are shown in Fig. A.2(b). These were obtained by recording the scattering yield from the narrow energy regions 1 and 2 of Fig. A.2(a) while tilting the ⟨111⟩ axis through the incident

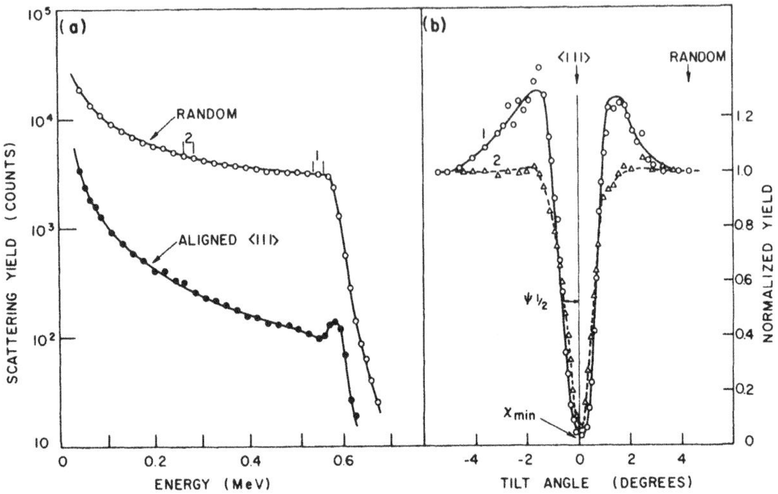

Fig. A.2. (a) Energy spectra for 1-MeV helium ions backscattered from a silicon crystal: ●, beam incident along the ⟨111⟩; O, beam incident in a random direction. The energy regions 1 and 2 indicate the two single-channel-analyzer settings used in (b). These correspond to depths of 900 and 6000 Å, respectively. (b) Orientation dependence of the normalized yield obtained from the energy regions 1 and 2 of (a). Tilt angles marked "⟨111⟩" and "random" show the orientations at which the energy spectra of (a) were taken. (From Picraux *et al.*[282])

beam direction. This figure illustrates the two main parameters of interest, viz. (1) the critical angle for channeling, defined as the angular half-width ($\psi_{1/2}$) at a level midway between the aligned and random levels, and (2) the minimum yield (χ_{min}), defined as the ratio of the yield in the perfectly aligned direction to that in a random direction. $(1 - \chi_{min})$ is therefore a measure of the channeled fraction of the beam.

Figure A.2(b) illustrates also the magnitude of the depth effect. The shape of curve 2 differs from that of curve 1, particularly in the *shoulder* region, but even the $\psi_{1/2}$ and χ_{min} values are seen to depend somewhat on the depth beneath the surface at which the measurements are made. The depth dependence of $(1 - \chi_{min})$ is illustrated in Fig. A.3. At a depth of 8000 Å in silicon, slightly more than 90% of the helium beam is still channeled. Germanium shows a somewhat larger *dechanneling* rate, due presumably to its larger vibrational amplitude relative to the screening distance *a*. In silicon, a careful investigation of dechanneling as a function of depth and temperature has recently been reported by Foti *et al.*[285a]

These channeling parameters, particularly the depth effect, depend also upon substrate temperature. In certain applications, it may be necessary to take this into account.

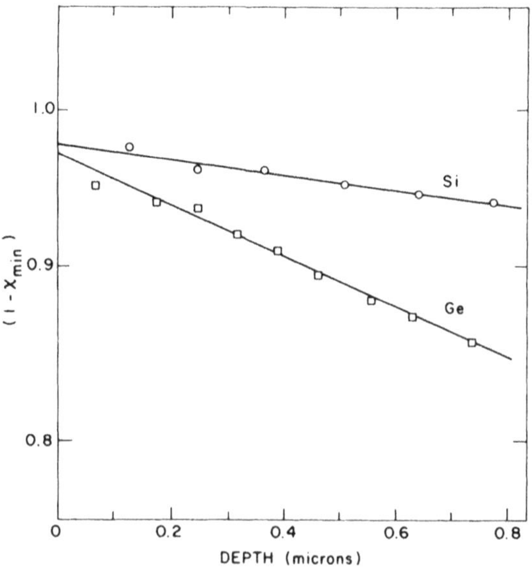

Fig. A.3. Depth dependence of the minimum yield χ_{min} for 1-MeV helium ions incident along the $\langle 111 \rangle$ axis in silicon and germanium. (From Picraux *et al.*[282])

A.3 Theory

The critical angle for axial channeling has been treated in Lindhard's theory[57] on directional effects and is predicted to be

$$(\psi_{1/2})_{axial} = \alpha\psi_1 \tag{A.1}$$

where ψ_1 is a characteristic angle expressing the functional dependence on energy, atomic number, and lattice spacing, and α is a constant that depends only on the vibrational amplitude of the lattice atoms.[286] Typical values of α range from 0.8 to 1.2.

In Lindhard's treatment, the axial rows of atoms that steer the channeled particles are treated in terms of an average potential of the atomic row or "string," and the characteristic angle ψ_1 is determined from the maximum transverse energy for which a particle may still be steered. For a monatomic row of atoms with uniform spacing, the resulting expression for ψ_1 is

$$\psi_1 = (2Z_1 Z_2 e^2/Ed)^{1/2} \tag{A.2}$$

provided $\psi_1 \lesssim a/d$,† where Z_1 and Z_2 are the atomic numbers of the projectile and target atoms respectively, e the electronic charge, a the Thomas–Fermi

† For $\psi_1 > a/d$, a less accurate expression must be used [cf. Eq. (2.11)]. For low-Z particles at MeV energies, however, the condition $\psi_1 < a/d$ is usually fulfilled, and so Eq. (A.2) is applicable.

screening distance [see Eq. (2.3) and Table 4.1], d the lattice spacing along the row, and E the projectile energy.

For rows containing more than one type of atom or for nonuniform spacing along the row (for example the ⟨111⟩ of Fig. A.1), the use of an averaged potential requires Z_2/d in Eq. (A.2) to be replaced by $\overline{Z_2/d}$, the average nuclear charge density along the row, i.e., the average atomic number per unit length.

Planar channeling may be treated in a similar manner. The use of an average sheet potential for the case of uniformly spaced planes leads to a planar critical angle of the form

$$(\psi_{1/2})_{\text{planar}} = \beta(Z_1 Z_2 e^2 N d_p a/E)^{1/2} \tag{A.3}$$

where N is the atomic density and d_p is the spacing between planes. Calculations by Picraux and Andersen[287] show that the proportionality constant β is only mildly temperature-dependent; however, it does depend somewhat on the atomic density of the chosen set of planes. Typical values of β range from 1.5 to 2.0. In the case of a plane containing more than one type of atom, Z_2 must of course be replaced by the average nuclear charge density, i.e., the average atomic number per unit area in the plane.

An estimate of the minimum yield for axial channeling is given by

$$\chi_{\min} = \pi N d\, r_{\min}^2 \tag{A.4}$$

where $r_{\min}$ is the closest distance of approach between a channeled particle and an aligned row. Rough estimates for $\chi_{\min}$ can be obtained (1) for a nonvibrating lattice, by setting $r_{\min} \simeq a$, and (2) for a lattice in which the mean-square vibrational amplitude $\overline{\rho_r^2}$ (measured perpendicular to the row) is large, by setting $r_{\min}^2 \simeq \overline{\rho_r^2}$. A better procedure, suggested by Andersen and Feldman,[288] enables the contribution from thermal vibrations to be expressed in terms of the experimentally determined ratio of $\psi_{1/2}$ to ψ_1:

$$\chi_{\min} = \pi N d(\pi a^2/4)(\psi_1/\psi_{1/2})^4 \tag{A.5}$$

Analytical expressions for depth effects (i.e., dechanneling) do not yet exist, but computer calculations have been made by several authors. These calculations are in qualitative agreement with the experiments, but much work remains to be done before accurate interpretation of depth effects will be possible. Dechanneling effects, due to the lattice disorder introduced during an implantation, were discussed briefly in Chapter 3.

A.4 Comparison between Experiment and Theory

A comparison of $\psi_{1/2}$ values for the ⟨110⟩ and ⟨111⟩ directions in a given monatomic diamond-type crystal enables the change from uniform to nonuniform spacing (Fig. A.1) to be studied, while keeping other parameters fixed. Picraux *et al.*[282] found good agreement with Eq. (A.2) for the ratio between ⟨110⟩ and ⟨111⟩ directions, using the average spacing $\bar{d}$ for the ⟨111⟩.

Absolute values of the critical angles in several diamond-type lattices are compared in Fig. A.4(a). Here the predicted dependence on energy, atomic number, and average lattice spacing has been included in the abscissa, thus enabling all the ⟨111⟩ and ⟨110⟩ measurements to be placed on the same graph. The systematic deviation observed between the diamond, silicon, and germanium data is due to different α values [Eq. (A.1)]. However, when the effect of vibrational amplitude is taken into account, using the parameter α from Andersen's calculations,[286] the agreement among the three crystals is greatly improved [Fig. A.4(b)]. (Vibrational amplitudes can be calculated from existing measurements of the characteristic Debye temperature, using the procedure outlined by Lonsdale.[289])

While the axial critical angles of Fig. A.4(b) exhibit the predicted functional dependence on Z_1, Z_2, and $\bar{d}$, they are about 20% lower in absolute magnitude than the predicted relationship (broken line). A similar discrepancy between experiment and theory has been observed in silicon at somewhat higher energies[215] and also in UO_2.[290] Nevertheless, for most channeling-effect applications, this agreement between theory and experiment is quite adequate.

In *diatomic* lattices there are two different situations (Fig. A.1): mixed rows containing both atomic species, and monatomic rows. For a mixed row such as the ⟨111⟩, the steering of the channeled particles is due to the average effect of both types of atoms in the lattice; hence, the critical angle is obtained using the average atomic number $\bar{Z} = (Z_A + Z_B)/2$ and the average spacing $\bar{d}$ of the ⟨111⟩ row. Along the ⟨110⟩ direction, however, there are individual monatomic rows of the two types of atoms in the lattice. Each of these two sets of rows is characterized by a critical angle within which it can channel the incident beam. This critical angle in turn is determined by the average potential of the individual row. Thus, near the surface, two separate critical angles—one corresponding to each of the two types of atomic rows—should exist. At larger depths in the crystal, the separation between the two critical angles becomes less distinct. Particles that are just within the critical angle of the Z_A row ($Z_A > Z_B$) have an angle greater than the critical angle of the Z_B row, and hence undergo normal multiple scattering by close encounters with the Z_B atoms; this scattering will rapidly cause them to become dechanneled with respect to the Z_A rows also. (This dechanneling effect should not strongly influence channeling-effect studies based on backscattering, since distances within 1000 Å of the surface can easily be probed.)

The existence of two separate critical angles along the ⟨110⟩ and their pronounced depth dependence are supported by the measurements of Picraux *et al.*[282] on gallium phosphide and gallium antimonide. They used back-

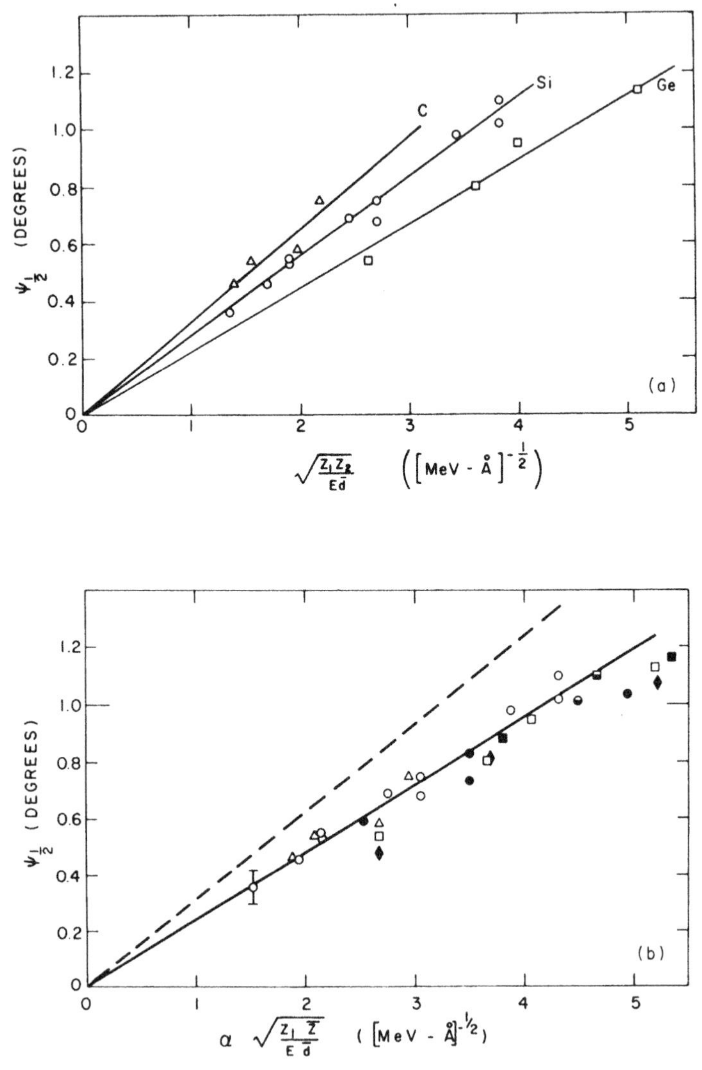

Fig. A.4. (a) The observed functional dependence of the axial critical angle ($\psi_{1/2}$) on energy (E), atomic number (Z_1, Z_2), and mean lattice spacing ($\bar{d}$) in diamond, silicon, and germanium. (From Picraux *et al.*[282]) (b) As in (a), but with Andersen's[286] calculated dependence on lattice vibrations included. Experimental data for some of the compound semiconductors are also shown: GaP—●, ⟨111⟩ and ◒, ⟨110⟩ (Ga rows); GaAs—◆, ⟨111⟩; GaSb—■, ⟨111⟩ and ⬓, ⟨110⟩ (Sb rows). The broken line is that predicted from Eqs. (A.1) and (A.2). (From Picraux *et al.*[282])

scattering *and* X-ray yield measurements in order to investigate simultaneously the interaction of an incident helium beam with both atomic species of the lattice. The data for backscattering have been included in Fig. A.4(b).

We shall not discuss here planar $\psi_{1/2}$ measurements, since they are of less importance for the applications described in Chapters 3 and 4. One can conclude, however, that the theory [Eq. (A.3)] predicts the correct functional dependence also for planar channeling.

Measured and estimated values of the minimum yield for the ⟨110⟩ direction are compared in Table A.1. For the cases of silicon, germanium,

TABLE A.1

RANDOM FRACTION (χ_{min}) ALONG THE ⟨110⟩ AXIS[a,b]

Crystal	Projectile	Rough estimates of χ_{min}: $Nd\pi\overline{\rho_r^2}$	Rough estimates of χ_{min}: $Nd\pi a^2$	χ_{min} (calculated)[c]	χ_{min} (observed)
C (diamond)	H^+	0.004	0.07	0.03	0.04
	He^+	0.004	0.06	0.03	0.05
Si	He^+	0.005	0.02	0.02	0.03
Ge	He^+	0.005	0.01	0.02	0.03
GaP	He^+	0.004_5	0.01_5	0.02	0.01_5
GaAs	He^+	0.005	0.01	0.01	0.04
GaSb	He^+	0.006	0.008	0.02	0.08

[a] Using 1-MeV H^+ or He^+ beams.
[b] Taken from Picraux *et al.*[282]
[c] Calculated from Eq. (A.5).

gallium phosphide, and diamond, the observed values of χ_{min} agree reasonably well with the values given by Eq. (A.5). In the other crystals studied (gallium antimonide, gallium arsenide), the χ_{min} values are significantly larger than predicted. It should be pointed out that surface contamination, lattice defects, etc. would always tend to *increase* the observed value of χ_{min}; such effects may perhaps account for the experimental discrepancy in gallium arsenide and gallium antimonide.

The minimum yield is a very important parameter, since it serves as a measure of the quality of a crystal. For example, it would be inadvisable to use the channeling effect to study implanted crystals, if the same crystals before implantation had shown minimum yields much larger than those predicted in Table A.1. Severe deviations from theory indicate poor crystal quality, excessive lattice disorder, surface contamination, or poor beam alignment. As was described in Chapter 3, χ_{min} measurements form the basis for the determination of lattice disorder by the channeling-effect technique. They also form the basis for locating the position of foreign atoms within the unit cell of a crystal (Chapter 4.)

References

1. J. Lindhard and M. Scharff, *Phys. Rev.* **124**, 128 (1961).
2. J. Lindhard, M. Scharff, and H. E. Schiøtt, *Kgl. Danske Videnskab Selskab. Mat. Fys. Medd.* **33**, No. 14 (1963).
3. O. S. Oen, D. K. Holmes, and M. T. Robinson, *J. Appl. Phys.* **34**, 302 (1963).
4. J. A. Davies, J. Friesen, and J. D. McIntyre, *Can. J. Chem.* **38**, 1526 (1960).
5. J. A. Davies, G. C. Ball, F. Brown, and B. Domeij, *Can. J. Phys.* **42**, 1070 (1964).
6. J. L. Whitton, *J. Appl. Phys.* **36**, 3917 (1965).
7. T. Andersen and G. Sørensen, *Can. J. Phys.* **46**, 483 (1968).
8. H. Lutz and R. Sizmann, *Z. Naturforsch.* **19a**, 1079 (1964).
9. B. G. Harvey, *Ann. Rev. Nucl. Sci.* **10**, 235 (1960).
10. Aarhus Conf. Proc. Isotope Separators and Their Application to Physics (J. Koch and K. O. Nielsen eds), *Nucl. Instr. Meth.* **38** (1965).
11. Chalk River Conf. Proc. Atomic Collision and Penetration Studies, *Can. J. Phys.* **46**, 449–782 (1968).
12. I. Bergström and B. Domeij, *Nucl. Instr. Methods* **43**, 146 (1966).
12a. G. L. Dearnaley, *Rept. Prog. Phys.* **32**, 405 (1969).
13. D. K. Nichols and V. A. Van Lint, *Solid State Phys.* **18**, 1 (1966).
14. J. F. Gibbons, *Proc. IEEE* **56**, 295 (1968).
15. P. Sigmund and J. B. Sanders, *Proc. Intern. Conf. Appl. Ion Beams Semiconductor Techn.* (P. Glotin, ed.), p. 215. Editions Ophrys, Grenoble, 1967.
16. J. B. Sanders, *Can. J. Phys.* **46**, 445 (1968).
17. L. Eriksson, J. A. Davies and P. Jespersgård, *Phys. Rev.* **161**, 219 (1967).
18. G. Dearnaley, J. H. Freeman, G. A. Gard, and M. A. Wilkins, *Can. J. Phys.* **46**, 587 (1968).
19. M. McCargo, J. A. Davies, and F. Brown, *Can. J. Phys.* **41**, 1231 (1963).
19a. J. Fairfield and B. Crowder, *Trans. AIME* **245**, 469 (1969).
20. E. Tannenbaum, *Solid State Electron.* **2**, 123 (1961).
20a. E. F. Duffek, C. Mybroic, and E. A. Benjamin, *J. Electrochem. Soc.* **111**, 1042 (1964).
20b. W. Przyborski, J. Roed, J. Lippert, and L. Sarholt-Kristensen, *Rad. Effects* **1**, 33 (1969).
21. J. L. Whitton and J. A. Davies, *J. Electrochem. Soc.* **111**, 1347 (1964).
22. R. Kelly and E. Ruedl, *Phys. Status Solidi* **13**, 55 (1966).
23. B. Domeij, F. Brown, J. A. Davies and M. McCargo, *Can. J. Phys.* **42**, 1624 (1964).
24. P. Jespersgård and J. A. Davies, *Can. J. Phys.* **45**, 2983 (1967).
25. J. P. S. Pringle, Private communication (1969), published in part in the *Ann. Conf. Mass Spectrometry, 15th, Denver, Colo.*, p. 240 (1967).
26. D. Powers and W. Whaling, *Phys. Rev.* **126**, 61 (1962).
27. D. Powers, W. K. Chu, and P. D. Bourland, *Phys. Rev.* **165**, 376 (1968).
28. W. K. Chu, P. D. Bourland, K. H. Wang, and D. Powers, *Phys. Rev.* **175**, 342 (1968).
29. E. Bøgh, Unpublished report from Aarhus Univ., Denmark (1963).
30. C. O. Thomas, D. Kahng, and R. C. Manz, *J. Electrochem. Soc.* **109**, 1055 (1962).
31. D. E. Davies, *Appl. Phys. Letters* **13**, 243 (1968).
32. J. W. Mayer, O. J. Marsh, R. Mankarious, and R. Bower, *J. Appl. Phys.* **38**, 1975 (1967).
33. R. W. Bower, R. Baron, J. W. Mayer, and O. J. Marsh, *Appl. Phys. Letters* **9**, 203 (1966).
34. C. van Opdorp, *Solid State Electron.* **11**, 397 (1968).
35. D. P. Kennedy, P. C. Murley, and W. Kleinfelder, *IBM J. Res. Develop.* **12**, 399 (1968).
36. J. A. Davies and G. A. Sims, *Can. J. Chem.* **39**, 601 (1961).

37. I. Bergström, J. A. Davies, B. Domeij, and J. Uhler, *Arkiv Fysik* **24**, 389 (1963).
38. J. A. Davies, J. D. McIntyre, R. L. Cushing, and M. Lounsbury, *Can. J. Chem.* **38**, 1535 (1960).
39. J. A. Davies, F. Brown, and M. McCargo, *Can. J. Phys.* **41**, 829 (1963).
40. N. Bohr, *Kgl. Danske Videnskab. Selskab, Mat. Fys. Medd.* **18**, No. 8 (1948).
41. K. B. Winterbon, AECL-3194, Unpublished rept. from Chalk River Nucl. Lab. (1968).
42. K. O. Nielsen, *in* "Electromagnetically Enriched Isotopes and Mass Spectroscopy" (M. L. Smith, ed.), p. 68. Academic Press, New York, 1956.
43. J. H. Ormrod and H. E. Duckworth, *Can. J. Phys.* **41**, 1424 (1963).
44. J. H. Ormrod, J. R. Macdonald, and H. E. Duckworth, *Can. J. Phys.* **43**, 275 (1965).
45. B. Fastrup, P. Hvelplund, and C. A. Sautter, *Nucl. Instr. Methods* **38**, 260 (1965).
46. B. Fastrup, P. Hvelplund, and C. A. Sautter, *Kgl. Danske Videnskab. Selskab, Mat. Fys. Medd.* **35**, No. 10 (1966).
47. B. Fastrup, A. Borup, and P. Hvelplund, *Can. J. Phys.* **46**, 489 (1968).
48. P. Hvelplund and B. Fastrup, *Phys. Rev.* **165**, 408 (1968).
49. F. Eisen, North American Rockwell Science Centre, Private communication (1969).
50. H. E. Schiøtt, *Kgl. Danske Videnskab. Selskab, Mat. Fys. Medd.* **35**, No. 9 (1966).
51. H. E. Schiøtt, *Can. J. Phys.* **46**, 449 (1968).
52. H. E. Schiøtt, Private communication (1969).
53. K. B. Winterbon, P. Sigmund, and J. B. Sanders, Submitted to *Kgl. Danske Videnskab. Selskab, Mat. Fys. Medd.* (1969).
54. C. Jech and R. Kelly, *J. Phys. Chem. Solids* **30**, 465 (1969).
55. S. T. Picraux, N. G. E. Johansson, and J. W. Mayer, *in* "Semiconductor Silicon" (R. R. Haberecht and E. L. Kern, eds.), p. 422. Electrochem. Soc., Inc., New York, 1969.
56. L. N. Large, H. Hill, and M. P. Ball, *Intern. J. Electron.* **22**, 153 (1967).
56a. A. G. Foyt, W. T. Lindley, C. M. Wolfe, and J. P. Donnelly, *Solid State Electron.* **12**, 209 (1969).
56b. C. Williamson and J. P. Boujot, CEA Rept. R-3042 (1966).
56c. J. F. Janni, Rept. No. AFWL-TR-65-150 (1966).
57. J. Lindhard, *Kgl. Danske Videnskab. Selskab, Mat. Fys. Medd.* **34**, No. 14 (1965).
58. J. W. Mayer and O. J. Marsh, *Appl. Solid State Sci.* **1**, 239 (1969).
59. J. L. Whitton, *Can. J. Phys.* **46**, 581 (1968).
60. L. Eriksson, *Phys. Rev.* **161**, 235 (1967).
61. J. F. Gibbons, A. El-Hoshy, K. E. Manchester, and F. L. Vogel, *Appl. Phys. Letters* **8**, 46 (1966).
62. W. J. Kleinfelder, W. S. Johnson, and J. F. Gibbons, *Can. J. Phys.* **46**, 597 (1968).
63. J. L. Whitton, G. Carter, J. H. Freeman, and G. A. Gard, *J. Materials Sci.* **4**, 208 (1969).
64. G. Dearnaley and M. A. Wilkins, *J. Sci. Instr.* **44**, 880 (1967).
65. J. U. Andersen, J. A. Davies, K. O. Nielsen, and S. L. Andersen, *Nucl. Instr. Methods* **38**, 210 (1965).
66. J. A. Davies, L. Eriksson, and J. L. Whitton, *Can. J. Phys.* **46**, 573 (1968).
67. J. A. Davies, L. Eriksson, and P. Jespersgård, *Nucl. Instr. Methods* **38**, 245 (1965).
68. F. H. Eisen, *Can. J. Phys.* **46**, 561 (1968).
69. O. B. Firsov, *Soviet Phys.-JETP (English Transl.)* **9**, 1076 (1959).
70. C. P. Bhalla and J. N. Bradford, *Phys. Letters* **27A**, 318 (1968).
71. I. M. Cheshire, G. Dearnaley, and J. M. Poate, *Phys. Letters* **27A**, 304 (1968).
72. A. H. El-Hoshy and J. F. Gibbons, *Phys. Rev.* **173**, 454 (1968).
72a. D. E. Harrison, *Appl. Phys. Letters* **13**, 277 (1968).
73. K. B. Winterbon, *Can. J. Phys.* **46**, 2429 (1968).

74. J. Lindhard and J. Finnemann, Unpublished report from Aarhus Univ., Denmark (1968).
75. R. P. Ruth and F. H. Eisen, *in* "Applications of Ion Beams to Semiconductor Technology," p. 539. Grenoble, 1967.
76. I. A. Galaktionova, V. M. Gusev, V. G. Naumenko, and V. V. Titov, *Soviet Phys.-Semiconductors* **2**, 656 (1968).
76a. G. Dearnaley, M. A. Wilkins, P. D. Goode, J. H. Freeman, and G. A. Gard, AERE-R6197 (1969).
76b. B. L. Crowder and J. M. Fairfield, Submitted to *J. Electrochem. Soc.*
77. K. E. Manchester, *J. Electrochem. Soc.* **115**, 656 (1968).
78. G. R. Piercy, M. McCargo, F. Brown, and J. A. Davies, *Can. J. Phys.* **42**, 1116 (1964).
79. H. Lutz and R. Sizmann, *Phys. Letters* **5**, 113 (1963).
80. J. L. Whitton, *Can. J. Phys.* **45**, 1947 (1967).
81. S. Roosild, R. Dolan, and B. Buchanan, *J. Electrochem. Soc.* **115**, 307 (1968).
82. D. E. Davies, *Can. J. Phys.* **47**, 1750 (1969).
83. D. E. Davies, *Solid State Electron.* **13**, 229 (1970).
84. P. Baruch, C. Constantin, J. C. Pfister, and R. Saintesprit, *Disc. Faraday Soc.* **31**, 76 (1961).
85. P. Baruch and J. C. Pfister. *Radiation Damage Solids* (printed by I.A.E.A.) **3**, 43 (1963).
86. J. C. Pfister, *in* "Radiation Damage in Semiconductors," p. 281. Paris-Royaumont, Dunod, 1965.
86a. D. G. Nelson, J. F. Gibbons, and W. S. Johnson, *Appl. Phys. Letters* **15**, 246 (1969).
86b. R. L. Minear, D. G. Nelson, and J. F. Gibbons, Stanford Electronics Laboratories, California, Private communication (1970).
87. A. Brelot, *in* "Radiation Effects in Semiconductors" (F. L. Vook, ed.), p. 460. Plenum Press, New York, 1968.
88. P. M. Glotin, *Can. J. Phys.* **46**, 705 (1968).
89. B. Domeij, F. Brown, J. A. Davies, G. R. Piercy, and E. V. Kornelsen, *Phys. Rev. Letters* **12**, 363 (1964).
90. E. V. Kornelsen, F. Brown, J. A. Davies, B. Domeij, and G. R. Piercy, *Phys. Rev.* **136**, A849 (1964).
91. J. A. Davies and P. Jespersgård, *Can. J. Phys.* **44**, 1631 (1966).
92. H. Hermann, H. Lutz, and R. Sizmann, *Z. Naturforsch.* **21a**, 365 (1966).
93. J. O. McCaldin, *Nucl. Instr. Methods* **38**, 153 (1965).
94. J. A. Davies, L. Eriksson, and J. W. Mayer. *Appl. Phys. Letters* **12**, 255 (1968).
95. J. O. McCaldin, *Progr. Solid State Chem.* **2**, 9 (1965).
96. H. H. Andersen and P. Sigmund, *Nucl. Instr. Methods* **38**, 238 (1965); also private communication.
96a. M. Sparks, *Phys. Rev.* **184**, 416 (1969).
97. T. Itoh, T. Inada, and K. Kanekawa, *Appl. Phys. Letters* **12**, 244 (1968).
97a. T. Itoh, and I. Ohdomari, *J. Appl. Phys.* **41**, 434 (1970).
98. O. J. Marsh, G. A. Shifrin, and R. Baron (to be published) (1970).
99. G. Carter and J. S. Colligon, "Ion Bombardment of Solids," Chapter 6, p. 204. American Elsevier, New York, 1968.
100. A. Sosin and W. Bauer, "Studies in Radiation Effects" (G. J. Dienes, ed.), Vol. 3. Gordon Breach, New York, 1969.
101. D. S. Billington and J. H. Crawford, "Radiation Damage in Solids." Princeton Univ. Press, Princeton, New Jersey, 1961.
102. J. Lindhard, V. Nielsen, M. Scharff and P. V. Thomsen, *Kgl. Danske Videnskab. Selskab, Mat. Fys. Medd.* **33**, No. 10 (1963).

103. A. R. Sattler, "Radiation Effects in Semiconductors" (F. L. Vook, ed.), p. 243. Plenum Press, New York, 1968
104. E. L. Haines and A. B. Whitehead, *Rev. Sci. Instr.* **37**, 190 (1966).
105. F. Seitz and J. S. Koehler, *Solid State Phys.* **2**, 305 (1956).
106. J. A. Brinkman, *Amer. J. Phys.* **24**, 246 (1956).
107. G. H. Kinchin and R. S. Pease, *Rept. Prog. Phys.* **18**, 1 (1955).
108. P. Sigmund, *Appl. Phys. Letters* **14**, 114 (1969).
109. R. Bauerlein, *in* "Radiation Damage in Solids" (D. S. Billington, ed.), p. 358. Academic Press, New York, 1962.
110. M. Yoshida, *J. Phys. Soc. Japan* **16**, 44 (1961).
111. J. R. Beeler, *Phys. Rev.* **150**, 470 (1966).
112. P. V. Pavlov, D. I. Tetel'baum, E. I. Zorin, and V. I. Alekseev, *Soviet Phys.-Solid State* **8**, 2141 (1967).
112a. D. K. Brice, *Appl. Phys. Letters* **16**, 103 (1970).
113. D. J. Mazey, R. S. Nelson, and R. S. Barnes, *Phil. Mag.* **17**, 1145 (1968).
114. R. S. Nelson and D. J. Mazey, *J. Mater. Sci.* **2**, 211 (1967).
115. R. S. Nelson and D. J. Mazey, *Can. J. Phys.* **46**, 689 (1968).
116. S. Kurtin, G. A. Shifrin, and T. C. McGill, *Appl. Phys. Letters* **14**, 223 (1969).
116a. T. C. McGill, S. L. Kurtin, and G. A. Shifrin, *J. Appl. Phys.* **41**, 246 (1970).
117. J. Tauc, A. Abraham, L. Pajasova, R. Grigorovici, and A. Vancu, *in Proc. Conf. Noncrystalline Solids Delft, The Netherlands*, 1964, p. 606. North-Holland Publ., Amsterdam, 1965.
118. T. M. Donovan and W. E. Spicer, *Phys. Rev. Letters* **21**, 1572 (1968).
119. R. Uyeda, ed., *Intern. Congr. Electron Microscopy 6th Kyoto* Maruzen, Tokyo, 1966.
120. R. L. Jacobson and G. K. Wehner, *J. Appl. Phys.* **36**, 2674 (1965).
121. J. J. Lander and J. Morrison, *J. Appl. Phys.* **34**, 3517 (1963)
122. A. U. MacRae and G. W. Gobeli, *J. Appl. Phys.* **35**, 1629 (1964).
123. R. E. Schlier and H. E. Farnsworth, *J. Chem. Phys.* **30**, 917 (1959).
124. J. R. Parsons, *Phil. Mag.* **12**, 1159 (1965).
125. D. J. Mazey, R. S. Barnes, and R. S. Nelson, in *Proc. Intern. Congr. Electron Microscopy 6th Kyoto* (R. Uyeda, ed.), p. 363. Maruzen, Tokyo, 1966.
126. L. N. Large and R. W. Bicknell, *J. Mater. Sci.* **2**, 589 (1967).
127. U. F. Gianola, *J. Appl. Phys.* **28**, 868 (1957).
128. P. B. Hirsch, A. Howie, R. B. Nicholson, D. W. Pashley, and M. J. Whelan, "Electron Microscopy of Thin Crystals," p. 453. Butterworths, London, 1965.
129. J. R. Parsons and R. W. Balluffi, *J. Phys. Chem. Solids* **25**, 263 (1964).
130. H. von Richter and G. Breitling. *Z. Naturforsch* **13a**, 988 (1958).
131. R. W. Bicknell, *Proc. Roy. Soc.* (*London*) Ser. A **311**, 75 (1969).
132. J. R. Parsons, Chalk River Nucl. Lab., Private communication (1968).
133. R. W. Bicknell, Plessy Co. Ltd., Private communication (1969).
134. G. Borrmann, *Z. Phys.* **127**, 297 (1950).
135. G. Borrmann, H. Hartwig, and H. Irmler, *Z. Naturforsch.* **13*a***, 423 (1958).
136. W. W. Webb, *in* "Direct Observation of Imperfections in Crystals," (J. B. Newkirk and J. H. Wernick, eds.), p. 29. Wiley (Interscience), New York, 1962.
137. F. W. Young, T. O. Baldwin, A. E. Merlini, and F. A. Sherrill, *in* "Advances in X-Ray Analysis," (G. R. Mallet, M. Fay, and W. M. Mueller, eds.), Vol. 9, p. 1. Plenum Press, New York, 1966.
138. G. H. Schwuttke, *J. Appl. Phys.* **33**, 2760 (1962).
139. G. H. Schwuttke, *J. Appl. Phys.* **36**, 2712 (1965).
140. T. O. Baldwin and J. E. Thomas, *J. Appl. Phys.* **39**, 4391 (1968).

141. G. H. Schwuttke, K. Brack, E. E. Gardner, and H. M. DeAngelis, *in* "Radiation Effects in Semiconductors" (F. L. Vook, ed.), p. 406. Plenum Press, New York, 1968
141a. U. Bonse, M. Hart, and G. H. Schwuttke, *Phys. Status Solidi* **33**, 361 (1969).
142. T. O. Baldwin, *Phys. Rev. Letters* **21**, 901 (1968).
143. G. den Ouden, *J. Appl. Phys.* **39**, 4509 (1968).
144. G. S. Anderson, G. K. Wehner, and H. J. Olin, *J. Appl. Phys.* **34**, 3492 (1963).
145. G. S. Anderson and G. K. Wehner, *Surface Sci.* **2**, 367 (1964).
146. G. S. Anderson, *J Appl. Phys.* **37**, 3455 (1966).
147. G. S. Anderson, *J. Appl. Phys.* **38**, 1607 (1967).
148. E. S. Mashkova and V. A. Molchanov, *Soviet, Phys.-Solid State* **8**, 1206 (1966).
149. E. S. Mashkova and V. A. Molchanov, *Can. J. Phys.* **46**, 713 (1968).
150. I. N. Evdokimov, E. S. Mashkova, and V. A. Molchanov, *Soviet Phys.-Solid State* **9**, 1434 (1967).
151. I. N. Evdokimov, E. S. Mashkova and V. A. Molchanov, *Phys. Letters* **25A**, 619 (1967).
152. F. L. Vook and H. J. Stein, *Rad. Effects* **2**, 23 (1969).
153. R. J. MacDonald and D. Haneman, *J. Appl. Phys.* **37**, 1609 (1966).
154. R. J. MacDonald and D. Haneman, *J. Appl. Phys.* **37**, 3048 (1966).
155. E. Bøgh, *Can. J. Phys.* **46**, 653 (1968).
156. J. A. Davies, J. Denhartog, L. Eriksson, and J. W. Mayer, *Can. J. Phys.* **45**, 4053 (1967).
157. J. W. Mayer, L. Eriksson, S. T. Picraux, and J. A. Davies, *Can. J. Phys.* **46**, 663 (1968).
158. A. C. Damask and G. J. Dienes, "Point Defects in Metals." Gordon and Breach, New York, 1963.
159. L. C. Feldman, Bell Telephone Lab., Private communication (1969).
159a. F. H. Eisen, B. Welch, J. E. Westmoreland, and J. W. Mayer, *Proc. Conf. Atomic Collision Phenomena in Solids, Sussex*, 1969.
160. M. F. Ashby and L. M. Brown, *Phil. Mag.* **8**, 1649 (1963).
160a. B. L. Crowder, R. S. Title, M. H. Brodsky, and G. D. Pettit, *Appl. Phys. Letters* **16**, 205 (1970).
161. I. Mitchell and D. Marsden, Chalk River Nucl. Lab., Private communication (1969).
162. S. T. Picraux, J. E. Westmoreland, J. W. Mayer, R. R. Hart, and O. J. Marsh, *Appl. Phys. Letters* **14**, 7 (1969).
163. G. Fladda, Res. Inst. Phys., Stockholm, Private communication (1969).
164. D. A. Marsden, G. R. Bellavance, J. A. Davies, M. Martini, and P. Sigmund, *Phys. Status Solidi* **35**, 269 (1969).
165. G. D. Watkins, *in* "Radiation Effects in Semiconductors" (F. L. Vook, ed.), p. 67. Plenum Press, New York, 1968.
165a. E. Bøgh, University of Aarhus, Private communication (1969).
166. D. Sigurd, G. Fladda, and K. Björkqvist, Res. Inst. for Phys., Stockholm. Unpublished work (1968).
167. R. S. Nelson, D. J. Mazey, M. D. Matthews, and D. F. Holloway, *Phys. Letters* **23**, 18 (1966).
168. W. M. Gibson, F. W. Martin, R. Stensgaard, F. Palmgren-Jensen, N. I. Meyer, G. Galster, A. Johansen, and J. S. Olsen, *Can. J. Phys.* **46**, 675 (1968).
169. K. Björkqvist and G. Fladda, Res. Inst. for Phys., Stockholm, Unpublished work (1969).
170. F. H. Eisen and J. E. Westmoreland, Calif. Inst. Tech., Private communication (1969).
170a. H. J. Stein, K. L. Brower, F. L. Vook, and J. A. Borders, *Appl. Phys. Letters* **16**, 106 and 108 (1970).
170b. K. L. Brower, F. L. Vook, and J. A. Borders, *Appl. Phys. Letters* **15**, 208 (1969).
171. L. Eriksson, G. R. Bellavance, and J. A. Davies, *Rad. Effects* **1**, 71 (1969).

172. L. Eriksson, J. A. Davies, J. Denhartog, J. W. Mayer, O. J. Marsh, and R. Mankarious, *Appl. Phys. Letters* **10**, 323 (1967).
173. J. W. Mayer, J. A. Davies, and L. Eriksson, *Appl. Phys. Letters* **11**, 365 (1967).
174. L. Eriksson, J. A. Davies, N. G. E. Johansson, and J. W. Mayer, *J. Appl. Phys.* **40**, 842 (1969).
174a. J. E. Westmoreland, J. W. Mayer, F. H. Eisen, and B. Welch, *Appl. Phys. Letters* **15**, 308 (1969).
175. R. S. Nelson, AERE Harwell, Private communication (1969).
176. R. R. Hart, Hughes Res. Lab., Private communication (1969).
176a. R. R. Hart and O. J. Marsh, *Appl. Phys. Letters* **15**, 206 (1969).
176b. D. Sigurd, G. Fladda, L. Eriksson, and K. Björkqvist, *Rad. Effects* (to be published).
177. H. J. Stein, F. L. Vook, and J. A. Borders, *Appl. Phys. Letters* **14**, 328 (1969).
178. L. J. Cheng and J. Lori, *Phys. Rev.* **171**, 856 (1968).
179. G. Hass and N. W. Scott, *J. Phys. Radium* **11**, 394 (1950).
180. R. N. Thomas and M. H. Francombe, *Appl. Phys. Letters* **13**, 270 (1968).
180a. E. V. Kornelsen and M. K. Sinha, *J. Appl. Phys.* **39**, 4546 (1968); **40**, 2888 (1969).
181. Hj. Matzke, *Can. J. Phys.* **46**, 621 (1968).
182. R. Kelly and Hj. Matzke, *J. Nucl. Mat.* **20**, 171 (1966).
183. R. Kelly, *Phys. Status Solidi* **21**, 451 (1967).
184. C. Jech and R. Kelly, Private communication (1969).
185. Hj. Matzke, Euratom, Ispra., Private communication (1969).
186. G. den Ouden, *Phil. Mag.* **19**, 321 (1969).
186a. D. F. Daley and K. A. Pickar, *Appl. Phys. Letters* **15**, 267 (1969).
187. J. M. Pankratz, J. A. Sprague, and M. L. Rudee, *J. Appl. Phys.* **39**, 101 (1968).
188. J. W. Mayer, *IEEE Trans.* **NS-15**, No. 6, 10 (1968).
189. L. J. Cheng, J. C. Corelli, J. W. Corbett, and G. D. Watkins, *Phys. Rev.* **152**, 761 (1966).
190. J. W. Corbett, "Electron Radiation Damage in Semiconductors and Metals." Academic Press, New York, 1966.
191. F. L. Vook, ed., *Proc. Santa Fe Conf. Rad. Effects Semiconductors.* Plenum Press, New York, 1968.
192. J. W. Cleland and J. H. Crawford, *J. Appl. Phys.* **30**, 1204 (1959).
193. B. R. Gossick, *J. Appl. Phys.* **30**, 1214 (1959).
194. H. J. Stein and R. Gereth, *J. Appl. Phys.* **39**, 2890 (1968).
195. F. L. Vook and H. J. Stein, "Radiation Effects in Semiconductors" (F. L. Vook, ed.), p. 99. Plenum Press, New York, 1968.
196. H. J. Stein, *Phys. Rev.* **163**, 801 (1967).
197. H. J. Stein, *J. Appl. Phys.* **38**, 204 (1967).
198. H. J. Stein, *IEEE Trans.* **NS-15**, No. 6, 69 (1968).
199. D. K. Wilson, *IEEE Trans.* **NS-15**, No. 6, 77 (1968).
200. R. E. Whan, *J. Appl. Phys.* **37**, 3378 (1966).
201. S. Datz, C. Erginsoy, G. Leibfried, and H. O. Lutz, *Ann. Rev. Nucl. Sci.* **17**, 129 (1967).
202. E. Bøgh and E. Uggerhøj, *Nucl. Instr. Methods* **38**, 216 (1965).
203. J. M. Khan, D. L. Potter, R. D. Worley, and H. P. Smith, *Phys. Rev.* **148**, 413 (1966); **163**, 81 (1967).
203a. E. Uggerhøj and A. Andreasen, University of Aarhus, Private communication (1969).
204. B. Domeij, *Arkiv Fysik* **32**, 179 (1966).
205. D. S. Gemmel and R. E. Holland, *Phys. Rev. Letters* **14**, 945 (1965).
206. A. F. Tulinov, V. S. Kulikauskas, and M. M. Malov, *Phys. Letters* **18**, 304 (1965).
207. B. Domeij and K. Björkqvist, *Phys. Letters* **14**, 127 (1965).

208. E. Bøgh and J. L. Whitton, *Phys. Rev. Letters* **19**, 553 (1967).
209. J. U. Andersen and E. Uggerhøj, *Can. J. Phys.* **46**, 517 (1968).
210. J. W. Mayer, *Nucl. Instr. Methods* **63**, 141 (1968).
211. J. U. Andersen, W. M. Gibson, and E. Uggerhøj, *Proc. Intern. Conf. Appl. Ion Beams Semiconductor Tech.* (P. Glotin, ed.), p. 153. Editions Orhrys, Grenoble, 1967.
212. E. Bøgh, "Interaction of Radiation with Solids," (A. Bishay, ed.), p. 361. Plenum Press, New York, 1967.
213. Hj. Matzke and J. A. Davies, *J. Appl. Phys.* **38**, 805 (1967).
213a. J. H. Patterson, A. L. Turkevich, and E. Franzgrote, *J. Geophys. Res.* **70**, 1311 (1965).
213b. D. A. Thompson, H. D. Barber, and W. D. Mackintosh, *Appl. Phys. Letters* **14**, 102 (1969).
214. W. Whaling, *Handbuch Phys.* **34**, 193 (1957).
215. J. A. Davies, J. Denhartog, and J. L. Whitton, *Phys. Rev.* **165**, 345 (1968).
215a. L. C. Feldman and B. R. Appleton, *Appl. Phys. Letters* **15**, 305 (1969).
215b. O. Meyer, J. Gyulai, and J. W. Mayer, *Surface Sci.* (to be published).
215c. J. Gyulai, O. Meyer, J. W. Mayer, and V. Rodriguez, *Appl. Phys. Letters* **16**, 232 (1970).
216. G. Fladda, P. Mazzoldi, E. Rimini, D. Sigurd, and L. Eriksson, *Rad. Effects* **1**, 249 (1969).
217. K. Björkqvist, J. A. Davies, B. Domeij, L. Eriksson, N. G. E. Johansson, and J. W. Mayer, Unpublished work (1968).
218. I. Bergström, K. Björkqvist, B. Domeij, G. Fladda, and S. Andersen, *Can. J. Phys.* **46**, 2679 (1968).
219. L. Eriksson, G. Fladda, and K. Björkqvist, *Appl. Phys. Letters* **14**, 195 (1969).
220. W. M. Gibson, J. U. Andersen, and E. Uggerhøj, *in Proc. Symp. Rad. Effects Semiconductor Components* Toulouse, France (1967).
221. J. A. Cairns and R. S. Nelson, *Phys. Letters* **27A** 14 (1968).
221a. W. F. van der Weg, and F. W. Saris, FOM Institute of Atomic and Molecular Physics, Amsterdam, Private communication (1969).
222. E. Uggerhøj and J. U. Andersen, *Can. J. Phys.* **46**, 543 (1968).
223. P. N. Tomlinson and A. Howie, *Phys. Letters* **27A**, 491 (1968).
224. G. Fladda, K. Björkqvist, and D. Sigurd, Res. Inst. for Phys., Stockholm, Unpublished work (1969).
225. I. Mitchell, Chalk River Nucl. Lab., Private communication (1969).
225a. L. Eriksson, J. A. Davies, and J. W. Mayer, "Radiation Effects in Semiconductors" (F. L. Vook, ed.), p. 398. Plenum Press, New York, 1968.
226. K. Björkqvist, B. Domeij, L. Eriksson, G. Fladda, A. Fontell, and J. W. Mayer, *Appl. Phys. Letters* **13**, 379 (1968).
227. G. D. Watkins, *in Proc. Symp. Rad. Effects Semiconductor Components* Toulouse, France (1967).
228. F. A. Trumbore, *Bell System Tech. J.* **39**, 205 (1960).
229. S. Fischler, *J. Appl. Phys.* **33**, 1615 (1962).
230. C. S. Fuller and J. A. Ditzenberger, *J. Appl. Phys.* **27**, 544 (1956).
231. N. I. Boltaks, "Diffusion in Semiconductors." Academic Press, New York, 1963.
231a. L. Eriksson, J. A. Davies, J. Denhartog, Hj. Matzke, and J. L. Whitton, *Can. Nucl. Techn.* **5**, 40 (1966).
231b. G. Fladda, K. Björkqvist, L. Eriksson, and D. Sigurd, *Appl. Phys. Letters* (to be published).
231c. J. C. North and W. M. Gibson, *Appl. Phys. Letters* **16**, 126 (1970).

231d. O. Meyer, N. G. E. Johansson, S. T. Picraux, and J. W. Mayer, *Solid State Comm.* (to be published).
231e. O. Meyer, Nuclear Research Center, Karlsruhe, Private communication (1969).
231f. G. L. Vick and K. M. Whittle, *J. Electrochem. Soc.* **116**, 1142 (1969).
232. V. I. Fistul, "Heavily Doped Semiconductors," Chapters 3, 5. Plenum Press, New York, 1969.
232a. L. N. Large, *Contemp. Phys.* **10**, 277, 505 (1969).
233. E. H. Putley, "The Hall Effect and Related Phenomena." Butterworths, London, 1960.
234. L. J. van der Pauw, *Philips Res. Repts.* **13**, 1 (1958).
235. N. G. E. Johansson, J. W. Mayer, and O. J. Marsh, *Solid State Electron.* **13**, 317 (1970).
235a. P. A. Schumann, *in* "Semiconductor Silicon" (R. R. Haberacht and E. L. Kern, eds.), p. 662. Electrochemical Society, Inc., 1969.
235b. W. R. Runyon, *in* "Silicon Semiconductor Technology," Chapter 8. McGraw-Hill, New York, 1965.
235c. K. B. Wolfstirn, *J. Phys. Chem. Solids* **16**, 279 (1960).
236. J. W. Mayer, O. J. Marsh, G. A. Shifrin, and R. Baron, *Can. J. Phys.* **45**, 4073 (1967).
237. R. Baron, G. A. Shifrin, O. J. Marsh, and J. W. Mayer, *J. Appl. Phys.* **40**, 3702 (1969).
238. A. H. Clark and K. E. Manchester, *Trans. AIME* **242**, 1173 (1968).
239. M. G. Buehler, Stanford Res. Rept. SEL-66-064 (1966).
239a. R. L. Petritz, *Phys. Rev.* **110**, 1254 (1958).
240. O. J. Marsh, R. Baron, G. A. Shifrin, and J. W. Mayer, *Appl. Phys. Letters* **13**, 199 (1968).
241. F. J. Morin and J. P. Maita, *Phys. Rev.* **96**, 28 (1959).
241a. M. D. Matthews, AERE, Harwell, Private communication [R-6047 (1969)].
241b. B. L. Crowder and J. Fairfield, *J. Electrochem. Soc.* **117**, No. 3, 363 (1970).
241c. S. M. Sze and J. C. Irvin, *Solid State Electron.* **11**, 599 (1968).
242. D. I. Tetel'baum, *Soviet Phys.-Semiconductors* **1**, 593 (1967).
243. T. E. Seidel and A. U. MacRae, *Trans. AIME* **245**, 491 (1969).
244. E. I. Zorin, P. V. Pavlov, and D. I. Tetel'baum, *Soviet Phys.-Semiconductors* **2**, 111 (1968).
245. J. Blakemore, "Semiconductor Statistics." Pergamon Press, New York, 1962.
246. N. G. E. Johansson and J. W. Mayer, *Solid State Electron.* **13**, 123 (1970).
247. P. V. Pavlov, E. I. Zorin, D. I. Tetel'baum, and E. K. Granitsyna, *Soviet Phys.-Solid State* **7**, 2386 (1966).
248. T. M. Golovner, V. V. Zadde, A. K. Zaitseva, M. M. Koltun, and A. P. Landsman, *Soviet Phys.-Semiconductors* **2**, 598 (1968).
249. J. P. Donnelly, A. G. Foyt, E. D. Hinkley, W. T. Lindley, and J. O. Dimmock, *Appl. Phys. Letters* **12**, 303 (1968).
250. R. G. Hunsperger and O. J. Marsh, *J. Electrochem. Soc.* **116**, 488 (1969).
251. A. G. Foyt, J. P. Donnelly, and W. T. Lindley, *Appl. Phys. Letters* **14**, 372 (1969).
252. V. K. Vasil'ev, E. I. Zorin, P. V. Pavlov, and D. I. Tetel'baum, *Soviet Phys.-Solid State* **9**, 1503 (1968).
253. V. M. Gusev, M. I. Guseva, V. I. Kurinny, V. G. Naumenko, V. V. Titov, and V. S. Tsyplenkov, *Proc. Conf. Rad. Effects Semiconductor Components* Toulouse, France (1967).
254. D. E. Davies, *Appl. Phys. Letters* **14**, 227 (1969).
255. R. F. Webber, R. S. Thorn, and L. N. Large, *Intern. J. Electron.* **26**, 163 (1969).
255a. B. L. Crowder and F. F. Morehead, *Appl. Phys. Letters* **14**, 313 (1969).
256. N. G. Blamires, M. D. Matthews, and R. S. Nelson, *Phys. Letters* **28A**, 178 (1968).
256a. B. L. Crowder, Private communication (1969).

257. E. I. Zorin, P. V. Pavlov, D. I. Tetel'baum, and Yu. N. Shutov, *Soviet Phys.-Semiconductors* **1**, 1051 (1968).
258. W. S. Johnson, Thesis, Stanford Univ. (1969).
258a. J. H. Shannon, R. Tree, and G. A. Gard, *Can. J. Phys.* **48**, 229 (1970).
259. K. Bulthuis, and R. Tree, *Phys. Letters* **28A**, 558 (1969).
259a. O. Meyer and J. W. Mayer, *J. Appl. Phys.* (to be published).
259b. O. Meyer and J. W. Mayer, *Solid State Electron.* (to be published).
260. P. V. Pavlov, V. A. Uskov, E. I. Zorin, D. I. Tetel'baum, and A. S. Baranova, *Soviet Phys.-Solid State* **8**, 2221 (1967).
261. A. S. Grove, "Physics and Technology of Semiconductor Devices." Wiley, New York, 1967.
262. O. J. Marsh, J. W. Mayer, G. A. Shifrin, and D. Jamba, *Appl. Phys. Letters* **11**, 92 (1967).
262a. E. Schibli, and A. G. Milnes, *Mat. Sci. Eng.* **2**, 173 (1967).
262b. E. Schibli, and A. G. Milnes, *Mat. Sci. Eng.* **2**, 229 (1967).
263. P. V. Pavlov, E. I. Zorin, D. I. Tetel'baum, and Yu. S. Popov, *Soviet Phys.-Doklady* (*English Transl.*) **10**, 786 (1966).
264. J. O. McCaldin, and A. E. Widmer, *J. Phys. Chem. Solids* **24**, 1073 (1963).
265. J. O. McCaldin, and A. E. Widmer, *Proc. IEEE* **52**, 301 (1964).
266. R. S. Ying, R. G. Mankarious, D. C. English, and R. W. Bower, *IEEE Trans. Semiconductors* **SC-3**, No. 3, 225 (1968).
266a. R. W. Bower, H. G. Dill, and R. MacPherson (to be published).
266b. R. W. Bower, To be published.
267. R. W. Bower, H. G. Dill, K. G. Aubuchon, and S. A. Thompson, *IEEE Trans. Electron Devices* **ED-15**, No. 10, 757 (1968).
268. N. J. Chou and B. L. Crowder, *J. Appl. Phys.* (to be published).
268a. R. W. Bower and H. G. Dill, Intern. Electron Device Meeting 16.6 (October 1966).
269. R. W. Bower and H. G. Dill (to be published).
270. N. E. Moyer, R. W. Bower and H. G. Dill, *Proc.* 1969 *GOMAC Conf.* (1969).
271. K. G. Aubuchon, Private communication (1968).
271a. K. G. Aubuchon, *Intern. Conf. Properties Use M.I.S. Structures* Grenoble, France (June 1969).
272. J. A. Kerr, Field Effects Device Conf., Westham Tech. College, London, England (May 1969).
272a. J. M. Shannon, J. Stephen, and J. H. Freeman, *Electronics* **42**, No. 3, 96 (1969).
273. H. G. Dill, *IEEE Trans. Electon Devices* **ED-15**, No. 10, 717 (1968).
273a. M. P. Lepselter, A. U. MacRae, and R. W. MacDonald, *Proc. IEEE* **57**, 812 (1969).
274. J. A. Kerr and L. N. Large, *Proc. Intern. Conf. Appl. Ion Beams Semiconductor-Techn.* (P. Glotin, ed.), p. 601. Editions Orhrys, Grenoble, 1969.
275. K. H. Nicholas, *Solid State Electron.* **9**, 35 (1966).
276. R. S. Ohl, *B.S.T.J.* **31**, 104 (1952).
276a. E. F. Kingsbury, and R. S. Ohl, *B.S.T.J.* **31**, 802 (1952).
277. P. Brook and C. S. Whitehead, *Electron. Letters* **4**, 16 (1968).
277a. G. F. Foxhall and R. A. Moline, To be published.
278. O. Meyer and G. Haushahn, *Nucl. Instr. Methods* **56**, 17 (1967).
278a. O. Meyer, *IEEE* **NS-51**, No. 3, 232 (1968).
278b. S. Kalbitzer, R. Bader, H. Herzer and K. Bethge, *Z. Physik* **203**, 117 (1967).
278c. S. Kalbitzer, R. Bader, W. Melzer, and W. Stumpfi, *Nucl. Instr. Methods* **54**, 223 (1967).
278d. E. Laegsgaard, F. W. Martin and W. M. Gibson, *Nucl. Instr. Methods* **60**, 24 (1968).

278e. O. Meyer, *Nucl. Instr. Methods* **64**, 279 (1969).
279. O. Meyer, *Nucl. Instr. Methods* **64**, 285 (1969).
280. G. Dearnaley, A. G. Hardacre, and B. D. Rogers, *Nucl. Instr. Methods* **71**, 86 (1969).
281. B. R. Appleton, L. C. Feldman, and W. L. Brown, Brookhaven Conf. Proc., Solid-State Physics Research with Accelerators, BNL 50083, p. 45 (1967).
282. S. T. Picraux, J. A. Davies, L. Eriksson, N. G. E. Johansson, and J. W. Mayer, *Phys. Rev.* **180**, 873 (1969).
283. A. R. Sattler and G. Dearnaley, *Phys. Rev.* **161**, 244 (1967).
284. B. R. Appleton, C. Erginsoy, and W. M. Gibson, *Phys. Rev.* **161**, 330 (1967).
285. A. R. Sattler and F. L. Vook, *Phys. Rev.* **175**, 526 (1968).
285a. G. Foti, F. Grasso, and E. Rimini, *Lettere Nuovo Cimento* **1**, 941 (1969).
286. J. U. Andersen, *Kgl. Danske Videnskab. Selskab, Mat. Fys. Medd.* **36**, No. 7 (1967).
287. S. T. Picraux and J. U. Andersen, *Phys. Rev.* **186**, 267 (1969).
288. J. U. Andersen and L. Feldman, Private communication (1968).
289. K. Lonsdale, *Acta Cryst.* **1**, 142 (1948).
290. L. Eriksson and J. A. Davies, *Arkiv Fysik* **39**, 439 (1969).

Author Index

Numbers in italics are reference numbers indicating first author and refer to the reference list beginning on p. 259. Numbers in parentheses show the pages on which citations appear. All other numbers are reference numbers and indicate that author is not the first author.

Abraham, A., 117
Alekseev, V. I., 112
Andersen, H. H., *96*(63)
Andersen, J. U., *65*(48, 128, 135), *209* (131), *211*(136), 220, 222, *286*(253, 255, 256), 287, *288*(254)
Andersen, S. L., 65, 218
Andersen, T., *7*(11, 15)
Anderson, G. S., *144*(86, 87, 107), *145*(86, 107), *146*(86, 107), *147*(86, 107)
Andreasen, A., 203a
Appleton, B. R., 215a, *281*(250), *284*(251)
Ashby, M. F., *160*(97)
Aubuchon, K. G., 267, *271*(239), *271a* (239)

Bader, R., 278b, 278c
Baldwin, T. O., 137, *140*(85, 86, 114, 119–122), *142*(86, 114, 119, 120, 122)
Ball, G. C., 5
Ball, M. P., 56
Balluffi, R. W., 129
Baranova, A. S., 260
Barber, H. D., 213b
Barnes, R. S., 113, 125
Baron, R., 33, 98, 236, *237*(185, 186, 190, 192–195, 200, 201–204, 206–213, 216, 217, 220, 221), 240
Baruch, P., *84*(59), *85*(59)
Bauer, W., 100
Bauerlein, R., *109*(72)
Beeler, J. R., *111*(73)
Bellavance, G. R., 164, 171
Benjamin, E. A., 20a
Bergström, I., *37*(20), *12*(11, 142, 143), *218*(142, 143)
Bethge, K., 278b
Bhalla, C. P., *70*(53)
Bicknell, R. W., 126, *131*(81, 83, 115, 121), *133*(83–85, 115)
Billington, D. S., *101*(66)
Björkqvist, K., 166, *169*(106, 109), 176b, 207, *217*(142, 143, 145, 171), 218, 219, 224, *226*(149, 166, 167, 170), 231b
Blakemore, J., *245*(197)
Blamires, N. G., *256*(203)
Bøgh, E., *29*(16), *155*(91, 93, 96, 97, 115, 140), *165a*(101, 138), *202*(128), *208*(131), *212*(136, 144, 149)
Bohr, N., *40*(21, 69)
Boltaks, N. I., *231*(158, 167, 206, 216)
Bonse, U., *141a*(85, 106)
Borders, J. A., 170a, 170b, 177
Borrmann, G., *134*(85), *135*(85)
Borup, A., 47
Boujot, J. P., 56b
Bourland, P. D., 27, 28

Bower, R., 32, *33*(18), 266, *266a*(229, 243), *266b*(229), *267*(232–234), *268a*(233), *269* (234–237), 270
Brack, K., 141
Bradford, J. N., 70
Breitling, G., 130
Brelot, A., *87*(59)
Brice, D. K., *112a*(75, 107)
Brinkman, J. A., *106*(71)
Brodsky, M. H., 160a
Brook, P., *277*(244)
Brower, K. L., 170a, *170b*(107, 122)
Brown, F., 5, 19, 23, 39, 78, 89, 90
Brown, L. M., 160
Brown, W. L., 281
Buchanan, B., 81
Buehler, M. G., *239*(190)
Bulthuis, K., *259*(215)

Cairns, J. A., *221*(148, 149)
Carter, G., 63, *99*(65, 66, 71)
Cheng, L. J., *178*(114, 121, 122), *189*(122)
Cheshire, I. M., *71*(53)
Chou, N. J., *268*(232)
Chu, W. K., 27, *28*(16, 20),
Clark, A. H., *238*(185, 186, 190, 197–199, 202, 203, 206)
Cleland, J. W., *192*(123)
Colligon, J. S., 99
Constantin, C., 84
Corbett, J. W., 189, *190*(122, 123)
Corelli, J. C., 189
Crawford, J. H., 101, 192
Crowder, B. L., 19a, *76b*(55, 194, 202), *160a*(99), *241b*(194, 202, 213), *255a* (202, 213), *256a*(204), 268
Cushing, R. L., 38

Daley, D. F., *186a*(121, 122)
Damask, A. C., *158*(91, 200)
Datz, S., *201*(126)
Davies, D. E., *31*(18, 57, 244), *82*(57), *83*(57, 58, 185, 219), *254*(185, 186, 202, 213, 214)
Davies, J. A., *4*(11, 15), *5*(11, 15, 42, 45, 47), 17, 19, 21, 23, 24, *36*(18, 20), 37, *38*(20), *39*(20, 91), 65, *66*(48, 60, 61), *67*(48), 78, 89, 90, *91*(60, 62), *94*(62, 149, 222), *156*(91, 98, 99, 106, 107, 110, 114, 136, 139, 143, 144, 149, 157, 171, 203, 216), 157, 164, 171, 172, 173, 174, 213, *215*(138, 250, 255), 217, 225a, 231a, 282, 290
De Angelis, H. M., 141
Dearnaley, G. L., *12a*(11), *18*(13, 42–48, 53, 54), *64*(48, 135), 71, *76a*(55, 62), *280*(247), 283
Denhartog, J., 156, 172, 215, 231a
Dienes, G. J., 158
Dill, H. G., 266a, 267, 268a, 269, 270, *273*(239)
Dimmock, J. O., 249
Ditzenberger, J. A., 230
Dolan, R., 81
Domeij, B., 5, 12, *23*(15, 31, 32), 37, *89*(60, 61), 90, *204*(130, 148), *207*(131), 217, 218, 226
Donnelly, J. P., 56a, *249*(200), 251
Donovan, T. M., *118*(77)
Duckworth, H. E., 43, 44
Duffek, E. F., *20a*(15)

Eisen, F. H., *49*(27, 53, 56), *68*(51–54, 251), 75, *159a*(94, 102, 106, 112), *170*(106, 109,) 174a
El-Hoshy, A., 61, *72*(53)
English, D. C., 266
Erginsoy, C., 201, 284
Eriksson, L., *17*(13, 42–44, 48–50), *60*(43, 44, 46, 48), 66, 67, 94, 156, 157, *171*(106, 149, 161, 200), *172*(107, 149), 173, *174*(107, 118, 141, 149–155, 158–163, 166, 170, 213), 176b, 216, 217, *219*(145, 149, 169, 170), *225a*(149,) 226, *231a*(149, 163), 231b, 282, *290*(255)
Evdokimov, I. N., *150*(87, 88, 107), *151* (87, 88, 89, 107)

Fairfield, J., *19a*(15, 215), 76b
Farnsworth, H. E., 123
Fastrup, B., *45*(27), *46*(27), *47*(27), 48
Feldman, L. C., *159*(94), *215a*(140), 281, 288
Finnemann, J., 74
Firsov, O. B., *69*(50, 52)
Fischler, S., *229*(158, 167)
Fistul, V. I., *232*(188, 221, 222)
Fladda, G., *163*(100), 166, 169, 176b, *216*(143, 149, 157, 162, 172–175), 218, 219, *224*(149, 166, 167, 168, 169, 170), 226, *231b*(149, 164, 165, 173, 175, 206)

Fontell, A., 226
Foti, G., *285a*(252)
Foxhall, G. F., *277a*(244)
Foyt, A. G., *56a*(37), 249, *251*(200)
Francombe, M. H., 180
Franzgrote, E., 213a,
Freeman, J. H., 18, 63, 76a, 272a
Friesen, J., 4
Fuller, C. S., *230*(158)

Galaktionova, I. A., *76*(53, 54, 185, 194, 203)
Galster, G., 168
Gard, G. A., 18, 63, 76a, 258a
Gardner, E. E., 141
Gemmel, D. S., *205*(130)
Gereth, R., 194
Gianola, U. F., *127*(80)
Gibbons, J. F., *14*(11, 17, 28), *61*(45, 57), 62, 72, 86a, 86b
Gibson, W. M., *168*(106, 109, 148, 149, 160, 163, 185, 186, 194, 198, 200, 201, 203, 205, 215), 211, *220*(146, 147), 231c, 278d, 284
Glotin, P. M., *88*(59, 60, 185, 200, 216)
Gobeli, G. W., 122
Golovner, T. M., *248*(199)
Goode, P. D., 76a
Gossick, B. R., *193*(123)
Granitsyna, E. K., 247
Grasso, F., 285a
Grigorovici, R., 117
Grove, A. S., *261*(218)
Gusev, V. M., 76, *253*(185, 201, 202, 205)
Guseva, M. I., 253
Gyulai, J., 215b, *215c*(140)

Haines, E. L., *104*(69, 70)
Haneman, D., 153, 154
Harrison, D. E., *72a*(53)
Hardacre, A. G., 280
Hart, M., 141a
Hart, R. R., 162, *176*(111, 215), *176a*(111, 112, 203, 215)
Hartwig, H., 135
Harvey, B. G., *9*(11)
Hass, G., *179*(114)
Haushahn, G., 278
Hermann, H., *92*(60, 62)
Herzer, H., 278b

Hill, H., 56
Hinkley, E. D., 249
Hirsch, P. B., *128*(80)
Holland, R. E., 205
Holloway, D. F., 167
Holmes, D. K., 3
Howie, A., 128, 223
Hunsperger, R. G., *250*(200)
Hvelplund, P., 45, 46, 47, *48*(27, 52)

Inada, T., 97
Irmler, H., 135
Irvin, J. C., 241c
Itoh, T., *97*(64, 80, 82, 186, 216), *97a*(64)

Jacobson, R. L., *120*(79)
Jamba, D., 262
Janni, J. F., *56c*(38)
Jech, C., *54*(36, 117, 216,) *184*(117)
Jespersgård, P., 17, *24*(15, 18, 19, 31, 32, 33, 34), 67, 91
Johansen, A., 168
Johansson, N. G. E., 55, 174, 217, 231d, *235*(186–190, 199, 215, 220), *246*(185, 186, 198, 200, 202–204, 213), 282
Johnson, W. S., 62, 86a, *258*(205)

Kahng, D., 30
Kalbitzer, S., *278b*(245), *278c*(245, 247)
Kanekawa, K., 97
Kelly, R., *22*(15), 54, *182*(117), *183*(117), 184
Kennedy, D. P., *35*(18)
Kerr, J. A., *272*(239), *274*(241, 242)
Khan, J. M., *203*(128)
Kinchin, G. H., *107*(71)
Kingsbury, E. F., *276a*(244)
Kleinfelder, W. J., 35, *62*(45, 55, 56)
Koehler, J. S., 105
Koltun, M. M., 248
Kornelsen, E. V., 89, *90*(60), *180a*(115)
Kulikauskas, V. S., 206
Kurinny, V. I., 253
Kurtin, S. L., *116*(76–78, 99), 116a

Laegsgård, E., *278d*(245)
Lander, J. J., *121*(79)
Landsman, A. P., 248
Large, L. N., *56*(37), *126*(80, 81, 83, 99, 115, 121, 182), *232a*(182), 255, 274

Leibfried, G., 201
Lepselter, M. P., *273a*(241)
Lindhard, J., *1*(11, 21), *2*(11, 21–34, 38, 52, 55, 57, 74, 211, 217), *57*(40, 126, 131, 251, 253, 254), *74*(53), *102*(67–69)
Lindley, W. T., 56a, 249, 251
Lippert, J., 20b
Lonsdale, K., *289*(255)
Lori, J., 178
Lounsbury, M., 38
Lutz, H., *8*(11, 15), *79*(55), 92, 201

MacDonald, J. R., 44
MacDonald, R. J., *153*(90, 104, 105), *154* (90, 91, 104, 105)
MacDonald, R. W., 273a
Mackintosh, W. D., 213b
MacPherson, R., 266a
MacRae, A. U., *122*(79), 243, 273a
Maita, J. P., 241
Malov, M. M., 206
Manchester, K. E., 61, *77*(55), 238
Mankarious, R., 32, 172, 266
Manz, R. C., 30
Marsden, D., 161, *164*(101, 104)
Marsh, O. J., 32, 33, 58, *98*(63, 64), 162, 172, 176a, 235, 236, 237, *240*(185, 191, 198, 200, 202, 204), 250, *262*(186)
Martin, F. W., 168, 278d
Martini, M., 164
Mashkova, E. S., *148*(87, 88, 107), *149* (87, 88, 107), 150, 151
Matthews, M. D., 167, *241a*(194, 202), 256
Matzke, Hj., *181*(117), 182, *185*(117), *213* (136), 231a
Mayer, J. W., *32*(18, 244), 33, 55, *58*(41, 54, 182, 186, 206), 94, 156, *157*(91–100, 112–116, 120, 121, 149, 154, 168, 178), 159a, 162, 172, *173*(107, 149), 174, 174a, *188*(122), *210*(135), 215b, 215c, 217, 225a, 226, 231d, 235, *236*(185, 186, 190, 192, 202, 205, 213, 219), 237, 240, 246, 259a, 259b, 262, 282
Mazey, D. J., *113*(76, 80, 81, 83, 110, 111, 115, 121), 114, 115, *125*(80, 83), 167
Mazzoldi, P., 216
McCaldin, J. O., *93*(62), *95*(62), *264*(185, 222) *265*(185, 222)
McCargo, M., *19*(14, 15, 20, 21), 23, 39, 78
McGill, T. C., 116, *116a*(76, 99)
McIntyre, J. D., 4, 38
Melzer, W., 278c
Merlini, A. E., 137
Meyer, N. I., 168
Meyer, O., *215b*(140), 215c, *231d*(149, 166, 167, 210, 223), *231e*(166, 247), *259a* (216), *259b*(216), *278*(245), *278a*(245), *278e*(247), *279*(247)
Milnes, A. G., 262a, 262b
Minear, R. L., *86b*(59)
Mitchell, I., *161*(100), *225*(149, 160, 173)
Molchanov, V. A., 148, 149, 150, 151
Moline, R. A., 277a
Morehead, F. F., 255a
Morin, F. J., *241*(192, 196, 197)
Morrison, J., 121
Moyer, N. E., *270*(236)
Murley, P. C., 35
Mybroic, C., 20a

Naumenko, V. G., 76, 253
Nelson, D. G., *86a*(59), 86b
Nelson, R. S., 113, *114*(76, 77, 102, 103), *115*(76, 102, 110), 125, *167*(102), *175* (111), 221, 256
Nicholas, K. H., *275*(241)
Nichols, D. K., *13*(11)
Nicholson, R. B., 128
Nielsen, K. O., *42*(22, 25), 65
Nielsen, V., 102
North, J. C., *231c*(149, 164, 206, 207)

Oen, O. S., *3*(11)
Ohdomari, I., 97a
Ohl, R. S., *276*(244), 276a
Olin, H. J., 144
Olsen, J. S., 168
Ormrod, J. H., *43*(27), *44*(27)
den Ouden, G., *143*(86, 119, 120), *186* (120)

Pajasova, L., 117
Palmgren-Jensen, F., 168
Pankratz, J. M., *187*(121)
Parsons, J. R., *124*(80, 81, 83, 97–99, 115, 120), *129*(80), *132*(83)
Pashley, D. W., 128
Patterson, J. H., *213a*(137)
Pavlov, P. V., *112*(73, 74, 76), 244, *247* (186, 199, 201), 252, 257, *260*(217, 218), *263*(222)

Pease, R. S., 107
Petritz, R. L., *239a*(190)
Pettit, G. D., 160a
Pfister, J. C., 84, 85, *86*(59)
Pickar, K. A., 186a
Picraux, S. T., *55*(36, 118, 149, 166, 185, 209, 210, 222, 223), 157, *162*(100, 107, 110), 231d, *282*(250–257), *287*(254)
Piercy, G. R., *78*(55), 89, 90
Poate, J. M., 71
Popov, Yu. S., 263
Potter, D. L., 203
Powers, D., *26*(16), *27*(16, 20, 33), 28
Pringle, J. P. S., *25*(15, 31)
Przyborski, W., *20b*(15)
Putley, E. H., *233*(186, 187, 190, 197)

Rimini, E., 216, 285a
Rogers, B. D., 280
Robinson, M. T., 3
Rodriguez, V., 215c
Roed, J., 20b
Roosild, S., *81*(55, 228)
Rudee, M. L., 187
Ruedl, E., 22
Runyon, W. R., *235b*(188, 216)
Ruth, R. P., *75*(53, 54, 199, 200)

Saintesprit, R., 84
Sanders, J. B., 15, *16*(12, 31), 53
Saris, F. W., 221a
Sarholt-Kristensen, L., 20b
Sattler, A. R., *103*(69), *283*(251), *285*(251)
Sautter, C. A., 45, 46
Scharff, M., 1, 2, 102
Schibli, E., *262a*(222), *262b*(222)
Schiøtt, H. E., 2, *50*(28, 29, 32, 38), *51*(28–31, 36), *52*(29, 30, 34, 35)
Schlier, R. E., *123*(79)
Schumann, P. A., *235a*(188)
Schwuttke, G. H., 141a, *138*(85), *139*(85), *141*(85, 105, 185)
Scott, N. W., 179
Seidel, T. E., *243*(186, 194, 196, 197, 201, 205, 206)
Seitz, F., *105*(71)
Shannon, J. H., *258a*(206), *272a*(239)
Sherrill, F. A., 137
Shifrin, G. A., 98, 116, 116a, 236, 237, 240, 262
Shutov, Yu. N., 257
Sigmund, P., *15*(12, 32, 34–36, 73, 75, 104, 105, 228), 53, 96, *108*(72), 164
Sigurd, D., *166*(101, 102), *176b*(111, 112), 216, 224, 231b
Sims, G. A., 36
Sinha, M. K., 180a
Sizmann, R., 8, 79, 92
Smith, H. P., 203
Sørensen, G., 7
Sosin, A., *100*(66, 71, 73)
Sparks, M., *96a*(63)
Spicer, W. E., 118
Sprague, J. A., 187
Stein, H. J., 152, *170a*(106, 122), *177*(114, 121, 122), *194*(123), 195, *196*(123), *197*(123), *198*(123)
Stensgaard, R., 168
Stephens, J., 272a
Stumpfi, W., 278c
Sze, S. M., 241c(226)

Tannenbaum, E., *20*(15)
Tauc, J., *117*(77)
Tetel'baum, D. I., 112, *242*(194, 198, 205), 244, 247, 252, 257, 260, 263
Thomas, C. O., *30*(18)
Thomas, J. E., 140
Thomas, R. N., *180*(114)
Thompson, D. A., *213b*(137)
Thompson, S. A., 267
Thomsen, P. V., 102
Thorn, R. S., 255
Title, R. S., 160a
Titov, V. V., 76, 253
Tomlinson, P. N., *223*(148)
Tree, R., 258a, 259
Trumbore, F. A., *228*(158, 167)
Tsyplenkov, V. S., 253
Tulinov, A. F., *206*(130)
Turkevich, A. L., 213a

Uggerhøj, E., 202, *203a*(129, 149), 209, 211, 220, *222*(148, 149)
Uhler, J., 37
Uskov, V. A., 260

Vancu, A., 117
Van Der Pauw, L. J., *234*(186, 188)
Van Der Weg, W. F., *221a*(148, 149)

Van Lint, V. A., 13
Van Opdorp, C., *34*(18)
Vasil'ev, V. K., *252*(185, 186, 201, 202, 205)
Vick, G. L., *231f*(178, 206)
Vogel, F. L., 61
Von Richter, H., *130*(80)
Vook, F. L., *152*(89, 111) 170a, 170b, 177, *195*(123), 285

Wang, K. H., 28
Watkins, G. D., *165*(101, 157), 189, *227* (157, 164)
Webb, W. W., *136*(85)
Webber, R. F., *255*(186, 202, 203, 206)
Wehner, G. K., 120, 144, 145
Welch, B., 159a, 174a
Westmoreland, J. E., 159a, 162, 170, *174a* (110–112, 114)
Whaling, W., 26, *214*(138)
Whan, R. E., *200*(123, 124)
Whelan, M. J., 128
Whitehead, A. B., 104
Whitehead, C. S., 277
Whittle, K. M., 231f
Whitton, J. L., *6*(11, 15), *21*(15), *59*(42, 43), *63*(46, 53), 66, *80*(55), 208, 215, 231a
Widmer, A. E., 264, 265
Wilkins, M. A., 18, 64, 76a
Williamson, C., *56b*(38)
Wilson, D. K., *199*(123)
Winterbon, K. B., *41*(21), *53*(31), *73*(53)
Wolfe, C. M., 56a
Wolfstirn, K. B., *235c*(188)
Worley, R. D., 203

Ying, R. S., *266*(227, 245)
Yoshida, M., *110*(73)
Young, F. W., *137*(85)

Zadde, V. V., 248
Zaitseva, A. K., 248
Zorin, E. I., 112, *244*(185, 194, 222), 247, 252, *257*(205), 260, 263

Subject Index

A

Alignment of single crystals, 135
Aluminum, ranges in 20, 33, 43
Aluminum–implanted Si
  disorder, 82
  electrical behavior, 183, 201–202, 210–213, 216–217, 220
  enhanced diffusion during anneal, 64, 216–217
Al_2O_3
  range straggling in, 34
  ranges in, 18, 19, 31, 32
Amorphous layer
  anneal of, 113, 114
  formation dose, various ions, 100
  influence on electrical behavior, 183, 201–205
    on lattice location, 153–155, 156
    on out-diffusion (gas release), 115, 117
Amorphous targets, *see* Range distributions, amorphous targets
Anodic oxidation, 15
Antimony–implanted Ge
  location, 167–168
Antimony–implanted Si
  disorder, 77, 98, 101, 107, 113, 150, 159
  electrical behavior, 183, 203–204, 219–220
  location, 150, 154, 158, 159
Argon–implanted Ge
  disorder (LEED), 79
  disorder distribution, low energy, 90, 104
  double scattering on Ge, 88
Arsenic–implanted Si
  disorder, 99, 113, 159
  electrical behavior, 202, 219–220
  location, 154, 158, 159
  ranges in, 55

B

Backscattering
  measurements, 136 ff, *see also* Channeling effect measurements
    depth scale, 137, 138
    detection sensitivity, 141 ff
      high Z atoms, 146
      low Z atoms, 139
    mass resolution, 143 ff
    mass scale, 138
    scattering cross section, 137
    range distribution, use in, 16
Binding energy, 72
Bipolar transistors, 241 ff
Bismuth–implanted Ge, location, 167

Bismuth–implanted Si
carrier distribution, 191, 204
disorder, 98, 104, 159
electrical, 203–204, 220
location, 154, 158–161, 169
mobility distribution, 191
Blocking, 130; *see also* Channeling behavior
Bohr radius, 21
Boron–implanted Si
disorder
depth distribution, 73, 106
dislocation loops, 84, 115
implantation temperature dependence, 100, 111
distribution (computed), 73
dopant distribution, 56, 58
dopant ionization energy, 194
electrical behavior, 194, 203, 206–207, 214–215, 219–220
location, 163–165, 173, 207
radiation sensitivity, 173, 175
ranges, 55
Z_1 correction, 27

C

Cadium–implanted Si
electrical behavior, 209–210
location, 166
out-diffusion, 118
Capacitance-voltage techniques, range distribution, 18
Carbon–implanted Ge, disorder, 106
Carrier mobility
dependence on dopant concentration, 194–195
influence of disorder on, 192, 201
Cesium–implanted Si, electrical behavior, 222
Channeling behavior
axial, 40, 253, 255
heavy ions, 38 ff; *see also* Range distributions
critical angle, 40
dechanneling, 42 ff
light ions
critical angle, 127, 252 ff
dechanneling, 93, 138, 252
minimum yield, 254, 257
planar, 254
Channeling effect measurements, *see also* Backscattering measurements
disorder, 91 ff
correction for dechanneling, 93, 94
double alignment, 96
experimental techniques, 91 ff
influence of lattice distortion, 91, 101
lattice location, 126 ff
attenuation in yield, 128, 133
experimental techniques, 131 ff
lattice positions, 133, 134
Circuit speed, 234 ff
Concentration, average, 19, 36
Critical angle, *see* Channeling behavior

D

Diamond-type lattice, *see* Lattice
Diffusion effects, *see also* Enhanced diffusion, and Out-diffusion
influence on dopant distribution, 57 ff
on electrical behavior, 215–217
on lattice location, 156, 168
Diodes
breakdown voltage, 229, 243
diffused, 227, 228
hyperabrupt, 244
IMPATT, 245
point-contact, 244
Dislocations, 83, 115
Disorder, *see also* Lattice disorder
mean surface area (LEED), 79
ranges, influence on, 46
Displacement processes, lattice atoms
average number displaced, 71 ff
energy available, 67 ff
spatial distribution, 73 ff
threshold energy, 72
Displacement spike, 71
Divacancies, 106, 114, 122
Dopant ionization energy, 194–197

E

Ejection patterns, *see* Sputtering
Elastic scattering, low energy, 87
Electrical behavior, carriers/cm^2
anneal characteristics, 198 ff

influence, of amorphous layer, 183
of channeling, 215
of compensation, 197, 201
of implantation temperature, 212–215
interstitial components, 208–212
maximum concentration values, 218–221
survey, 185–186
Electron diffraction, 78 ff
Electron emission, secondary, 86 ff
Electron transmission microscopy, 81
dislocations, 83, 84
isolated disordered regions, 83, 97
Electronic collisions, *see* Energy loss, electronic
Electronic stopping power, *see* Energy loss, electronic
Emitter-push, 241–242
Energy loss, 21
electronic, 23
Z_1 oscillations, 27
stopping power, 23, 138
nuclear, 21
lattice atom displacement, 67 ff
maximum values of, 23
range distribution, influence on, 21 ff
Energy partition, 67 ff
Enhanced diffusion, 56 ff; *see also* Diffusion effects
gas release, 116, 117
interstitial diffusion, 60
postbombardment anneal, 64, 115 ff, 216–217
radiation-enhanced, 59

F

Fast neutron irradiation
comparison with ion implantation, 118 ff
defect clusters, 123
X-ray transmission, 85
Four-point probe measurements, 188

G

Gallium–implanted Si
disorder, 98, 113
electrical behavior, 183, 188, 199, 213, 218, 220
influence of channeling, 215
location, 155, 158, 162–163, 169
Gas release, *see* Enhanced diffusion
Germanium
ejection patterns, sputtering, 87
enhanced diffusion, proton induced, 59
ions, ionization produced in Ge, 69
Gold–implanted Si, location, 166

H

Hall-effect measurements, 186 ff
Hall coefficient, 187
Hall mobility, 188
influence of carrier distribution, 190–191
layer removal studies, 192–193
range study, use in, 16
ratio of Hall to conductivity mobility, 188
technique, 186–190
Helium-ion ranges, 37, 38

I

IMPATT diodes, 227, 245
Indium–implanted Ge
disorder, 96, 99, 113
location, 167, 168, 169
Indium–implanted Si
electrical behavior, 188–189
location, 155, 158, 162–163, 169
Integrated circuits, 234 ff
Interstitial diffusion, 60

J

Junction characteristics
implanted diodes, 227–229
influence of masking, 228
passivation, 231–233
Junction location, use in range measurements, 17
Junction–staining method, 17

K

Krypton
out-diffusion in Si, Ge (gas release), 115, 117
range in Al_2O_3, 18

L

Lattice, diamond-type
atomic configuration, 38, 251
diatomic, 255
lattice positions in, 133, 134, 152
Lattice disorder, *see also* Displacement processes
amorphous layer, formation of, 98
anneal behavior, 112 ff
channeling, influence on, 46, 101 ff
depth distribution, 104 ff
dislocations, 83–85
dose dependence, 98 ff
energy dependence, 104
experimental techniques, 76 ff
implant temperatures, 107 ff
isolated regions, 81, 83, 97
measurement of, *see* Channeling effect measurements
optical indication of, 76 ff
Lattice location, *see also* Channeling effect measurements
charge state effects, 157, 170, 179–180
factors influencing location, 156–157, 176–180
germanium, Group III and Group V elements in, 167–169
implantation reactions, 177–180
influence of lattice disorder, 150, 156, 160
mixed Group III and Group V implants, 169–170
postbombardment effects, 170–173
replacement reaction, 157, 164, 173, 178
silicon
Group III elements in, 152–154, 161–165
Group V elements in, 152–154, 158–161
other ion species, 152–154, 165–167
species dependence, *see* individual ions
summary of observations, 152–157
survey, 149
Lattice parameter changes, fast neutron irradiation, 86, 119 ff
Layer-removal, 14, 15, 193
Lead-implanted Ge, location, 167
Lithium-implanted Si
disorder, 106
lattice location, nuclear reaction, 146
Low-energy electron diffraction, lattice disorder, 78 ff
LSS range-energy, 21 ff, 28

M

Masking
application to MOSFET, 233, 234
influence on junction characteristics, 227–229
Mercury-implanted Ge, Si, location, 166, 167
MOSFET, 233 ff, 239 ff
Multiplexer, 238

N

Neon-implanted Si, disorder in
influence of channeling, 103
optical indication, 77, 110
transmission microscopy, 83
Nitrogen-implanted Si
disorder, 105
dopant distribution, 56
dopant ionization energy, 194, 222
electrical behavior, 222
Nonplanar process, application of implantation to, 243 ff
Nuclear collisions, *see* Energy loss
Nuclear reactions, determination of lattice location in, 146–148

O

Operational amplifier, 239
Optical reflectivity (as indication of disorder), 76, 102, 110, 111
Oscillator, 236–237
Out-diffusion, *see also* Enhanced diffusion
gas release, 115 ff
reordering of amorphous layer, 115
Oxygen-implanted Ge, disorder, 80, 97
Oxygen-implanted Si, disorder, 81, 106
Oxygen-vacancy complex, 123

P

Passivation, 231, 232
Path-length correction, 19, 28 ff

Phosphorus-implanted Si
disorder, 98, 100, 106, 115
dopant distribution, 56, 58
dopant ionization energy, 194
electrical behavior, 198, 201–203, 205, 219–220
enhanced diffusion during implanation, 59, 60
location, 160
range distribution, 45 ff, 53
supertail, 62
Planar process, 229 ff
application of ion-implantation to, 231–232, 242–243
Polyatomic targets, *see* Range distributions
Potassium, ranges, 42, 44, 46, 61
Projected range, 19, 28 ff; *see also* Range distributions, amorphous targets
Proton ranges, 37, 38
Pulse-height defect, 69

R

Radioactive emission (blocking), 130, 131
Radio-tracer techniques, 14–16
Radon, alpha emission (blocking), in tungsten, 131
Range distributions
amorphous targets, 19 ff
average dopant concentration, 19, 36
correction factors, 27 ff
projected range, 19, 28 ff
range-energy parameters, 21, 23
range straggling, 19, 32 ff
rules of thumb, 25, 29, 34
total range, 25
variable energy, 37
measurement technique, 14–18
electrically active centers, 16–18
implanted atoms, 14–16
oxide layers, 37
polyatomic targets, 30, 31
polycrystalline targets, 18, 20, 32
Si, 53 ff
single crystals, 38 ff
channeling, 39 ff
implantation parameters, 43 ff
bombardment dose, 44–46
misalignment, 46, 47
substrate temperature, 44
surface disorder, 46, 47
maximum range, 42, 48 ff
Z_1 oscillations, 52
Range-energy relationship, 24
dimensionless parameters, 21
Range straggling, *see* Range distributions, amorphous targets
Resistors (implanted), 242–243

S

Scattering, *see* Backscattering
Secondary electron emission, lattice disorder, measurement of, 87
Selenium-implanted Si
electrical behavior, 223
location, 166
Semiconductor detectors, resolution, 142, 144
implanted, 245 ff
Sheet resistivity, 187–188
influence of dose on, 226–227
Shift register, 237–238
Silicon
electron emission (ion bombardment), 89
ionization produced in Si, 69
range distribution in, 53 ff
Z_1 oscillations in, 52
SiO_2
composition analysis, 139, 140
implantation through, 231, 232
ranges in, 37
Sodium-implanted Si
electrical behavior, 222
enhanced diffusion, 62
Spin resonance, 99, 106, 122
Sputtering
ejection patterns, 86, 87
lattice disorder, measurement of, 89, 90, 105
Stopping power, 25, 138; *see also* Energy loss
Supertail, 60, *see also* Enhanced diffusion

T

Ta_2O_3, ranges in, 31, 32
Tellerium-implanted Si
electrical behavior, 222–223
location, 166

Thallium–implanted Ge, location, 167–168, 169–170, 172
Thallium–implanted Si
electrical behavior, 208–209, 220
lattice location, 141, 155, 158, 161–163, 169–170, 172–175
Thermal equilibrium solubility, 219–220
Thomas–Fermi screening length, 21, 91, 127
Tin–implanted Si, location, 166
Total range, *see* Range distributions, amorphous targets
Tungsten
enhanced penetration in, 60, 61
ranges in, 42, 44 ff
Z_1 oscillations in, 52

V

Variable energy implants, 38
Vibrational amplitude of lattice atoms, 254, 255

X

Xenon
enhanced penetration, 61
ranges in crystals, 43, 46, 49
Xenon–implanted Si
disorder, 101
range in, 46
X-ray techniques, 85 ff
anomalous transmission (Borrmann), 85, 119 ff
small angle scattering, 86, 119 ff

Z

Z_1 oscillations, 27, 52 *see also* Energy loss, electronic
Zinc–implanted Si, location, 166

Printed in Dunstable, United Kingdom

91343006R00167